Requirements Practice in Conceptual Design

A Practical Guide

Second Edition

For a complete listing of titles in the
Artech House Technology Management and Professional Development Library,
turn to the back of this book.

Requirements Practice in Conceptual Design

A Practical Guide

Second Edition

Michael Ryan

ARTECH HOUSE

BOSTON | LONDON

artechhouse.com

Library of Congress Cataloging-in-Publication Data
A catalog record for this book is available from the US Library of Congress.

British Library Cataloguing in Publication Data
A catalogue record for this book is available from the British Library.

ISBN 13: 978-1-68569-135-6

Cover design by Iain Hill

Artech House
685 Canton Street
Norwood, MA 02062

10 9 8 7 6 5 4 3 2 1

Contents

Preface

It may not always be obvious why formal requirements are necessary, particularly since much of our decision making in our personal and business lives is not conducted using formal methodologies. For projects, however, an agreed-upon set of system requirements is essential from a number of different perspectives.

From the point of view of business management, the set of system requirements provides a mechanism whereby the organization's resources can be allocated knowingly to the project that has been established to bring the system into being. Unless the requirement set is well-defined, the project has no firm base and will have inadequate resources allocated to it, because the required work will not have been properly quantified in scope, cost, schedule, or quality. Worse still, without sufficient oversight of the formulation of a set of requirements, the business may be held ransom by users who insist on the inclusion of unrealistic requirements that business management has not approved.

From the project manager's perspective, the set of system requirements is an essential part of the definition of the scope of the project—it is from this scope that suitable estimates will be developed for schedule and budget. Scope management is one of the major functions of a project manager; a well-defined system scope (based on a formal set of system requirements) is necessary to be able to justify any expenditure of funds or effort within the project, to report adequately on the progress of work, and, eventually, to determine when the project is complete. The lack of adequate requirements processes is one of the major reasons why projects fail. Clearly then, if the project manager does not begin with a well-defined, universally agreed-on set of system requirements for a fixed-price contract, the project is doomed to fail.

From the systems engineering perspective, the formal set of system requirements represents the transition from the business world into the systems engineering and engineering worlds. Once the set has been established for the next time period (project, phase, build, increment, spiral, or sprint), it is not possible to rectify a poor set of requirements with good design or good engineering, nor with good production/process control, particularly when that set of requirements is the basis of a fixed-price contract between the acquirer and the contractor. Consequently, a project can rarely recover from poor definition, regardless of how much good work is performed subsequently.

An appropriate requirements-engineering process allows us to begin a project with a set of requirements that is complete, consistent, comprehensible, feasible, and able to be validated. A useful process ensures that all stakeholders have had input and that the various points of view have been reconciled. To do that, we must, therefore, be able to trade off functionality for cost—which implies that we understand the required functionality, priorities, and costs. We also need to be able to use the requirements-engineering process to manage changes in requirements for a wide variety of reasons.

The cost of developing requirements is estimated to be approximately 10–15% of the total system cost. Yet it is estimated that it costs some 200 times more to rectify an error in requirements during utilization than it does during early system design. Poor requirements definition cannot be rectified by good design or good engineering, so rigorous development of requirements is essential for successful system development. Requirements engineering provides this rigor through the application of a formal methodology. When requirements-engineering processes are ad-hoc, the end result is, at best, an unsatisfactory product or, at worst, a canceled project.

Note that the preceding comments regarding a set of requirements are true regardless of the development methodology—that is, they are as true for the start point for a monolithic waterfall project, a project phase, an increment in incremental development, a build in evolutionary development, a spiral in spiral development, or a sprint in Agile development.

Despite the considerable effort expended in systems engineering and requirements-engineering standards, there are surprisingly few methodologies available for the actual conduct of requirements engineering. At one level, this is not so surprising because standards such as ISO/IEC/IEEE 15288 (*Systems and Software Engineering—System Life Cycle Processes*), ANSI/EIA-632, IEEE-STD-1220 (*IEEE Standard for Application and Management of the Systems Engineering Process*), and ISO/IEC/IEEE 29148 (*Systems and Software Engineering—Life Cycle Processes—Requirements Engineering*) have a true systems engineering background; they are therefore focused on what must be done and do not prescribe how it should be done. Consequently, all of these standards are extremely useful in the guidance that they provide to an organization, but a requirements engineer cannot "do" them, because they (deliberately) provide little or no guidance on how to perform the tasks associated with good requirements practice.

Similarly, much has been written about requirements engineering, by both academics and practitioners. There are thousands of conference and journal papers addressing the topic and at least a hundred books directly addressing requirements engineering as well as several hundred more books (on business analysis, project management, systems engineering, and software engineering) that make significant mention of it. Yet, in all these descriptions, the focus is on what to do, rather than how to do it. Most of what is written focuses either on the very high-level consideration of requirements frameworks and elicitation techniques, or on very low-level activities such as the syntax of good requirement statements. It is important to understand the upper level issues such as the appropriate life cycle models and requirements-engineering processes, and it is essential to understand the templates for documents and the language used (vocabulary, syntax, and structure) of requirements, but it is also important to address what are arguably the most important activities—those that are undertaken to articulate the purpose of the system; gather the relevant information; analyze, negotiate, and synthesize the needs and requirements; and then present them in such a manner that describes the system's mission in order to meet the business needs.

This text focuses on those latter issues for good requirements practice. While both upper level frameworks and lower level requirements-writing skills are addressed, it focuses principally on how to carry out requirements engineering—that is, the fundamentally important aspect of how to conduct system-level logical design such that the resulting artifacts not only have the right shape but also have the right content.

We should make clear at the outset which types of systems are relevant to the practices presented here. We are focused on one type of system and on one way to represent requirements. There are many types of systems based on the various combinations of characteristics—open or closed; natural or human-made or human-modified; and precedented or unprecedented. It is important to recognize that this text and most of the standards discussed (such as ISO/IEC/IEEE 15288) refer to open, physical systems that are human-made/modified from largely precedented elements. It should also be noted at the outset that we are focused here on text-based requirements.

The text assumes an organization that is undertaking requirements engineering in systems acquisition within a framework that conforms to ISO/

IEC/IEEE 15288; in this assumed organization, the chief engineer is required to develop a systems engineering process that conforms to a standard such as IEEE-STD-1220 and, as part of the development of a suitable systems engineering process, is required to develop a requirements-engineering methodology that conforms to ISO/IEC/IEEE 29148 (or similar). The methodology proposed here, which provides a framework for requirements-engineering activities associated with the development of the logical (functional) architecture (the conduct of logical design in the conceptual design phase of the system life cycle), is called, for convenience, the conceptual design methodology (CDM).

The book is written to serve as a text for undergraduate and postgraduate courses in requirements engineering, as well as a manual for practitioners in business analysis, project management, systems engineering, and requirements engineering. For students, review questions are provided at the end of each chapter to assist in the confirmation of understanding.

We begin in Chapter 1 by setting the context for requirements engineering, with a description of systems, system life cycles, project management, systems engineering, and requirements engineering to provide the context for the detail in Chapters 2 to 9. Chapter 2 provides an overview of a number of the important aspects of the extant requirements-engineering body of knowledge. Chapter 3 describes the key features that underpin the design of CDM and outlines and further details the upper level processes. Subsequently, Chapters 4 to 9 present and illustrate CDM's major processes, activities, steps, and tasks. Chapters 8 and 9 focus, respectively, on defining system requirements and requirements writing. Throughout the book, a number of unifying examples are used to illustrate the application of CDM.

1

Requirements Practice Overview

1.1 Introduction

The principal role of project management is to meet project requirements—the project scope—within a given schedule and budget (to the desired quality). Despite significant efforts over the last two decades to define the discipline of project management and to improve the competency of project managers, projects continue to fail with alarming regularity. When examining the causes, it becomes apparent that project failure is rarely because business analysts do not understand the problem or opportunity to be pursued, or that project managers cannot manage their schedule or their budget, or that contract managers can't manage contracts, or that engineers can't design adequately, or that production processes cannot produce a quality product. Rather, projects fail principally because the project requirements—upon which the project's scope (and consequently the schedule and budget) is based—are poorly defined and incomplete and are often allowed to change, uncontrolled, throughout the course of the acquisition. It is clear that a major factor in reducing the number of project failures is to focus on the methodologies and processes through which requirements are better defined.

In defining requirements—defining scope—project management is assisted by the disciplines of business analysis, systems engineering, and requirements engineering. Business analysts work to understand the structure, policies, and operations of an organization and to recommend solutions that enable the organization to achieve its goals [1]. Business analysis therefore involves understanding the organization (defining business needs, goals, and objectives) and its internal and external stakeholders; determining the courses of action required to achieve those goals; and defining the capabilities required to provide products, services, and results to external stakeholders. Here, we focus on the systems that are fielded by the organization to provide such capabilities. That is, we are interested in systems and, in particular, in capability systems. In business it is common to refer to the system at this level as an asset and to refer to its management as asset management [2].

Once business needs, goals, and objectives are understood, systems engineering [3] provides the framework within which complex systems can be adequately defined, analyzed, specified, procured, manufactured, tested, operated, supported, and, ultimately, retired. Systems engineering methodologies and practices emerged from experience gained in the US Department of Defense acquisition programs of the 1950s. These programs often involved complex and challenging user requirements, and most entailed high technical risk because they involved a large number of different technical disciplines and the use of emerging (unproven) technologies. Requirements engineering, as a significant function of systems engineering, plays an essential role in making sure that a rigorous approach is taken to the development of a complete set of unambiguous requirements; it ensures that all relevant requirements are included, and all irrelevant requirements are excluded.

Essentially, requirements engineering is synonymous with business analysis, and the bodies of knowledge of both disciplines are in harmony. Here, to avoid any perceived bias that may come from use of the word engineering in the term requirements engineering or the word business in the term business analysis, we refer to both generically as requirements practice. Since we are focused on that practice during the early phases of the system life cycle, principally during conceptual design, this book's interest is in requirements practice in conceptual design.

Although requirements practice is the principal focus of this book, this introductory chapter provides an overview of systems, system life cycles, project management, systems engineering, and requirements engineering in order to provide the context for the detail in Chapters 2 to 9. The chapter concludes with some qualifications regarding the focus and coverage of these remaining chapters.

1.2 Definition of a System

ISO/IEC/IEEE 15288 defines a system in its most general form as an arrangement of parts or elements that together exhibit behaviors or meaning that the individual constituents do not [4]. We will see shortly that, from our perspective of the design of an engineered system, it is more useful to use a slightly narrower definition that a system is a combination of interacting elements organized to achieve one or more stated purposes.

In the broadest sense then, an engineered system is something that provides a solution to a complex problem. It has been engineered to meet some defined purpose, which must be clearly stated by the business it serves (to become an asset to the organization), and which represents the starting point of the design process. This purpose statement also provides the basis for the ultimate test of the system's fitness for purpose once it has been fielded.

From further examination of the definition, we can see that an engineered system is comprised of internal system elements with interconnections/interactions between those elements contained within an external boundary and of interfaces (inputs/outputs) across that boundary to external elements (perhaps in a related system) that exist in an operating environment. As illustrated in Figure 1.1, when we draw the boundary around selected system elements, we define the system of interest (SOI), which consists of those system elements and their interconnections that exist within the defined boundary (as well as the external interfaces that cross the boundary). As illustrated in Figure 1.2, an SOI might be considered as part of a wider SOI (WSOI) within an operating environment, which can be conceived as being part of a wider environment [5].

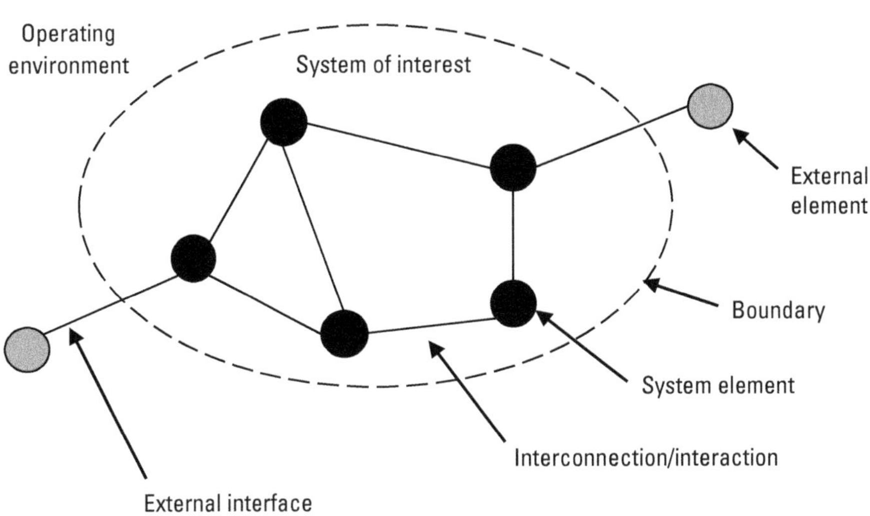

Figure 1.1 An SOI: its elements, interconnections, boundary, and interfaces to external elements in the operating environment.

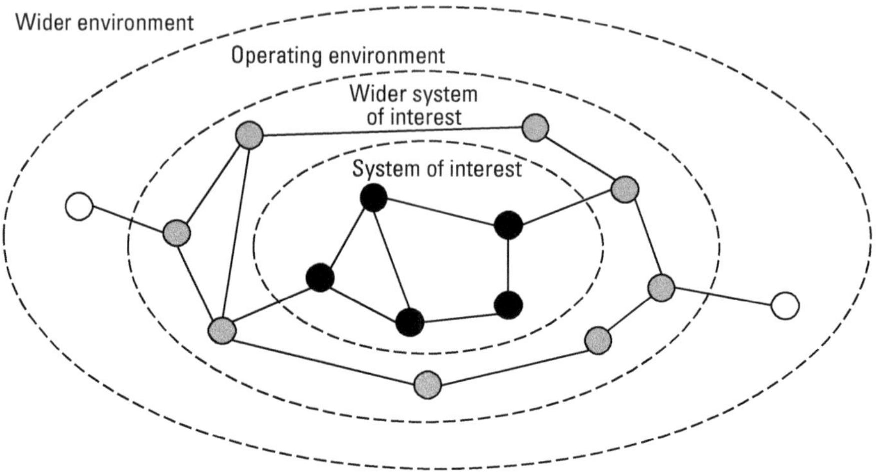

Figure 1.2 An SOI, which might be considered as part of wider SOI, which can be conceived as being part of an operating environment, as part of a wider environment.

Throughout the book, we see that (as illustrated in Figure 1.3) systems engineering is responsible, inter alia, for the following:

- Describing the SOI functionally as part of its environment;
- Extracting during conceptual design a stand-alone description of the SOI in a manner that can be contracted with a developer;
- Developing the SOI through the acquisition project;
- Introducing the SOI into service back in its environment as a physical solution.

Figure 1.3 demonstrates that our articulation of the SOI with its boundary and interfaces is a critical step in bringing the system into being. Even with a perfect acquisition process, any deficiencies in the initial description will result in delivering a system that is not fit-for-purpose, because it was not sufficiently described in the first place.

In a physical sense the term system is sometimes considered to be synonymous with product—that is, we say that a project is delivering a system or is delivering a product. A system is normally, however, considered to comprise a number of products. Figure 1.4 shows that EIA-632 [6] defines a system as comprising operational products (end products) and enabling products (such as test, training, and disposal). In this case, the product might also include a service or a result.

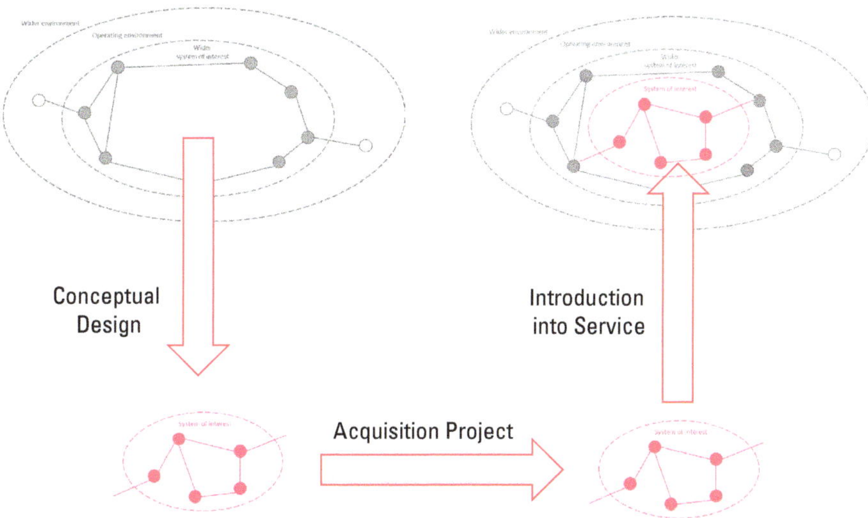

Figure 1.3 An SOI described functionally as part of its environment, extracted as a stand-alone description during conceptual design, developed through the acquisition project, and then introduced into service back in its environment as a physical solution.

1.2.1 Types of Systems

There are numerous ways to classify systems. The four main classifications are described as follows:

- *Closed or open systems:* An open system interacts with its operating environment; it accepts inputs from that environment across its boundary and returns outputs across the same boundary to the external environment. A closed system is isolated from its external environment. We are only interested in useful and, thus, open systems.

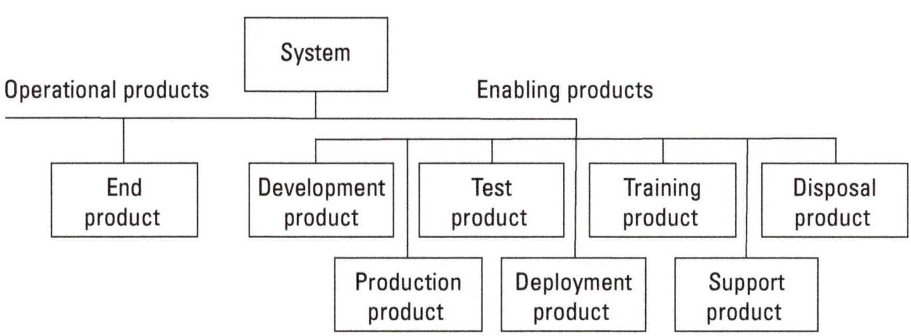

Figure 1.4 EIA-632 building block concept of a system comprising operational products and enabling products.

- *Natural or human-made/human-modified systems:* Natural systems contain natural elements and are the result of natural processes; human-made systems come into existence through the efforts of humans and may contain human-made elements or natural elements adapted to human-designed purposes. Natural systems that have been modified for human purposes are called human-modified systems. The requirements engineering for natural systems is certainly not conducted by humans, so we are only interested in human-made systems and the human contribution to human-modified systems.
- *Physical or conceptual systems:* Physical systems exist in a physical form; conceptual systems, such as economic, political, and religious systems, do not have a physical form. We focus here on physical systems comprising combinations of integrated hardware and software elements.
- *Precedented or unprecedented systems:* In a precedented system, similar systems (or, at least, the majority of system elements) have been produced before. An unprecedented system is one that has not been previously produced, nor have any of the system elements. Systems that comprise mostly unprecedented elements are the result of scientific research and development and, thus, do not lend themselves to be part of a project that can be managed with a robust scope, schedule, and budget. Here we focus on systems that comprise largely precedented elements—that is, those to which engineering, systems engineering, and project management are appropriate.

The wide variety of combinations of the above (and other) characteristics can lead to a large number of types of systems, each of which has markedly difficult properties. It is important to recognize that this book and most of the standards discussed herein (such as ISO/IEC/IEEE 15288) refer to engineered systems: open, physical systems that are human-made/human-modified from largely precedented elements.

1.2.2 A Systems Approach

When considering the system, either from a logical or a physical perspective, a systems approach focuses on the system as a whole, linking the value judgements of what is desired to solve the problem to the design decisions, which are based on what is feasible to implement the solution. Systems approaches draw on the wide range of knowledge appropriate to the SOI and its wider context to produce desired outcomes without adverse, unintended consequences.

Systems approaches are either soft or hard, defined as follows:

- Soft systems approaches are suited to addressing problems involving incomplete data and unclear goals drawing on subjective and humanist philosophies with constructivist and interpretivist foundations. Soft systems approaches comprise the following:
  - Creative thinking tools including brainstorming, morphological analysis, de Bono's six thinking hats, and Van Grundy's creative problem-solving techniques;
  - Systems thinking methodologies and techniques including cognitive mapping, influence diagramming, soft systems methodology, strategic choice approach, iterative planning, strategic assumptions surfacing and testing, the viable systems model (method), system dynamics modeling, rich pictures, systems thinking, causal loop diagramming, Actifeld, and field anomaly relaxation.
- Hard systems approaches are suited to solving well-defined problems with reliable data and clear goals, using analytical methods and quantitative techniques. Hard systems approaches comprise systems engineering and engineering.

Soft systems approaches and hard systems approaches are complementary. As Figure 1.5 implies, soft systems approaches are useful to assist in gaining an understanding of the problem space early on; they focus on creative thinking, human perspectives, and systemic interrelationships. These methodologies focus on the understanding of problems in ill-structured, ambiguous, or socially complex situations. They involve stakeholder engagement, dialogue, and iterative problem structuring, making them particularly useful in organizational change, policy development, and social systems. Techniques such as rich pictures, mind mapping, and causal loop diagrams help visualize multiple perspectives and emergent properties of systems, leading to adaptive and context-sensitive solutions.

In contrast, hard systems approaches rely more on structured, quantitative, and analytical techniques rooted in the systems engineering and engineering disciplines. These approaches focus on designing, optimizing, and managing well-defined systems with measurable parameters. They employ mathematical analysis, simulations, rigorous methodologies, decision analysis, and modeling.

Together, these systems approaches enhance problem-solving capabilities. Soft systems thinking fosters creativity and adaptability, while hard systems thinking provides rigor and precision. Integrating both approaches enables a holistic understanding of complex challenges, balancing technical feasibility

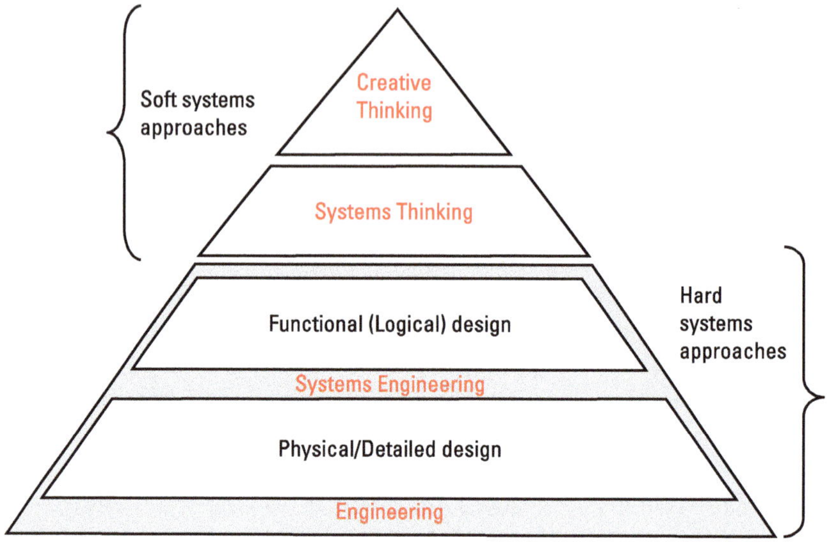

Figure 1.5 The hierarchy of soft and hard systems approaches: creative thinking, systems thinking, systems engineering, and engineering.

with social and organizational dynamics. A systems approach therefore means that the design process remains focused on the problem/opportunity as well as the solution.

1.2.3 Hierarchical Descriptions of a System

Recall that a system can be defined as a combination of interacting system elements organized to achieve some purpose. Each of these system elements can be implemented to fulfill its respective specified requirements [7], and we can consider the system to be a hierarchical composition of system elements, as illustrated in Figure 1.6.

The system elements in Figure 1.6 can be logical elements or physical elements, which supports the concepts of a logical (hierarchy) architecture and a physical (hierarchy) architecture, discussed briefly in Sections 1.2.3.1 and 1.2.3.2.

1.2.3.1 Logical Hierarchy

In a logical description of a system (see Figure 1.7), the system's mission is broken down into a hierarchical structure of its major functions. The logical description or architecture is therefore often called a functional hierarchy, or a functional (or logical) architecture.

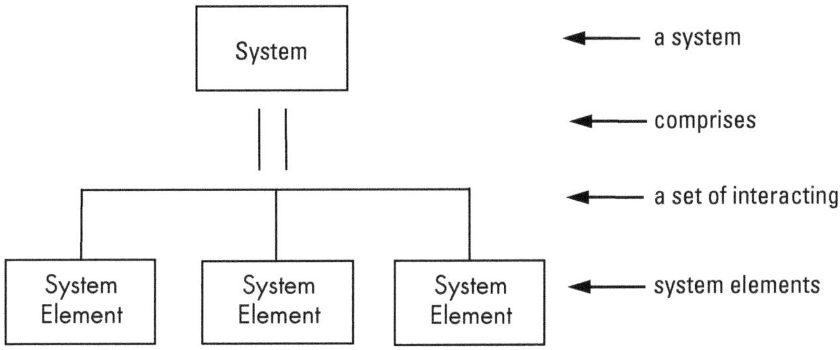

Figure 1.6 A system is composed of a set of interacting system elements.

1.2.3.2 Physical Hierarchy

In a physical sense, as we have described, a system can be considered to consist of two types of products: operational (end) products and enabling products. The end products of a system are also normally described in a hierarchy; here we use a four-layer hierarchy. We describe a top-level entity known as the system that comprises a number of subsystems that comprise a number of assemblies that comprise a number of components. Although these terms are perhaps the most common, there are others in use. For example, Figure 1.8 illustrates what is perhaps the most complete hierarchy of system elements as proposed by IEEE-STD-1220 [8]; this configuration adds elements such as products, subassemblies, subcomponents, and parts to our simple four-layer taxonomy.

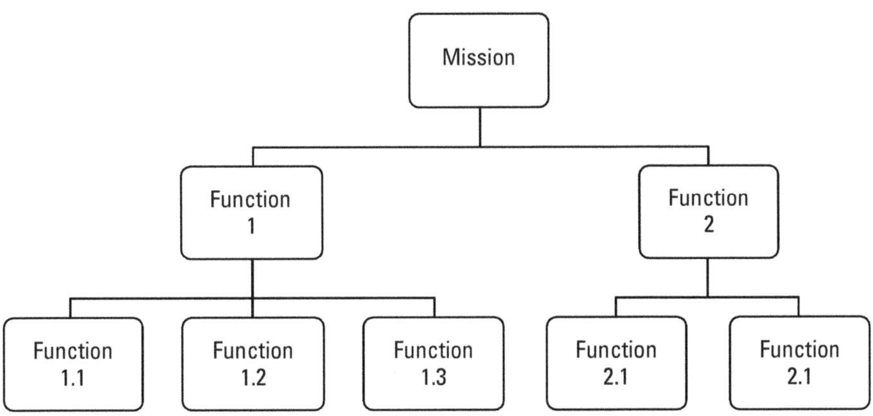

Figure 1.7 A logical (or functional) hierarchy of system functions.

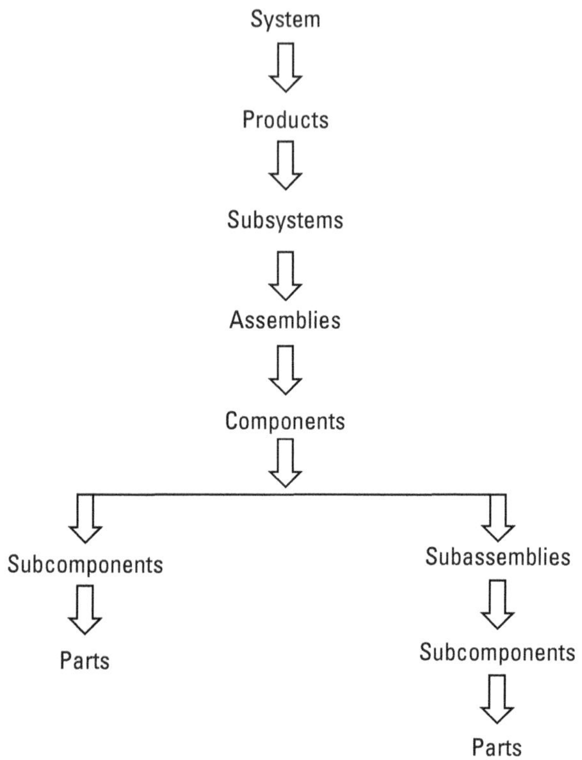

Figure 1.8 Hierarchy of physical elements of a system from IEEE-STD-1220.

ISO/IEC/IEEE 15288 [9] considers an SOI to comprise a combination of interacting elements, some of which, as illustrated in Figure 1.9, may be systems in their own right. When the SOI consists only of system elements that are systems in their own right—the SOI is called a system-of-systems (SoS).

Figure 1.10 illustrates how the hierarchy of Figure 1.9 can be delineated into groupings of systems and SoS. However, to have a useful view, we need a better understanding of the ways in which system elements can be grouped.

1.2.4 Collections of Systems

If a system comprises a set of interacting elements, there are a number of ways in which those elements can be collected. Unfortunately, we are not always very careful about the language we use to describe those connections. First, we are not always careful to make the distinction between a system and a subsystem. The difference matters because we simply cannot refer to

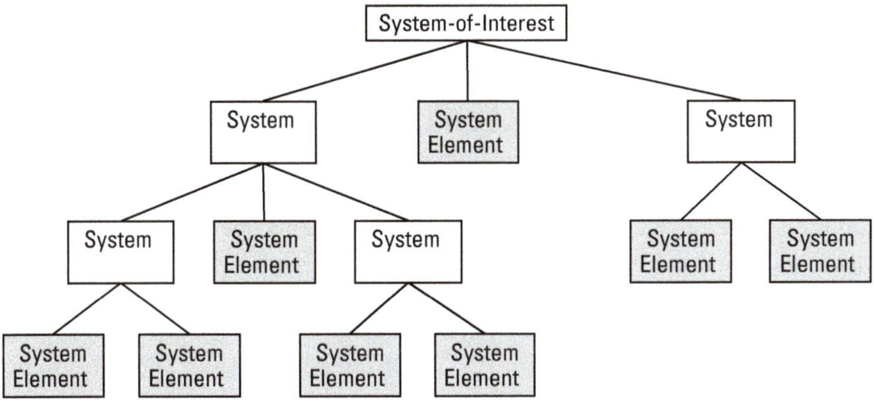

Figure 1.9 Hierarchy of system elements of an SOI in which some elements may be systems in their own right. (*After:* ISO/IEC/IEEE 15288.)

any grouping of elements as a system. For example, an aircraft system may contain, among other elements, the engine, which may consist of assemblies such as fuel tanks, pumps and lines, turbines, compressors, gear boxes, and hydraulic pumps. Commonly, the engine manufacturer will refer to the engine as the engine system. If we continue such use of the term to the lowest level of components, a nut and bolt could be considered a fastening system, because those two elements interact for that purpose. Using the term in that manner would make almost everything a "system" and, in the process, render the term almost meaningless. In fact, it would also dilute the meaning of SoS. Consider, for example, that the term SoS could even then be applied to two

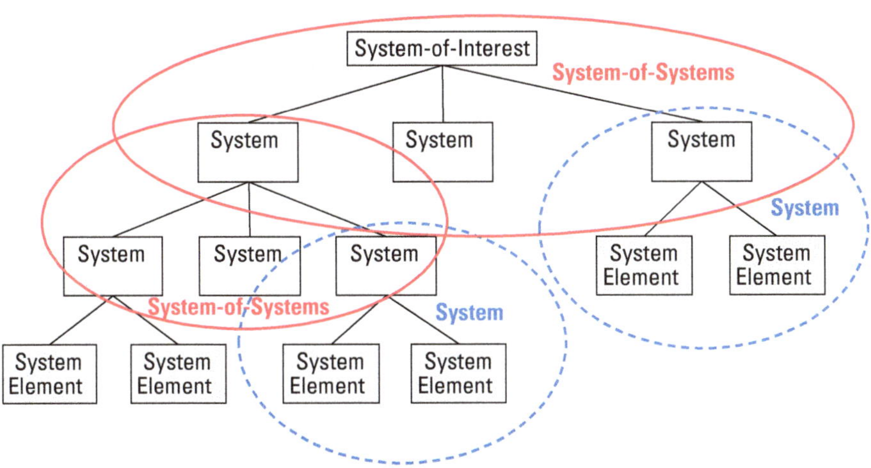

Figure 1.10 Hierarchy of elements of an SOI. (*After:* ISO/IEC/IEEE 15288.)

bolts and two nuts (that is, to two "systems") used to join two plates, the end result of which would be another SoS [10].

So, although we should always take a systems approach to design at all levels, not all combinations of elements are usefully considered "systems." To use the term more effectively in our consideration of systems engineering, an implicit part of the definition of a system is that it must be able to stand alone in its own right—that is, be managerially and operationally independent. With that qualification, an engine is not a system—it is only ever useful as an element of a system (that is, as a subsystem) and has no purpose other than to be a good servant of the parent system. Thus, it is optimized to that effect. To be more careful about the use of the terms system and subsystem, it is helpful to keep in mind the following:

- A system is a set of tightly coupled, codependent elements (subsystems) optimized at the system level for the permanent purpose of the system. Systems are managerially and operationally independent and will no doubt have independent life cycles. (They will almost certainly have been procured separately.)
- A subsystem is a set of tightly coupled, codependent elements (components) optimized at the system level for the permanent purpose of the system; they remain permanently part of the system. Subsystems are not independent and only exist to serve the system—they will have been designed by the system designer to be part of the system. Further, since the system is to be optimized, the subsystems are invariably suboptimal.

Note again that, although we state that a subsystem is not a system, we should take a systems approach to its design. IEEE-STD-1220 calls the systems approach the systems paradigm.

Since a system element can be a system in its own right, there are three types of ways in which systems can be grouped [11]; these are described as follows:

- *Portfolio-of-systems (PoS):* A PoS is a collection of systems managed jointly with a unified budget to fulfill one or more missions (that may not be related). Member systems interact to share (or perhaps compete for) resources.
- *SoS:* As noted in Section 1.2.3, an SoS is a collection of independent systems that interact for a common purpose. The constituent systems of an SoS are loosely coupled, and, since they are independent and

can join or leave the SoS over time, they have been optimized for their own purpose before joining the SoS. An individual system may be a member of more than one SoS.

- *Family-of-systems (FoS):* A collection of systems (also commonly called a product line) that are grouped because they are jointly designed, developed, and manufactured—even though they may not interact in operations.

In this book we are particularly interested in three levels of combination: subsystem, system, and SoS. As illustrated simplistically in Figure 1.11, the three groupings have similar architectures in that each comprises elements that are interconnected.

This similarity in upper level architecture can lead to the misapprehension that the three groupings are the same. However, there are differences, described as follows:

- *SoS.* SoS elements are systems in their own right, so that they are managerially independent and operationally independent and have been optimized for their own purpose before contributing to the purpose of the SoS. Further, since the constituent systems must be allowed to be optimized for their own purpose, the resultant SoS will invariably be suboptimal. Individual systems in an SoS can come and go and can be replaced readily, as they are loosely coupled, and can belong to one or more higher level SoS, even at the same time.

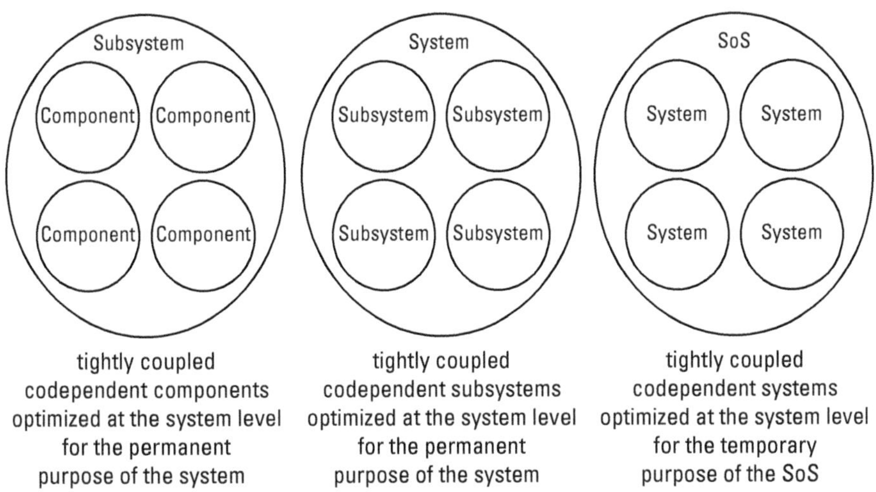

Figure 1.11 Elements of a subsystem, a system, and a SoS.

- *System:* On the other hand, a system comprises subsystems that are not independent and that only exist to provide the level of performance that serves the parent system. Subsystems are therefore invariably suboptimal since it is the system (the combination of subsystems) that is to be optimized, not the constituent subsystems. An individual system can be a member of one or more SoS, even at the same time.
- *Subsystem:* Subsystems comprise components that are not independent and that only exist to serve the system as good servants of the subsystem—again, it is the system that is to be optimized, not the constituent subsystems. An individual subsystem is only ever a member of the one system and cannot readily be removed from the operational system.

Table 1.1 summarizes the attributes of subsystems, systems, and SoS.

An important observation from Table 1.1 (particularly the last row) is the central position of system design; note that it is the system that is optimized at each level of design: subsystem, system, and SoS. Here, therefore, we focus on the system level.

Again, however, we should emphasize that, although we have been pedantic about the definition of a system, we do recognize that design at all levels should make use of a systems approach, even for those combinations of elements that are not systems per se.

1.2.5 Logical and Physical Descriptions of a System

A system can be described in two broad ways—in logical terms and in physical terms. A logical description (historically often referred to as a functional description) of a system articulates what the system will do, how well it will

Table 1.1
Attributes of a Subsystem, a System, and an SoS

Attribute	Subsystem	System	SoS
Purpose	System	System	SoS
Element Type	Component	Subsystem	System
Integration	Tightly Coupled	Tightly Coupled	Loosely Coupled
Element Interaction	Codependent	Codependent	Independent
Period	Permanent	Permanent	Temporary
Optimization Level	System	System	System

do it, how it will be verified, under what conditions it will perform, and what other systems will be involved with its operation. A physical description relates to the system elements and explains what the elements are, how they look, and how they are to be manufactured, integrated, and verified. The logical description contains the "whats" of the system, and the physical description contains the "hows."

It is worth noting that in the context of systems engineering, the term "physical" is used to differentiate this hierarchy from the functional hierarchy described earlier and that physical elements within a system may include elements of software that have been designed to operate on hardware elements to deliver prescribed functionality. The term "functional" is related to what the system is to do (and how well it is to do it and how it is proposed to be validated) and the term "physical" is related to how the system operates and is supported.

The two descriptions are valid independent descriptions of a system, and it is very important that a system is described both logically and physically, focusing first on the logical description for reasons described as follows:

- It is axiomatic that we develop the logical description first. To determine whether any particular physical implementation (i.e., how we are going to implement the elements of the system) is appropriate, we first must understand (from the logical architecture) what we want the system to do (that is, what purpose it serves). We therefore need to focus on the logical description (what) first, from which a series of candidate physical descriptions (how) can be developed, one of which can be selected as the preferred physical solution.
- Also, we must not allow the way in which we implement current physical systems to color unnecessarily the way in which we might describe future systems. An initial focus on the logical description therefore allows us to provide novel solutions to new (or even old) problems. Focusing on the physical description initially would lead to the tendency to solve new problems with old physical implementations.
- Upper level trade-offs and feasibility analyses must be conducted at the logical level before deciding on the physical implementation—if not, significant waste may result from the selection of physical solutions that either perform unnecessary functions or do not possess critical functionality.
- A logical description is ideally suited to the interface between the solution and the business case that drove its implementation. While it is

often possible for the business case to be met directly by an obvious physical solution, it is better for business management to transition from the business case into a more detailed logical description of what is required before considering how to achieve it in a physical sense. The definition of the logical description before the development of an appropriate physical description therefore moves from the business case to the final physical solution in controlled verifiable steps.
- The logical description changes slowly; the physical description changes much faster, particularly as the pace of technological change quickens. Arguably, for example, the need for, and upper level logical description of, an internal combustion engine has changed little (other than in performance requirements) over the last two centuries, while the physical implementations have changed dramatically. That is, while the purpose of the engine subsystem, as part of the car system, has not changed over the years, the physical implementation is obviously very different.

In the development of a system, therefore, there are at least two architectural views: a system logical architecture and a system physical architecture. Of course, these two descriptions are of the same system so they must be related. Section 5.3.5 details how the logical architecture, as outlined in the requirements breakdown structure (RBS), is mapped onto the physical architecture as represented by the hardware and software configuration items contained in the systems breakdown structure (SBS) in the project's work breakdown structure (WBS).

1.2.6 Problem Domain and Solution Domain

Since a system can be considered to be the solution to a problem, it is common to consider the activities being undertaken throughout the life of the system to be in either the problem domain (problem space) where we use predominantly logical descriptions, or the solution domain (solution space) where we use predominantly physical descriptions.

Activities in the problem domain (including production of the logical architecture) are generally considered to be the responsibility of the customer (the acquirer or tasking activity); activities in the solution domain (including the physical architecture) are generally considered to be the responsibility of the organization implementing the system (the developer or the performing activity).

1.2.7 A System as a Capability—A Capability System

In viewing a system as an organization's solution to a problem regarding the provision of products, services, and results, we need to recognize that a system is much more than an aggregation of hardware or software products; it must also be described in terms of all of its constituent elements, such as the major equipment solution (hardware and software), organization, personnel, collective training, facilities, data, support system (including supplies), operating procedures, and organizational policies. The system is fully defined by the combination of these resources operating in an environment that defines the context within which the system must operate and dictates the requirements to be exhibited by the system. In that sense then, a system is defined as delivering an operational capability. It is common, therefore, particularly in defense environments, to refer to the system at this level as a capability system. In business it is common to refer to the system at this level as an asset and to refer to its management as asset management [12].

Having acknowledged all of the elements of a capability system, it must also be recognized that each of the elements will most probably have a different acquisition cycle—for example, people are "acquired" in a different manner to that in which the major equipment are developed—and each element of the capability may even be acquired through a different acquisition means. In the remainder of this text, for ease of description, we focus on the acquisition of the major equipment element (often called the materiel system) of the capability system. We must always keep in mind, however, that this acquisition is being undertaken in parallel with the acquisitions of the other elements of the desired capability and that all the elements must be brought back together prior to introduction into service to field an operational capability.

1.3 A Generic System Life Cycle

There is no single accepted agreement on how many phases exist in a system life cycle or what those phases are called. In fact, there are a considerable number of generic life cycle stages in various life cycle viewpoints—for example, the International Council on Systems Engineering (INCOSE) *Systems Engineering Handbook* [13] provides six example life cycles, each of which has utility in a particular organizational context. For the discussion in this book, we use the system life cycle phases illustrated in Figure 1.12, and for further detail during the acquisition phase, we use the life cycle activities defined by Blanchard and Fabrycky [14] and MIL-STD-499B [15].

Preacquisition Phase	Acquisition Phase	Utilization Phase	Retirement Phase

Figure 1.12 Phases of a generic system life cycle.

We use this model shown in Figure 1.12 for a number of reasons. It emphasizes that a system begins with a perceived need and finishes upon retirement—the so-called cradle-to-grave approach. There is a clear delineation between the acquisition and in-service (utilization) phases of a system, allowing the differing application of systems engineering during those two phases to be investigated and recorded. In addition, it shows sufficient detail in the early activities of the acquisition where systems engineering methodologies and practices have the potential to make the most significant contribution.

As shown in Figure 1.12, a generic system life cycle can be divided into four very broad phases, described as follows:

- *Preacquisition phase:* The life cycle begins in the preacquisition phase with an idea for a system being generated as a result of business planning. Early consideration of the possible options results in the confirmation of the early business needs for the system, which are elaborated by a business case that justifies expenditure of organizational resources on acquisition of the system. In some instances, the preacquisition phase may determine that it is not feasible or cost-effective to proceed to acquisition (due to technology limitations or funding shortfalls, for example). In that sense then, the preacquisition phase is where organizations expend research and development funds to ensure that only feasible, cost-effective projects are taken forward to acquisition.
- *Acquisition phase:* The business needs for the system provide the input for the acquisition phase, which is focused on bringing the system into being and into service in the organization. This would normally involve defining the system in terms of business needs and requirements, stakeholder needs and requirements, and system requirements and then engaging a contractor to develop/deliver the system.
- *Utilization phase:* The system remains in service during the utilization phase until the business has no further need for it, or it no longer can meet the functions required of it by the organization, or it is no longer cost-effective to keep it in service. During utilization, the system may undergo a number of modifications and upgrades to rectify

performance shortfalls, to meet changing operational requirements or external environments, to enable ongoing support for system maintenance, or to enhance current performance or reliability.

- *Retirement phase:* Following operational use and system support, the system is eventually phased out and retired from service. The system life cycle concludes with the retirement phase. If the business need for the capability still exists in the organization, the conclusion of one system life cycle marks the start of another, and the process begins again. If the system has reached the end of its economic life, the life cycle ends in disposal.

1.3.1 Parties Involved

There are a number of parties involved throughout the system life cycle, and there is not any standard taxonomy for their names. This section introduces the nomenclature we use throughout these chapters.

The customer organization is managed by *enterprise management* who set the direction for the organization and for *business management* who are responsible for managing the activities conducted by the operations element of the organization (undertaken by the *operators*—often called *users*).

The systems employed within the organization are acquired by the *acquisition* element (also called the acquirer [16] or tasking activity [17]) of the organization under the auspices of a *project* (managed by a *project manager*). Project managers are supported by a number of related disciplines including *business analysis, systems engineering, requirements engineering, specialist engineering disciplines, quality assurance,* and *integrated logistic support.* Operators are aided in their operation of the system by the *support* element of the organization, which supports, sustains, and maintains the system throughout its life. In addition to the operational, acquisition, and support staff, there are many others within the customer organization who have a stake in the successful implementation of the project. These *stakeholders* can include representatives from the management, financial, operations, supply, maintenance, and facilities areas of the organization.

The system is obtained from a *supplier* [18] (also called the *performing activity* [19]) who may deliver the system off-the-shelf (OTS) or may develop it, in which case they are often called the *developer* [20]. The supplier (developer or performing activity) may be an internal part of the customer (acquirer) organization. If the development of the system is undertaken in house, the acquisition element of the organization (the acquirer) will engage with the development element (the developer) to develop the system. It is increasingly common these days for the supply or development to be undertaken by an outside organization called a *contractor*, which is the entity responsible for supplying the system (perhaps by designing and developing it) to meet the customer requirements. The relationship between the customer and the contractor varies with each project, but, for each project, the relationship is formally defined by the terms and conditions of the contract between the two parties. In many cases the contractor is not able to perform all of the work required and devolves packages of work to a number of *subcontractors*. The terms and conditions relating to this work are described in the relevant *subcontract*.

Responsibility for the various phases of the system life cycle is spread across the enterprise (or organization) within which the eventual system will operate. The initial preacquisition phase is the responsibility of enterprise management, who conduct business planning and establish the business case for the projects required to support an organization. A project is then established with a project charter providing authority to a project manager to expend organizational resources on the acquisition of the system. Systems engineering is an important discipline that is responsible for performing the technical management of the project throughout acquisition and then to the capability (product) manager during utilization. Once acquisition is complete, and the system is introduced into service, it is operated by the users and supported by the support element. Note that, rather than considering that each party has a role to play in a particular stage, it is best to consider that all parties are involved at all stages in the life cycle, with the roles and responsibilities of each party shifting in emphasis between stages.

1.4 Acquisition Phase and Utilization Phase

Systems engineering (and therefore requirements engineering) is predominantly related to the acquisition phase and, to a lesser extent, the utilization phase of the system life cycle. For these two major phases, we use the life cycle activities in Figure 1.13, which are based on those defined by Blanchard and Fabrycky [21]. It should be noted that there are a considerable number of other generic life cycle stages in other life cycle viewpoints. (See the INCOSE *Systems Engineering Handbook* [22] for six examples.)

In the acquisition phase, the activities are *conceptual design, preliminary design, detailed design and development*, and *construction and/or production*. In the utilization phase, the activities are *operational use* and *system support*. While there is no universally agreed-on taxonomy for these activities, we choose these

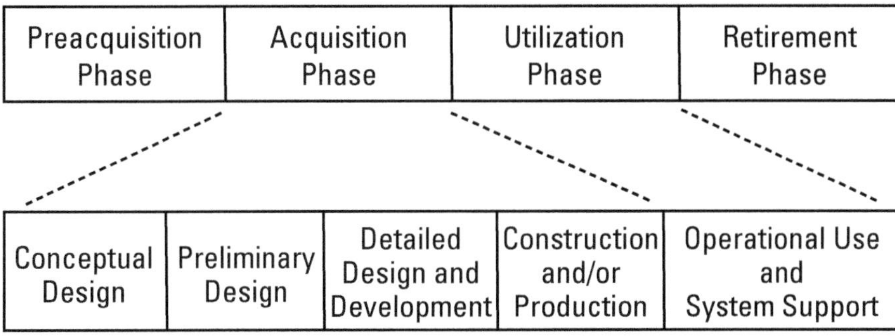

Figure 1.13 Activities in the acquisition phase and utilization phase of the system life cycle. (*After:* Blanchard and Fabrycky).

particular activities as a framework here, because they are generally accepted in the systems engineering community; we also consider the following reasons:

- These activities emphasize that a system begins with perceived business needs and finishes upon retirement and, ultimately, with disposal at end of life—the so-called cradle-to-grave approach.
- There is a clear delineation between the acquisition phase and the utilization phase of a system, which recognizes that there is a period of transition required as the system is transferred from the acquirers to the operators/users.
- The activities show sufficient detail in the early stages of the acquisition phase (particularly in conceptual design and preliminary design) where the application of systems engineering methodologies and practices have the potential to make the most significant contribution.
- Importantly, within the acquisition phase, the activities also differentiate clearly between the problem domain, which contains the logical description of the system (the product of conceptual design), and the solution domain, which contains the physical description of the system (the products of preliminary design and detailed design and development).
- Additionally, the separation of the early system design into conceptual (what and why) design and preliminary (how) design is very important, since the responsibility in most programs transitions from the customer for conceptual design to the contractor for preliminary design.

The significance of focusing on the system life cycle is that decisions made early in the acquisition phase are informed by the proposed and intended activities in the utilization phase. For example, the design of an aircraft airframe must take into account the maintenance and operation of that airframe during the utilization phase. It would be pointless to design the best airframe in the world if it did not have the necessary access points to allow maintenance personnel to service it, nor operators to operate it in the intended environment.

1.4.1 Acquisition Phase

Figure 1.13 shows that the acquisition phase comprises the four main activities of conceptual design, preliminary design, detailed design and development, and construction and/or production. Sections 1.4.1.1 to 1.4.1.4 describe each of these activities in more detail, outlining the major tasks undertaken, the main artifacts produced, and the major baseline developed during each. (See Figure 1.14 for an overview of artifacts, baselines, and reviews.)

1.4.1.1 Conceptual Design

Conceptual design focuses on the development of a set of clearly defined requirements, at the system level, and in logical terms. Although clearly defining the requirements of the system would seem an obvious (and essential) first step, it is often poorly done and is commonly the direct cause of problems later in the development process. Business managers and stakeholders sometimes, understandably, prefer to describe their needs and requirements in loose and ambiguous terms to protect themselves from subsequent changes in their business environment. The conceptual design process aims to avoid this ambiguity by providing a formal process by which the *business needs and requirements* (BNR) [23] are articulated and confirmed by business management, and then elaborated by stakeholders at the business operations level into a set of stakeholder needs and requirements (SNR) [24], which are further elaborated by business analysts and requirements engineers into a set of system requirements in the *system requirement specification* (SyRS) [25]. There may be one SyRS for

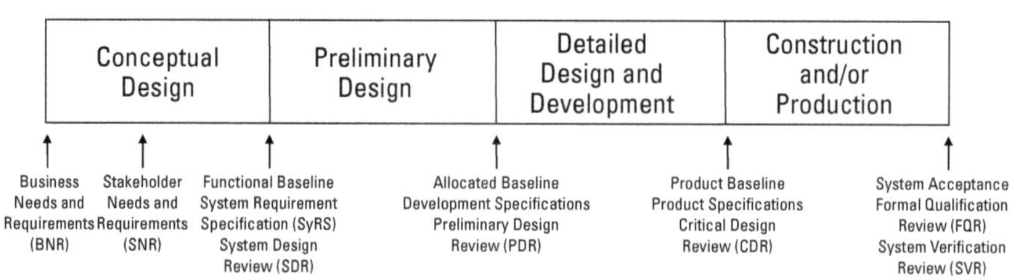

Figure 1.14 Acquisition phase activities and the major artifacts, baselines, and reviews associated with each.

the entire capability system, but it is more likely that there is one SyRS for each of the constituent elements of capability—that is, one for each of the major materiel system, personnel, support, training, facilities, and so on. As noted earlier, each of these constituent capability elements may be developed independently, perhaps through separate contracts. Chapter 2 discusses this aspect in more detail and illustrates the concept in Figure 2.3.

The BNR, SNR, and the SyRS are key elements of what is called the *functional baseline* (FBL), which describes the whats and whys of the system. The FBL therefore represents a system-level logical architecture that meets the business and stakeholder needs and requirements.

Conceptual design ends with the system design review (SDR), which finalizes the initial FBL. The SDR provides a formalized check of the logical design; communicates that design to the major stakeholders; confirms external interface and interoperability issues; confirms the BNR, SNR, and SyRS; and provides a formal record of design decisions and design acceptance.

1.4.1.2 Preliminary Design

The aim of preliminary design is to convert the FBL into an upper level physical definition of the system configuration or architecture (the hows of the system). Preliminary design is therefore the stage where logical design is translated into physical design, where focus shifts from the problem domain into the solution domain. The result of preliminary design is a subsystem-level design known as the allocated baseline (ABL) in which the logical groupings defined in the FBL are defined in more detail and then regrouped and allocated to subsystem-level physical groupings [elements of which are called configuration items (CIs)] that combine to form the upper level physical design of the system. At the center of the ABL are a series of development specifications that contain the subsystem-level requirements grouped by subsystem/CI as well as the interface requirements among those subsystems/CIs.

The ABL is so-called because the system-level requirements that are logically grouped in the FBL are allocated at this next baseline into physical groupings (subsystems/CIs). The ABL therefore represents a subsystem-level architecture (couched in physical terms) that meets the requirements of the system-level architecture (couched in logical terms) contained in the FBL.

The initial ABL is formalized at the preliminary design review (PDR). The PDR ensures the adequacy of the preliminary design effort prior to focusing on detailed design. PDR assesses the technical adequacy of the proposed solution in terms of technical risk and the likely satisfaction of the FBL. PDR also confirms the identification of subsystem/CI interfaces and the compatibility of each of the subsystems/CIs.

1.4.1.3 Detailed Design and Development

The ABL developed during preliminary design is used in the detailed design and development process to complete development of the individual subsystems, assemblies, and components in the system. Prototyping may occur and the system design is confirmed by test and evaluation. The result of the detailed design and development process is the initial establishment of the product baseline (PBL) as the system is now defined by the numerous products (subsystems, assemblies, and components) making up the total system (as well as the requisite materials and processes for manufacturing and construction). The definition of the system at this stage should be sufficiently detailed to support the commencement of the construction and/or production activities.

The PBL is established at the critical design review (CDR). The CDR is the final design review resulting in the official acceptance of the design and the subsequent commencement of construction and/or production activities. CDR evaluates the detailed design; determines readiness for production/construction; and ensures design compatibility, including a detailed understanding of all external and internal interfaces.

1.4.1.4 Construction and/or Production

During the final activity within the acquisition phase, construction and/or production. System elements are produced in accordance with the detailed design specifications in the PBL, and the system is ultimately constructed in its final form. In addition, formal test and evaluation (T&E) activities (acceptance tests) are conducted to ensure that the final system configuration meets the requirements in the SyRS.

Construction and/or production, and the acquisition phase, ends with the formal qualification review (FQR) or system verification review (SVR), or equivalent upon which the acquirer accepts the system from the contractor. This provides the basis upon which the customer accepts the system from the contractor. The FQR is informed by the results of acceptance T&E (AT&E).

1.4.2 Utilization Phase

On acceptance from the supplier, the system moves into the utilization phase. Further testing in conjunction with all elements of the capability (i.e., people, processes, training, facilities, and support)—or operational T&E (OT&E)—validates that the system meets its original business needs and that the system is considered suitable to enter operational service. The major activities during this phase are operational use and system support as well as continuous OT&E activities to validate that the capability system still meets the business needs. Systems engineering activities may continue during the utilization phase to support any modification that may be required to rectify performance shortfalls, to meet changing operational requirements or external environments, to enable ongoing support for the system to be maintained, or to enhance current performance or reliability.

1.4.3 Retirement Phase

The system life cycle ends with retirement of the system in the retirement phase, which may well overlap with the introduction into service of the replacement system should the organization still require the capability. Functions associated with retirement include transportation and handling, decomposition, and processing of the retiring system. If the system is at the end of its useful life at the end of the life cycle, retirement will mean disposal because the system cannot be reused in any way. A retirement concept should be developed during the early stages of the acquisition phase. If considered early, disposal and phase-out issues will form some of the criteria against which the system is designed (often called "design for disposability"). It is important, however, that system designers focus on retirement, as well as the more limiting issues of disposal; planning for disposal is important, but a system may retire from several life cycles before it is ultimately disposed of at the end of life.

1.5 The Role of Systems Engineering

In the modern world the authoritative definition of systems engineering should come from INCOSE, which describes it as follows:

> ". . . a transdisciplinary and integrative approach to enable the successful realization, use and retirement of engineered systems, using systems principles and concepts, and scientific, technological and management methods" [26, 27].

However, there is a wide range of systems engineering definitions [28–32], each of which tends to reflect the particular focus of its source as well as being a product of its time. Although each of these definitions has a slightly different focus, a number of common themes are evident, so we focus on those in Sections 1.5.1 to 1.5.6.

1.5.1 Top-Down Approach

Traditional engineering design methods are based on a bottom-up approach in which known components are combined into assemblies and then into the subsystems from which the system is then constructed. Once assembled, the system is tested for the desired properties, and the design is modified in an iterative manner until the system meets the desired needs. This approach is valid and extremely useful for relatively straightforward problems that are well-defined. Unfortunately, complicated problems cannot be solved with the bottom-up approach.

Systems engineering begins by addressing the system as a whole, which facilitates the initial allocation of requirements as well as the subsequent analysis of the system and its interfaces. Once system-level requirements are understood, the system is broken down into subsystems and the subsystems further broken down into assemblies and components, until a complete understanding is achieved of the system from top to bottom. This top-down approach is a very important aspect of managing the development of complex systems. By viewing the system initially as a whole and then progressively breaking it into smaller elements, the interaction between the components can be understood more thoroughly, which assists in identifying and designing the necessary interfaces between system elements (internal interfaces) and between this and other systems (external interfaces). For example, Figure 1.15 illustrates the EIA-632 approach to top-down development [33].

It must be recognized, however, that while design is conducted top-down, the system is implemented using a bottom-up approach as illustrated in Figure 1.16 for a simple four-tier system. That is, one of the aims of system engineering is to provide a rigorous, reproducible process by which the complex system can be broken down into a series of simpler elements that can then be designed and built using the traditional bottom-up engineering approach. Importantly, the second principal facet of systems engineering therefore is to provide a process by which the elements—the components, assemblies, and subsystems—can be integrated and verified as achieving the desired system purpose.

1.5.2 Focus on Life Cycle

Systems engineering is focused on the entire system life cycle and takes this life cycle into consideration during decision making at any point in the life cycle. Unfortunately it is still too common to consider solutions only in the light of the issues associated with the acquisition phase and to pay little attention to through-life support aspects. It is proper for project managers and their teams to focus on the acquisition phase of the project and on the development of a system that meets its requirements while minimizing cost and schedule. However, a lack of attention to whole-of-life considerations can often lead to larger-than-expected costs in the utilization phase that must be met from budgets that are insufficient to keep systems in service. A life cycle focus is

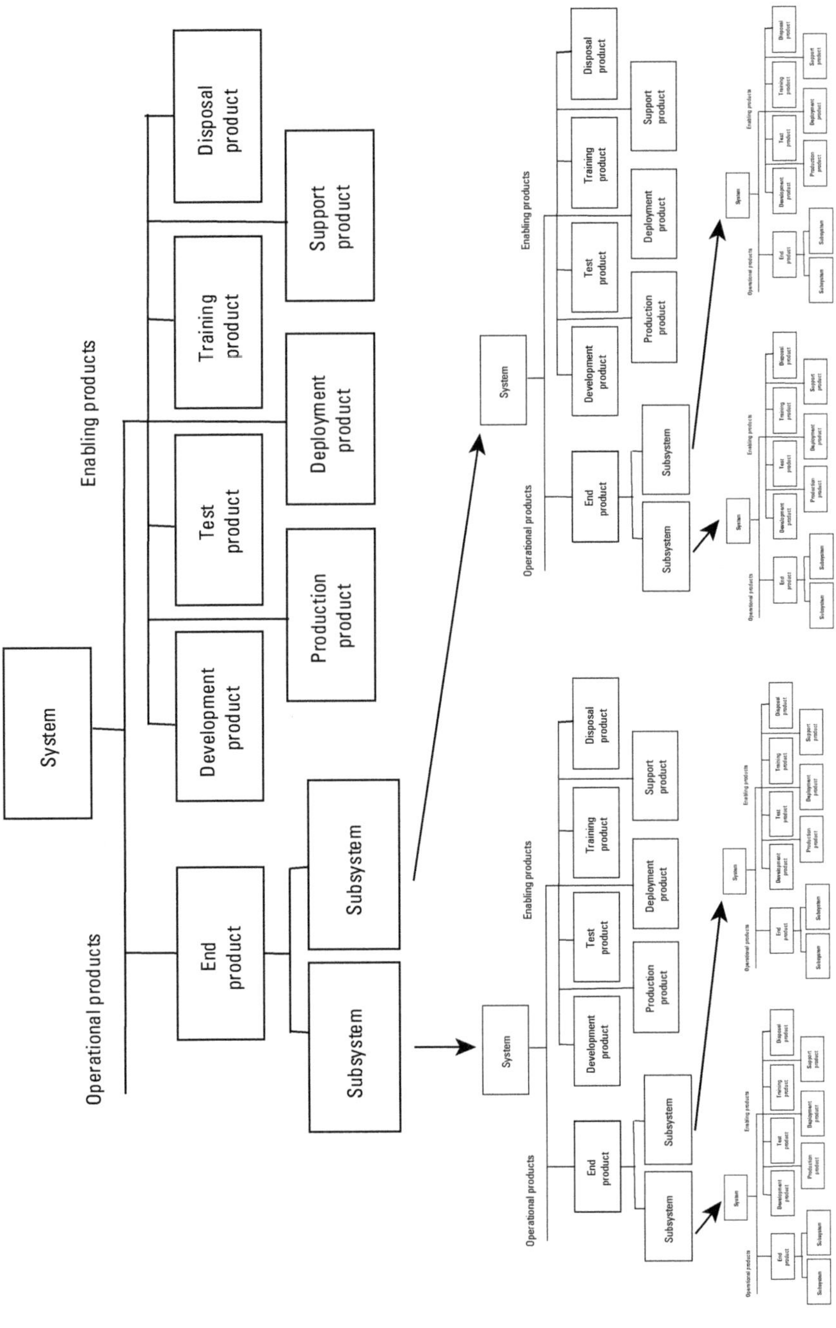

Figure 1.15 EIA-632 building block concept for top-down development of a system.

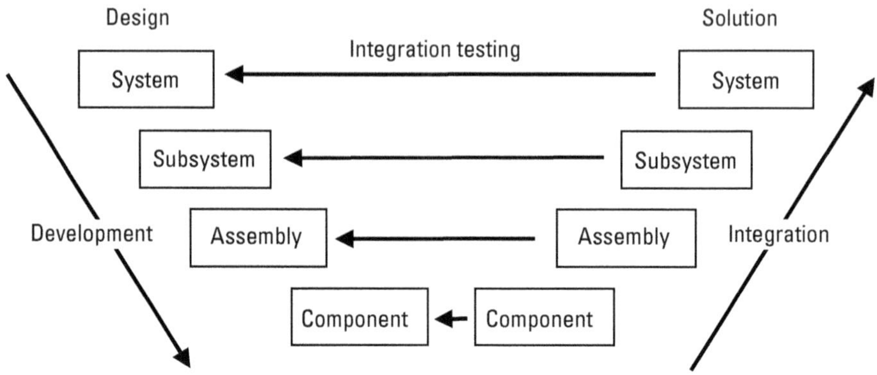

Figure 1.16 Bottom-up integration of a system.

on the capability system throughout its life cycle, not just on the constituent products throughout acquisition.

The significance of focusing on the system life cycle is that decisions made early in conceptual design are informed by the activities proposed to be conducted later in the acquisition phase and particularly in the utilization phase. For example, the design of an aircraft airframe must take into account the maintenance and operation of that airframe during the utilization phase—it would be pointless to design the best airframe in the world if it did not have the necessary access points to allow maintenance personnel to service it, nor operators to operate it in the intended environment.

Other examples of utilization phase requirements that impact on equipment design or selection during the acquisition phase include *reliability* and *availability*. Reliability normally refers to the ability of the equipment to operate without failure for a given period. Availability is a measure of the degree to which a system is in an operable condition when required at some random point in time. Again, a superior design is pointless unless the system can meet specified minimum levels of reliability and availability and, for that matter, all the other "-ilities."

Economic factors provide arguably the most compelling evidence to support the focus on life cycle as opposed to an emphasis on the end product itself. In short, a life cycle focus can save money in the long term. History has shown that a large proportion of total life cycle cost for a given system stems from decisions made early in the acquisition phase of the project. To that end, the maximum opportunity to reduce the total life cycle cost of a system is presented in the acquisition phase. Poor requirements cannot be rectified by good design; rigorous development of requirements is essential for the acquisition to be successful. Put slightly differently, it is impossible for a project manager to manage scope, time, and quality to achieve project success if the scope of the project is fundamentally flawed—success or failure is defined early in a project's life.

1.5.3 System Optimization and Balance

A system architecture must represent a balance among the large number of requirements and constraints that cover a wide range of factors such as human factors and environmental, economic, moral, ethical, social, cultural, and psychological factors. A simple, but essential, example is the balance between just two design factors: cost and performance—if either is allowed to dominate, it will generally be at the expense of the other.

It does not necessarily follow that the combination of optimized system elements leads to an optimized system. Additionally, the system architecture must represent a balance of a large number of requirements that, as well as the technical considerations, cover a wide range of factors, including human factors and environmental, economic, moral, ethical, social, cultural, and psychological factors. It is not normally useful, therefore, to allow the designers of subsystems to optimize their part of the system in isolation of system-level considerations—put another way, it is generally necessary for system-level considerations to constrain subsystem-level design in order for the system to be optimized.

It follows that a number of subsystems may need to be suboptimal (or at least constrained in some manner) to allow their combination (the system) to be optimal. A further advantage, therefore, of the top-down approach in systems engineering is that system optimization and balance can be achieved as a by-product of the design process, something that cannot be guaranteed in a bottom-up design method.

A balance must also be struck across the life cycle. Metrics such as cost effectiveness must be measured across all life cycle phases, not just acquisition. It is often the case that savings made in acquiring the system are then countered by significant maintenance and repair costs in service. Although, as a corollary, it doesn't necessarily follow that paying more for a system during acquisition will guarantee that the support costs are lower.

1.5.4 Integration of Disciplines and Specialties

The focus of systems engineering is on the system as a whole, and the maintenance of a strong interdisciplinary approach. Project management, requirements engineering, quality assurance, integrated logistics support, and traditional design disciplines such as hardware and software engineering are but a few of the many disciplines that are part of a coordinated systems engineering effort.

Systems engineering aims to manage and integrate the efforts of a multitude of technical disciplines and specialties to ensure that all user requirements are adequately addressed. Rarely is it possible for a large system to be designed by a single discipline. The systems engineering function focuses on breaking up the task into components that can be developed by these disparate disciplines and specialties and then providing the management to integrate their efforts to produce a system that meets the business needs and requirements. In modern system developments, this function is all the more important because of the complicated nature of large projects and their contracting mechanisms and the geographic dispersion of customer, contractor, and subcontractor personnel across the country and around the world. Commonly, complex systems can involve millions of hours of work by thousands of people from a wide range of disciplines and backgrounds spread across dozens of countries.

1.5.5 Management

While systems engineering clearly has a technical role and provides essential methodologies for systems development, it is not simply limited to technical issues and is not simply another engineering process to be adopted. Systems engineering has both a management and a technical role. Project management is responsible for ensuring that the system is delivered on time and within budget, with appropriate quality. The trade-offs and compromises implicit in those functions are informed by the products of systems engineering. Additionally, the scope of the project is defined by the WBS, which is the result of requirements engineering. Systems engineering, business analysis, requirements engineering, and project management are therefore inextricably linked.

1.5.6 Focus on Requirements Engineering

Requirements engineering involves the complete and accurate definition of system requirements, which is a primary focus of the early project management and systems engineering effort. A simple mission statement for the system must be translated into a large number of statements of requirement that form the basis for the logical design and that are subsequently elaborated further to form the physical architecture. These transitions must be managed by a rigorous process that guarantees that all relevant requirements are included and all irrelevant requirements excluded. The establishment of correct requirements is fundamental to the success of subsequent design activities.

Requirements engineering ensures that a rigorous approach is taken to the development of a complete set of unambiguous requirements through elicitation, analysis, negotiation, definition, verification, and validation of system requirements. Once requirements have been collected, the requirements-engineering process then focuses on the management of these requirements from the system level right down to the lowest constituent component.

Good requirements are essential to project and system success. Goldsmith observes that, "Defining business requirements is the most important and poorest performed part of system development" [34]. Before examining requirements engineering in more detail, however, let's look briefly at the related discipline of project management.

1.6 The Role of Project Management

The Project Management Institute's (PMI) *Project Management Body of Knowledge* [35] defines a project as, ". . . a temporary endeavor undertaken to create a unique product, service or result," and project management as, ". . . the application of knowledge, skills, tools and techniques to project activities to meet the project requirements." Although there are any number of unique products, services, or results that could be the output of a project, we are interested here in projects that bring systems into being in order to solve a business problem or to take up a business opportunity. Since a project is a temporary endeavor (that is, it has a defined starting point and a defined end point), a project can only exist for a portion of the system life cycle. Consequently, we should note that the principal responsibility of project management during the system life cycle is in the acquisition phase during which the capability system is brought into being and introduced into utilization. However, as illustrated in Figure 1.17, we will need to establish a number of projects throughout utilization in order to undertake major activities such as modifications to the system.

A useful way to investigate that relationship between systems (requirements) engineering and project management is through the 10 project management knowledge areas as defined in the PMI's *Project Management Body of Knowledge*, a summary of which is provided in the following paragraphs, including a brief note on the role of systems engineering in the activities defined within each knowledge area.

Project integration management requires that all elements of the project are properly planned and coordinated. The coordination involves making

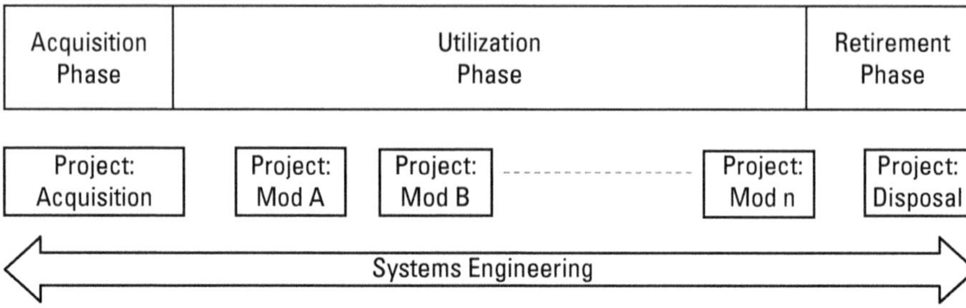

Figure 1.17 The role of project management and systems engineering throughout the system life cycle.

trade-offs between conflicting alternatives to meet that the needs and requirements of the business and of the key project stakeholders. The keys to integration management are the development of the project charter and then the project plan, which result from the project planning processes and form a governing approach to control and execution. The plan must be cognizant of the business constraints and organizational policies as well as the outputs from the project planning processes. Once the plan is in place, the project integration management function ensures that the plan is executed. Systems engineering contributes significantly to the development of the activities upon which the project plan is developed as well as to the subsequent integration effort.

Project scope management is primarily concerned with ensuring that the project includes all the work required to complete the project successfully without including any additional or unnecessary work. The first stage in scope management is to define a formal scope as the basis of the subsequent scope management effort. The *Guide to the Project Management Body of Knowledge* [36] references the use of systems engineering as fundamental to the completion of this task—almost all the activities we discuss in this book are focused on developing and managing the scope of the system. Once the scope has been developed and approved, the project must be broken into smaller, more manageable components. This breakdown is commonly performed using a WBS. The scope and the WBS must then be formalized through an acceptance process to ensure that all parties are committed to the same scope and the same work; systems engineering has an important role in defining the system requirements and then allocating them to the elements SBS within the project WBS. Once approved, the scope management activity must take control of any changes to the approved project scope.

Project schedule management is a key project management activity that is sometimes referred to as *time management.* To manage the project's schedule, the project manager must start with a comprehensive list of the work activities that make up the project. Tools such as the WBS assist in this identification, and, again, systems engineering provides an essential contribution. These activities will normally be performed in a particular sequence that is dictated often by the systems engineering aspects of the project. For example, the activity "hardware/software integration" can only be performed following the activities "hardware test and acceptance" and "software test and acceptance." Other complications with respect to time management eventuate when activities have specific start and end dates that cannot be altered. Once the activities and their sequences have been determined, the durations for each of the activities need to be estimated. Based on these pieces of information, the project manager can determine the schedule for the entire project. It then becomes an exercise in schedule control as activities are added, deleted, or modified throughout the project. The project manager is assisted in the time management function by tools ranging from sophisticated, automated software packages to a simple reliance on expert judgment to predict activity duration. Systems engineering activities are the major source of information regarding WBS activities, in terms of their formation, progress, and completion.

Project cost management ensures that the project is completed within the approved budget. The first task in determining the budget, called resource planning, determines the resources required to complete the project activities. Resources include personnel, materials, and equipment. Based on the resources required for each activity, the project manager can estimate the costs associated with each activity and the total cost of the entire project. Changes in the project scope will invariably impact the cost aspects of the project—those impacts need to be monitored and assessed on an ongoing basis. Tools are available to assist with cost management, but a great deal of the estimation and control effort relies on expert judgment and experience. Systems engineering provides critical input for the development of the WBS but also provides input to cost-benefit trade-offs as well as investigations to determine the effect of any proposed change on project cost and schedule.

Project quality management. The project manager must lead in the quality assurance effort to ensure that quality is taken seriously by all involved. The first stage in the process is to plan the quality management effort by identifying which quality standards are to be applied to the project and how these will be satisfied. The overall project performance must be evaluated regularly to provide users with confidence that the selected standards are being satisfied. The systems engineering efforts in establishing work packages, as well as in review and audit (particularly configuration audits), provide essential input to project quality management.

Project resource management is not exclusive to project management but is, nonetheless, a vital part of the project management function to ensure the effective use of resources across the project. Human-resource management ensures the most effective use of the project personnel. One of the first tasks of the project manager is often to acquire and organize the personnel required to perform the activities associated with the project. Once this has been done, the project manager must maintain the unity and direction of those involved with the project, a job that involves sound leadership and management skills such as negotiation, communication, motivation, and team building. In addition to human resources, the project manager is responsible for all other resources used or consumed to complete the project. Again, the extent, nature, and timing of the use of these resources is largely defined by systems engineering endeavors.

Project communications management is a fundamental activity for project managers. Communication is required to ensure that relevant project information is collected and disseminated to the appropriate personnel in a timely manner. The project manager must determine the information requirements of business managers and the key stakeholders in the project and the preferred format for the information to be provided to them. The project manager then must ensure that the information is distributed as planned. A particularly important part of communications management is the report generation associated with status reporting and phase or project closure. The role of communications management is worthy of emphasis here as a project management to which systems engineering provides great value. The systems engineer is, in some respects, part management staff and part technical staff. Systems engineers, like business analysists, are therefore largely conversant with both technical and management languages and can act as very effective translators between the management layers within organizations. Systems engineers can also assist with horizontal communication between different design teams to assist with technical issues such as design space consideration, interfacing issues, and technical conflicts. Of course, the major systems engineering reviews also provide a significant contribution to the communication of project information.

Project risk management encompasses all aspects of project risk. The systems engineering effort focuses on assessment of the technical risks associated with the project, whereas the project manager broadens the coverage of risk to include areas such as schedule and cost and the other elements of project management as defined in sources such as the *Project Management Body of Knowledge*. Systems engineering supplies input to the risk management process in concert with the project manager. Project risk management involves the same steps as technical risk management. The main source of information for the task of risk management is intuition and experience.

Project procurement management ensures the timely supply of goods and services from outside the immediate project environment in support of the project. The initial step is to plan what products and services are required to complete the project and when they will be needed. These details should then be recorded. The project manager should seek quotes or offers from the suppliers and select the best of the offers. In some cases, the procurement will be a one-time purchase but often one or more standing offers or contracts will result. The project manager will be responsible for managing the resulting contracts. Once all aspects of the procurement are completed, the project manager is responsible for the closure of any extant contracts.

Project stakeholder management is the process by which stakeholders are identified and an appropriate management plan is developed, implemented, and controlled. The PMI's *Project Management Body of Knowledge* defines a stakeholder as any individual, group, or organization that is affected by or can affect the system of interest. Since stakeholders can influence the system development, it is essential that they are identified and that appropriate mechanisms are established to engage with them throughout the project to ensure that their needs are met. From a systems engineering perspective, engagement with stakeholders is essential to the establishment of appropriate business and stakeholder needs as well as their transformation into system requirements.

1.7 What Is Requirements Engineering?

Requirements engineering can be described most simply as the process by which we identify a problem's context, locate the business and stakeholder needs and requirements within that context, and deliver system requirements that meet those needs [37]. As illustrated in Figure 1.18, that process requires a considerable reduction of the complexity of the real world, and a subset within which the customer organization operates, into a set of requirements that define the scope of the project. Therefore, considerable judgement is required to identify BNR and SNR and then convert them into a SyRS. Further, as Andriole points out, most requirements-engineering methodologies have been created by people who like precision, diagnosticity, and rigor—yet, requirements management is fundamentally a political process, not a technical one [38].

Dorfman and Thayer describe requirements engineering as the "science and discipline concerned with analyzing and documenting requirements" [39].

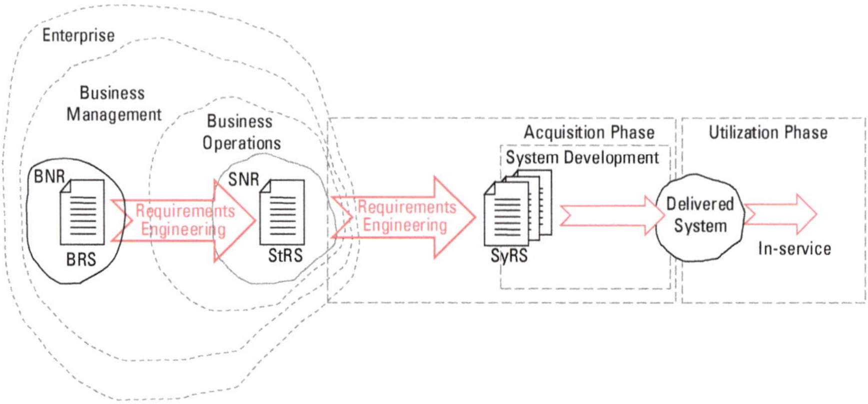

Figure 1.18 The role of requirements engineering in translating BNR into SNR and then into system requirements in one or more SyRS that define the capability system.

Kotonya and Sommerville define requirements engineering as "the systematic process of eliciting, understanding, analyzing, documenting, and managing requirements" [40]. ISO/IEC/IEEE 29148 defines requirements engineering as the "interdisciplinary function that mediates between the domains of the acquirer and supplier to establish and maintain the requirements to be met by the system . . . [and] . . . is concerned with discovering, eliciting, developing, analyzing, determining verification methods, validating communicating, documenting, and managing requirements" [41].

1.7.1 Why We Need Requirements

It may not always be obvious why requirements are necessary in the first place. An agreed-on set of requirements is essential, however, from a number of different perspectives.

From the point of view of the business management, the set of requirements provides a mechanism whereby they can agree that the organization's resources can be allocated to the project that has been established to bring the system into being. Unless the requirement set is well-defined, the project has no firm base, and the resources allocated to it will be inappropriate. Worse still, without sufficient oversight, the business may be held ransom by stakeholders insisting on unrealistic requirements that business management has not approved.

From the project manager's perspective, the set of requirements is an essential part of the definition of the scope of the project—it is from this scope that a suitable system will be developed. Scope management is one of the 10 major functions of a project manager [42], and a well-defined scope (set of requirements) is necessary to able to justify any expenditure of funds or effort within the project, to be able to report adequately on the progress of work, and to be able, eventually, to determine when the project is complete. Clearly then, a well-defined, universally agreed-on set of requirements is necessary at the outset of the project phase, increment, iteration, spiral, or sprint.

From the systems engineering perspective, the requirements set instantiated in the SyRS represents the transition from the business world into the systems engineering and engineering worlds. In this way, the SyRS defines what an acceptable technical solution needs to be capable of and communicates this information to both technical and nontechnical stakeholders. It should also be emphasized that it is not possible to rectify a poor set of requirements with good design, good engineering, or good production/process control—particularly if that set of requirements is the basis of a fixed-price contract between the acquirer and the contractor. Consequently, a project can rarely recover from poor definition, regardless of how much good design work is performed subsequently.

1.7.2 Why We Need Requirements Engineering

An appropriate requirements-engineering process allows us to guarantee that we begin the project with a set of requirements that is complete, consistent, comprehensible, and feasible and able to be validated. A useful process ensures that all stakeholders have had input and that the various points of view have been reconciled. To do that, we must therefore be able to trade off functionality for cost—which implies that we understand the required functionality, priorities, and costs. We also need to be able to use the process to manage changes in requirements for a wide variety of reasons.

Requirements engineering focuses on what is needed by the system, rather than how it is to be designed. The cost of developing requirements is estimated to be 10–15% of the total system cost, yet it is estimated that it costs some 200 times more to rectify an error in requirements during utilization than it does during early conceptual design [43]. As discussed in Section 1.7.1, poor requirements definition cannot be rectified by good design, so it invariably follows that rigorous development of requirements is essential for the successful system—requirements engineering provides this rigor through the application of a formal methodology. When requirements-engineering processes are ad hoc, the end result is almost always an unsatisfactory product or a canceled project [44]. Unfortunately, few organizations have a well-defined requirements-engineering process, and there are few, if any, standard methodologies that assist. Chapters 2 to 8 provide a useful methodology.

1.8 Benefits of Requirements Engineering

There are a number of potential benefits from the successful implementation of processes and methodologies associated with systems engineering and requirements engineering.

The first and most visible benefit is the potential for saving money during all phases of the system life cycle—the so-called life-cycle cost (LCC) savings. While some may argue that the additional effort required can increase costs, these increases are small (10–15%) and are generally felt in the very early design phases. If applied appropriately, systems engineering and requirements engineering can ensure that the savings achieved far outweigh the cost of implementing appropriate procedures and methodologies. Experience indicates that an early emphasis on systems engineering, and requirements engineering can result in significant cost savings later in the construction and/or production, operational use and system support, and disposal stages of the life cycle [45]. Systems engineering and requirements engineering should also assist in reducing the overall schedule associated with bringing the system into service.

Perhaps the most persuasive evidence is provided by data from NASA that shows that projects that expended the industry average of 2–3% of total project cost/effort on the requirements process throughout the life cycle experienced an 80–200% cost overrun, while projects that invested 8–14% of total project cost/effort in the requirements process had 0–50% cost overruns [46].

Requirements engineering ensures that the user requirements are accurately reflected in the design of the system, helping to minimize costly and time-consuming changes later in the life cycle. If changes are required, they can be incorporated early in the design and in a controlled manner. The rigorous consideration and evaluation of feasible design alternatives during the design phases of the project promote greater design maturity earlier. Approximately, 60–70% of errors in system development originate in the requirements analysis process [47], but some 80% are discovered at systems integration or later with a rework cost as high as 300–1,000 times the cost of in-phase correction [48]. To that end, the maximum opportunity to reduce the total life cycle cost of a system is presented in the acquisition phase. Poor requirements cannot be rectified by good design; rigorous development of requirements is essential for the acquisition to be successful.

System failures, cost overruns, and schedule problems are often the direct result of poor requirements-engineering practices. The systems engineering discipline aims to put in place a rigorous process of requirements engineering to produce well-defined requirements and adequate levels of traceability between the different levels of technical design documentation back to requirements that are unambiguous, complete, verifiable, and consistent.

Figure 1.19 illustrates the impact of systems engineering and requirements engineering on the system life cycle. For system developments that are more complex, the profile may be more complex than Figure 1.19, especially when the development spans many years and when early production systems are used to refine the functionality and performance of later versions. Note that systems engineering and requirements engineering have the greatest impact through the rigorous application of processes and methodologies during the early stages of the project where the ease of change and cost of modification is the lowest. In fact, the curve in Figure 1.19 could be relabeled as "the ease with which changes can be made" throughout the system life cycle. Consequently, requirements engineering provides the ideal opportunity to have the greatest impact on a project at a time when changes are easiest to make. Additionally, as illustrated in Figure 1.20, the greatest impact of requirements engineering comes at a time when the cost of implementing changes is the lowest.

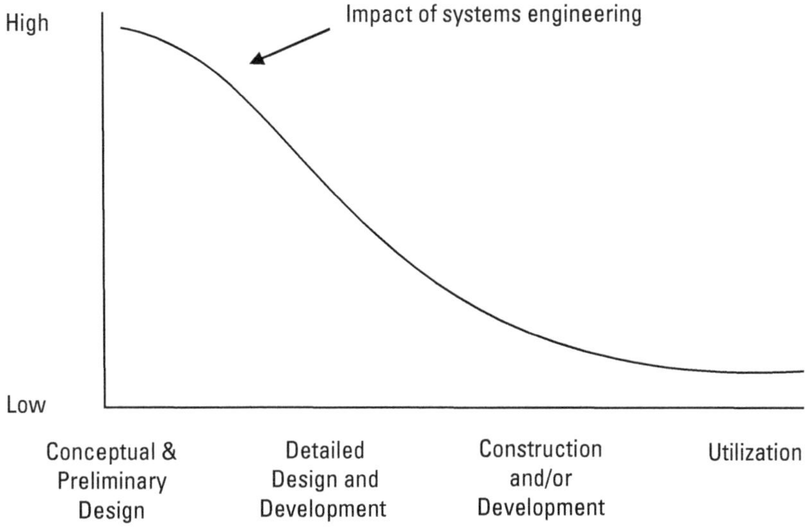

Figure 1.19 Impact of systems engineering and requirements engineering on the system life cycle.

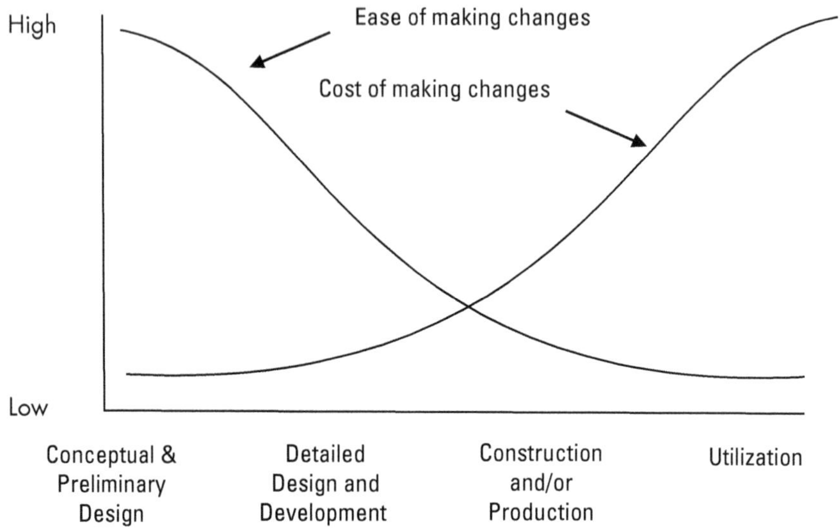

Figure 1.20 Ease of, and cost of, making changes throughout the system life cycle.

To quantify the cost of errors in requirements, Figure 1.21 (based on [49] and [50]) illustrates the relative cost to fix a defect in different phases of the life cycle of a software development project. A unit cost is assigned to the effort to fix an error in requirements during the requirements elicitation/generation phase. Note that the cost to fix an error in requirements during utilization is some 200 times greater than the cost of addressing the source of the same error in early conceptual design.

1.9 Development Approaches

It should be noted that the generic system life cycle illustrated in Figure 1.13 shows the phases and activities in sequence and is not intended to represent any particular development or acquisition model. Throughout this book we describe systems engineering and requirements engineering without discussing in great detail the development and acquisition context within which those endeavors might be undertaken. In order to explain the activities and artifacts of systems engineering, we have assumed that life cycle activities are undertaken sequentially. In doing so, we have assumed an approach that is generally referred to as the waterfall approach to system development. There are, however a number of other (more suitable) development approaches to implementing the activities of the system life cycle in Figure 1.13—such as the incremental, spiral, or evolutionary acquisition models [51], each of which has strengths and weaknesses depending on the nature of the system under development. The selection of a suitable development approach is a critical activity early in a system life cycle.

For efficiency in this text, however, the waterfall approach to system development is assumed in order to provide a logical, sequential flow of activities and deliverables, which simplifies the explanation of systems engineering and requirements engineering. Additionally, the waterfall approach is generally considered to be the basic building block upon which alternative approaches such as incremental, evolutionary, spiral, and Agile development are built [52, 53]. A solid understanding of waterfall development is therefore important, regardless of which other development methodology is employed.

1.10 Verification and Validation (V&V) and T&E

The entire systems engineering process aims to produce a system that is both verified against the descriptions produced during the systems engineering process and validated against the original needs, goals, and objectives that initiated the system development in the first case. Often these two associated aims are combined into the term V&V. However, this is misleading, and the practice is discouraged [54] since the two activities have quite different purposes. Verification ensures that we have "built the system right," and validation assures that we have also "built the right system" [55]. A well-managed approach to T&E aims to support the delivery of a system that is both verified and validated. Other systems engineering management functions, including

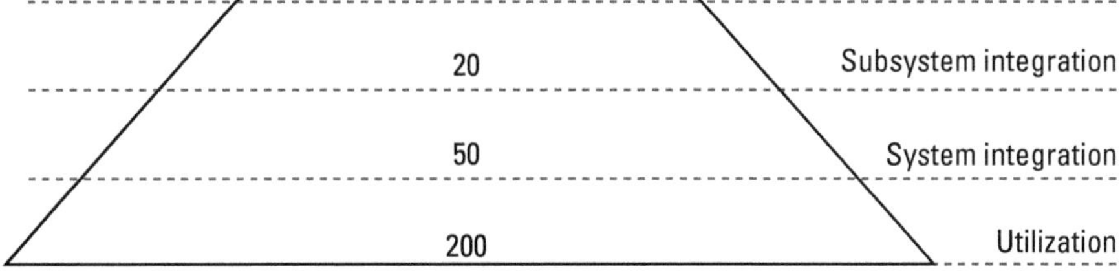

Figure 1.21 Relative cost to fix a defect at different phases in the life cycle of a software development project.

technical reviews and configuration audits, also support the overall objectives of verification and validation.

Three major categories of T&E that are applied to coincide broadly with the verification and validation required in the activities of the acquisition phase, the transition between the acquisition and utilization phases, and the retirement phase are described as follows.

- *Developmental test and evaluation (DT&E):* DT&E refers to the T&E activities undertaken during the acquisition phase to support the design and development effort. DT&E activities may also occur during the utilization phase to support activities such as the development of modifications. DT&E mostly supports verification activities to assure that the system element being developed is meeting, or has met, the design requirements of that system element.
- *AT&E:* AT&E represents the formal verification conducted to enable the acquirer to verify that the system meets the SyRS and to accept it from the developer. AT&E effectively forms the boundary or transition between the acquisition phase and the utilization phase and, as such, is an important contribution to FQR/SVR, which verifies that the delivered system meets the SyRS.
- *OT&E:* Once the system is accepted from the developer, OT&E may be conducted under realistic operational conditions by operational personnel in order to validate the capability system—that is, the major equipment combined with operational users, collective training, facilities, support, and so on. OT&E is normally conducted for a period following acceptance of the system by the acquirer from the supplier (that is, after AT&E) and to validate that the system meets its BNR/SNR before introduction into service. OT&E is also conducted continuously throughout the system life cycle in order to determine continued fitness for purpose and to identify the need for any modifications and upgrades.

1.11 The Purpose of This Book

Given the importance of requirements engineering during the early activities of the life cycle, particularly during conceptual design, this book introduces a requirements-engineering (business-analysis) methodology for the logical design (conceptual design). In particular, the focus is on the logical process required to develop the functional architecture (as contained in the BNR, SNR, and SyRS). While CDM is the focus of the remainder of the book, Chapter 2 provides a summary of the extant body of knowledge of requirements practice (requirements engineering and business analysis).

1.12 Some Qualifications Before Continuing

Now that this introductory chapter has provided useful background and explained the focus of the remainder of the book, before continuing we need to be clear about what the remaining chapters are *not* going to do (and why).

A project, by definition, produces a unique output. So, while there are general principles involved in the conduct of projects, the constituent activities of an individual project must be tailored into a relatively unique combination for that project to produce the unique output. There are, therefore, arguably as many ways to implement a particular project as there are unique systems to be developed. We cannot possibly address here the many combinations of factors associated with complex technical projects, and it would not be useful to attempt to do so, because any benefit of an introductory overview would be lost. So, to navigate the multidimensionality of the project space in a useful manner, we take a direct path and make the following simplifications, which should be kept in mind when reading Chapters 2 to 9:

- Most organizations involved in systems development will incorporate a number of development methodologies (perhaps at least a dozen) as part of their systems engineering and project management policies and procedures. Then, as part of the early project management activities, organizational practice requires a project manager to justify the selection of an appropriate methodology as part of their development of a project management plan. Here we use the linear sequential development methodology (commonly known as waterfall development) as a framework. We then provide a description of other major development methodologies (i.e., incremental, iterative, evolutionary, spiral, and Agile). There are two main reasons for taking this approach in an introductory text; they are described as follows:
  - First, the various elements of systems engineering and requirements practice are much easier to understand on first contact if they are explained sequentially—the following chapters are therefore structured that way. However, just because we have explained the processes sequentially does not mean at all that we suggest that the steps must be implemented that way. For each unique project

  output, an appropriate project development methodology must be chosen, tailored, and applied.
  - Second, even when a more elegant development methodology is applied, an understanding of the logical steps is essential—that is, it doesn't matter whether the project team is currently in a phase in waterfall, an increment in incremental, a build in evolutionary, a spiral in spiral, or a sprint in Agile; the steps we take the reader through here still have to be applied to the fundamental time period (phase, increment, build, spiral, or sprint) of the selected approach.
- In any system development, there will be a number of what most modern standards call acquirer-supplier arrangements. These are also called acquirer-developer, customer-contractor, and tasking-performing activity arrangements, among a number of other terms. It does not matter whether the development is conducted, as are most government acquisitions, through a major formal tendering/contracting process or, as many in-house developments are, undertaken between two or more groups internal to the organization, or in any other arrangement. In all cases (albeit slightly less formally for in-house development) it is essential that the acquirer-supplier arrangement is clear and unambiguous and that there are agreements for factors such as milestones, deliverables, budget, schedule, review and audit, and verification and validation. This remains true (albeit less formally), even for short development periods such as sprints in Agile development methodologies. Consequently, since there can be as many unique contracting arrangements as there are projects bringing unique systems into being, we focus our explanations on the most formal of those arrangements as may be implemented between a large government acquirer and a major supplier. Once the reader is familiar with the issues that drive formal developments, it is relatively straightforward to apply the general principles to less formal arrangements, such as those guiding in-house developments.
- One of the most dynamic areas in modern systems engineering is in digital engineering (DE) and the associated model-based systems engineering (MBSE). Here we focus on project artifacts that could be documents, spreadsheets, or databases and on tools that could be word processors, requirements management tools (RMTs), or models in MBSE. For example, we say that system requirements are recorded in the SyRS; while we use the word "specification," which might be

construed to imply document, we do not use the term in that way, and expect that the SyRS will be recorded in any form suitable to the organization—from a document to a spreadsheet, a database, a model, or any other digital form, or some combination. So, when we use the term "document" as a verb we mean to "record" as either a paper-based document, a database, or a model. We avoid direct reference to DE and MBSE in the remainder of the text for two reasons:

  - First, many projects still use paper-based documents to record project activities and will do so for years—or even decades—to come. In fact, we have worked with a number of major international companies that still use word processors to capture requirements, contracts, and design outputs. While digital engineering should be, and is being, embraced enthusiastically wherever possible, there is a substantial associated overhead that requires significant organizational investment. Additionally, with major government contracts that can span a decade or more, many projects will remain in the nondigital or part-digital world for some considerable time.
  - Second, although the form of the physical artifact changes between paper-based and digital environments, the logic of activities and relationships does not change—the same principles apply in both environments. This book, therefore, focuses on these principles and logical steps. For convenience, the book describes the resultant artifacts as a logical result of each stage, phase, and activity without proposing that those artifacts take any particular physical or digital form—that is a matter for the project and the organization it supports.

- Huge advances have been made in the 2023 revisions of ISO/IEC/IEEE 15288, the INCOSE *Systems Engineering Handbook* and its attendant manuals and guides to provide a common view of systems engineering in terms of life cycle processes, methodologies, tools, techniques, and terms. However, there are still many subtle (and sometimes not-so-subtle) differences in the way in which systems engineering is applied in organizations around the world. This book, therefore, aims to provide a balance to reflect the older terms and processes as well as the new, in the interest of making the content of the book as relevant to system developments that are just beginning as it is to those that have been in progress for a decade.
- In introducing CDM, the focus of the book is on the activities of the acquisition phase of the system life cycle; all life cycle phases and stages

are considered, but CDM concentrates on the activities in conceptual design that bring a system into being.

Review Questions

1. Briefly define/describe a system and what is meant by a SOI.
2. Requirements engineering applies to which of the following types of systems: open, closed, natural, human-made, human-modified, physical, conceptual, precedented, and unprecedented?
3. Briefly describe the differences between a subsystem, a system, and a SoS.
4. Briefly describe the differences between the problem domain (problem space) and the solution domain (solution space).
5. List the major elements that make up a capability system.
6. A system can be described in both logical and physical terms. Briefly explain why the two descriptions must coexist, what each description provides, and the relationship between the two descriptions.
7. Identify the major responsible parties (enterprise management, business management, project management, systems engineering, developer (contractor), and users/support) for the major phases of the life cycle.
8. Briefly describe the four major phases of the generic systems life cycle.
9. Briefly explain why the preacquisition phase is an essential precursor to the acquisition phase.
10. Briefly describe the activities in the acquisition phase and the utilization phase of the system life cycle (as proposed by Blanchard and Fabrycky).
11. List and briefly describe the major reviews that are held at the end of each of the activities in the acquisition phase.
12. Briefly describe the relevance to systems engineering of the following aspects: top-down approach, focus on life cycle, system optimization and balance, integration of disciplines and specialties, management, and focus on requirements engineering.
13. Briefly describe the three main categories of T&E: DT&E, AT&E, and OT&E.
14. List and briefly describe the 10 knowledge areas of the *Project Management Body of Knowledge.*
15. Briefly outline why we need requirements.

16. Briefly outline why we need requirements engineering.
17. Briefly outline the benefits of requirements engineering.

Endnotes

[1] International Institute of Business Analysis (IIBA), *A Guide to the Business Analysis Body of Knowledge® (BABOK® Guide)*, Toronto: IIBA, Version 3, 2015.

[2] ISO 55000, *Asset Management—Vocabulary, Overview and Principles*, 2024.

[3] Further reading on systems engineering can be obtained from the following major sources: Aslaksen, E. W., *Designing Complex Systems: Foundations of Design in the Functional Domain*, Boca Raton, FL: CRC Press, 2008.

Aslaksen, E. W., *The Systems Concept and Its Application to Engineering*, New York: Springer, 2013.

Aslaksen, E., and R. Belcher, *Systems Engineering*, Upper Saddle River, NJ: Prentice-Hall, 1992.

EIA-632, *Processes for Engineering a System*, Washington, DC: Electronic Industries Association (EIA), 2003.

Beam, W., *Systems Engineering: Architecture and Design*, New York: McGraw-Hill Book Co., 1990.

Blanchard, B. S., and J. E. Blyler, *Systems Engineering Management*, Hoboken, NJ: John Wiley & Sons, 2016.

Blanchard, B., and W. Fabrycky, *Systems Engineering and Analysis*, Upper Saddle River, NJ: Prentice-Hall, 2011.

Boardman, J., *Systems Engineering: An Introduction*, Upper Saddle River, NJ: Prentice-Hall, 1990.

Buede, D. M., and W. D. Miller, *The Engineering Design of Systems: Models and Methods*, New York: John Wiley & Sons, 2016.

Chestnut, H., *Systems Engineering Tools*, New York: John Wiley & Sons, 1965.

Chestnut, H., *Systems Engineering Methods*, New York: John Wiley & Sons, 1967.

Defense Systems Management College (DSMC), *Systems Engineering Management Guide*, Washington, DC: US Government Printing Office, 1990.

Dickerson, C. E., and D. N. Marvis, *Architecture and Principles of Systems Engineering*, Boca Raton, FL: CRC Press, 2010.

EIA/IS-632, *EIA Interim Standard—Systems Engineering*, Washington, DC: Electronic Industries Association, 1994.

De Weck, O. L., D. Roos, and C. L., Magee, *Engineering Systems*, Cambridge, MA: MIT Press, 2011.

Dori, D., *Model-Based Systems Engineering with OPM and SysML*, New York: Springer, 2016.

Douglass, B. P., *Agile Systems Engineering*, Birmingham, UK: Morgan Kaufmann, 2015.

Eisner, H., *Essentials of Project and Systems Engineering Management*, New York: John Wiley & Sons, 2008.

Faulconbridge, R. I., and M. J. Ryan, *Managing Complex Technical Projects*, Norwood, MA: Artech House, 2026.

Faulconbridge, R. I., and M. J. Ryan, *Applied Systems Engineering*, Norwood, MA: Artech House, 2026.

Forrester, J., *Principles of Systems*, Cambridge, MA: The MIT Press, 1968.

Gangl, M., et al., *Modern Methods of Systems Engineering: With an Introduction to Pattern and Model Based Methods*, Createspace Independent Publishing Platform, 2011.

Grady, J. O., *Systems Integration*, Boca Raton, FL: CRC Press, 1994.

Grady, J. O., *System Engineering Planning and Enterprise Identity*, Boca Raton, FL: CRC Press, 2010.

Haberfellner, R., et al., *Systems Engineering: Fundamentals and Applications*, Cham, Switzerland: Birkhauser, 2019.

Hall, A., *A Methodology for Systems Engineering*, Princeton, NJ: D. Van Nostrand, 1962.

Heap, G., J. Stanway, and A. Windsor, *A Structured Approach to Systems Development*, London: McGraw-Hill Book Company, 1992.

Hitchins, D., *Putting Systems to Work*, Chichester, UK: John Wiley & Sons, 1992.

Hitchins, D., *Advanced Systems Thinking, Engineering, and Management*, Norwood, MA: Artech House, 2003.

Hitchins, D., *Systems Engineering: A 21st Century Systems Methodology*, Chichester, UK: John Wiley & Sons, 2007.

Hubka, V., and W. Eder, *Theory of Technical Systems*, Berlin: Springer-Verlag, 1988.

Holt, J., *Systems Engineering Demystified: A Practitioner's Handbook for Developing Complex Systems Using a Model-Based Approach*, Birmingham, UK: Packt Publishing, 2021.

Holt, J., and S. Perry, *SysML for Systems Engineering: A Model-Based Approach*, London: Institution of Engineering and Technology, 2013.

Hunger, J., *Engineering the System Solution: A Practical Guide to Developing Systems*, Upper Saddle River, NJ: Prentice-Hall, 1995.

IEEE-STD-1220, *IEEE Standard for Application and Management of the Systems Engineering Process*, New York: IEEE Computer Society, 2005.

INCOSE, *Systems Engineering Handbook: A Guide for System Life cycle Processes and Activities,* D. D. Walden, et al. (eds.), INCOSE-TP-2003-002-05 2023, New York: John Wiley & Sons, 2023.

INCOSE, *Journal of the International Council on Systems Engineering*, Seattle, WA: International Council on Systems Engineering.

INCOSE, *Proceedings of Annual Conference*, Seattle, WA: International Council on Systems Engineering.

Jackson, S., *Systems Engineering for Commercial Aircraft*, Surrey, England: Ashgate, 2015.

Jamshidi, M., *Systems of Systems Engineering: Principles and Practice*, Boca Raton, FL: CRC Press, 2009.

Jenney, J., et al., *Modern Methods of Systems Engineering*, CreateSpace Independent Publishing Platform, 2011.

Kamrani, A. K., and M. Azimi, *Systems Engineering Tools and Methods*, Boca Raton, FL: CRC Press, 2011.

Khisty, C. J. and J. Mohammadi, *Fundamentals of Systems Engineering with Economics, Probability, and Statistics*, Upper Saddle River, NJ: Prentice-Hall, 2012.

Kossiakoff, A., et al., *Systems Engineering Principles and Practice*, Hoboken, NJ: John Wiley & Sons, 2020.

Lacy, J., *Systems Engineering Management: Achieving Total Quality*, New York: McGraw-Hill Book Co., 1992.

Larson, W., et al., *Applied Space Systems Engineering*, McGraw-Hill, 2009.

Machol, R., *System Engineering Handbook*, New York: McGraw-Hill, 1965.

Maier, M., and E. Rechtin, *The Art of Systems Architecting*, Boca Raton, FL: CRC Press, 2009.

Martin, J., *Systems Engineering Guidebook: A Process for Developing Systems and Products*, Boca Raton, FL: CRC Press, 1997.

MIL-STD-499B, *Military Standard—Systems Engineering—Draft*, Washington DC: US Department of Defense, 1994.

Micouin, P., *Model-Based Systems Engineering: Fundamentals and Methods*, London, John Wiley & Sons, 2014.

Parnell, G. S., P. J. Driscoll, and D. L. Henderson (eds.), *Decision Making in Systems Engineering and Management*, Hoboken, NJ: John Wiley & Sons, 2008.

Rechtin, E., *Systems Architecting: Creating and Building Complex Systems*, Upper Saddle River, NJ: Prentice-Hall, 1991.

Reilly, N., *Successful Systems Engineering for Engineers and Managers*, New York: Van Nostrand Reinhold, 1993.

Sage, A., *Decision Support Systems Engineering*, New York: John Wiley & Sons, 1991.

Sage, A., *Systems Engineering*, New York: John Wiley & Sons, 1992.

Sage, A., *Systems Management for Information Technology and Software Engineering*, New York: John Wiley & Sons, 1995.

Sage, A., and J. Armstrong, *Introduction to Systems Engineering*, New York: John Wiley & Sons, 2000.

Sage, A., and W. Rouse (eds.), *Handbook of Systems Engineering and Management*, New York: John Wiley & Sons, 2009.

Shishko, R., *ASA Systems Engineering Handbook*, Washington, DC: NASA, 1995.

Stevens, R., et al., *Systems Engineering: Coping with Complexity*, Hertfordshire, UK: Prentice Hall Europe, 1998.

Sydenham, P. H., *Systems Approach to Engineering Design*, Norwood, MA: Artech House, 2004.

Thome, B. (ed.), *Systems Engineering: Principles and Practice of Computer-Based Systems Engineering*, New York: John Wiley & Sons, 1993.

Truxal, J., *Introductory Systems Engineering Management*, New York: McGraw-Hill Book Co., 1972.

Wasson, C., *System Engineering Analysis, Design, and Development: Concepts, Principles, and Practices*, Hoboken, NJ: John Wiley & Sons, 2015.

Westerman, H., *Systems Engineering Principles and Practice*, Norwood, MA: Artech House, 2001.

Wymore, A., *Model-Based Systems Engineering*, Boca Raton, FL: CRC Press, 1993.

[4] ISO/IEC/IEEE 15288, *Systems and Software Engineering—System Life Cycle Processes*, 2023.

[5] Flood, R. L., and E. R. Carson, *Dealing With Complexity: An Introduction to the Theory and Application of Systems Science*, New York: Plenum Press, 1993.

[6] EIA-632, *Processes for Engineering a System*, Washington, DC: Electronic Industries Association (EIA), 2003.

[7] ISO/IEC/IEEE 15288, *Systems and Software Engineering—System Life Cycle Processes*, 2023.

[8] IEEE-STD-1220, *IEEE Standard for Application and Management of the Systems Engineering Process*, New York: IEEE Computer Society, 2005.

[9] ISO/IEC/IEEE 15288, *Systems and Software Engineering—System Life Cycle Processes*, 2023.

[10] This observation perhaps needs a little elaboration. When we have a nut and a bolt and need to join them in a useful way, we do need to take a "systems approach" to the mature of the two elements and their interface requirements (just as we would in considering the interconnection of a number of subsystems in a system, or a number of systems in an SoS. However, taking the same view at any level of engineering design does not mean that any combination at any level is usefully called a "system." That is not universally agreed on in the systems engineering community; hence, the term SOI is used to apply to any elements that are interconnected. That is a defendable perspective below system level; however, it does rather fall down when defining a SoS, which, therefore, becomes any combination of elements at the next level up. Nonetheless, the nuance of the debate is not important here, and the remainder of this text is applicable regardless of the reader's predilection.

[11] Maier, M. W., "Architecting a Portfolio of Systems," *Systems Engineering*, Vol. 1, No. 4, 2019, pp. 1–13.

[12] ISO 55000, *Asset Management—Vocabulary, Overview and Principles*, 2024.

[13] INCOSE, *Systems Engineering Handbook: A Guide for System Life Cycle Processes and Activities* D. D. Walden, et al. (eds.), INCOSE-TP-2003-002-05, John Wiley & Sons, 2023.

[14] Blanchard, B., and W. Fabrycky, *Systems Engineering and Analysis*, Upper Saddle River, NJ: Prentice-Hall, 2011.

[15] MIL-STD-499B, *Military Standard—Systems Engineering—Draft*, Washington DC: US Department of Defense, 1994.

[16] EIA-632, *Processes for Engineering a System*, Washington, DC: Electronic Industries Association (EIA), 2003.

[17] MIL-STD-499B, *Military Standard—Systems Engineering—Draft*, Washington, DC: US Department of Defense, 1994.

[18] EIA-632, *Processes for Engineering a System*, Washington, DC: Electronic Industries Association (EIA), 2003.

[19] MIL-STD-499B, *Military Standard—Systems Engineering—Draft*, Washington DC: US Department of Defense, 1994.

[20] EIA-632, *Processes for Engineering a System*, Washington, DC: Electronic Industries Association (EIA), 2003.

[21] Blanchard, B., and W. Fabrycky, *Systems Engineering and Analysis*, Upper Saddle River, NJ: Prentice-Hall, 2011.

[22] INCOSE, *Systems Engineering Handbook: A Guide for System Life cycle Processes and Activities* D. D. Walden, et al. (eds.), INCOSE-TP-2003-002-05, John Wiley & Sons, 2023.

[23] ISO/IEC/IEEE 15288, *Systems and Software Engineering—System Life Cycle Processes*, 2023.

[24] ISO/IEC/IEEE 15288, *Systems and Software Engineering—System Life Cycle Processes*, 2023.

[25] ISO/IEC/IEEE 29148, *Systems and Software Engineering—Life Cycle Processes—Requirements Engineering*, 2018.

[26] INCOSE Fellows, *Systems Engineering and Systems Definitions*, 2018.

[27] INCOSE, *Systems Engineering Handbook: A Guide for System Life cycle Processes and Activities* D. D. Walden, et al. (eds.), INCOSE-TP-2003-002-05, Hoboken, NJ: John Wiley & Sons, 2023.

[28] Defense Systems Management College (DSMC), *Systems Engineering Management Guide*, Washington, DC: US Government Printing Office, 1990.

[29] IEEE-STD-1220, *IEEE Standard for Application and Management of the Systems Engineering Process*, New York: IEEE Computer Society, 2005.

[30] EIA/IS-632, *EIA Interim Standard—Systems Engineering*, Washington DC: Electronic Industries Association, 1994.

[31] SECMM-95-01, *Systems Engineering Capability Maturity Model*, Version 1.1, Carnegie Mellon University, Pittsburgh, PA: Software Engineering Institute, 1995.

[32] Lake, J., "Unraveling the Systems Engineering Lexicon," *Proceedings of the INCOSE Symposium*, 1996.

[33] EIA-632, *Processes for Engineering a System*, Washington, DC: Electronic Industries Association (EIA), 2003.

[34] Goldsmith, R. F., *Discovering Real Business Requirements for Software Project Success*, Norwood, MA: Artech House, 2004, p. 1.

[35] Project Management Institute Standards Committee, *A Guide to the Project Management Body of Knowledge*, Upper Darby, PA: Project Management Institute, Version 6, 2017.

[36] Project Management Institute Standards Committee, *A Guide to the Project Management Body of Knowledge*, Upper Darby, PA: Project Management Institute, Version 6, 2017.

[37] Verner, J., et al., "Requirements Engineering and Software Project Success: An Industrial Survey in Australian and the U.S.," *Australasian Journal of Information Systems*, Vol. 13, No. 1, September 2005.

[38] Andriole, S., "The Politics of Requirements Management," *IEEE Software*, November/December 1998, pp. 82–84.

[39] Dorfman, M., and R. H. Thayer, *System and Software Requirements Engineering*, IEEE Computer Society Press, 1990.

[40] Kotonya, G., and I. Sommerville, *Requirements Engineering*, Chichester, UK: John Wiley & Sons, 2000.

[41] ISO/IEC/IEEE 29148, *Systems and Software Engineering—Lifecycle Processes—Requirements Engineering*, 2018.

[42] Project Management Institute Standards Committee, *A Guide to the Project Management Body of Knowledge*, Upper Darby, PA: Project Management Institute, Version 6, 2017.

[43] Davis, A.M., *Software Requirements: Objects, Functions and States*, Englewood Cliffs, NJ: Prentice Hall, 1993.

[44] Verner, J., et al., "Requirements Engineering and Software Project Success: An Industrial Survey in Australia and the U.S.," *Australasian Journal of Information Systems*, Vol. 13, No. 1, September 2005.

[45] Blanchard, B., and W. Fabrycky, *Systems Engineering and Analysis*, Upper Saddle River, NJ: Prentice-Hall, 2011.

[46] Hooks, I. F., and K. A. Farry, *Customer-Centered Products: Creating Successful Products Through Smart Requirements Management*, New York: AMACOM, 2001. (Reported in: Young, R., *The Requirements Engineering Handbook*, Norwood, MA, Artech House, 2004.)

Goldenson, D., and J. Elm, *The Business Case for Systems Engineering Study: Results of the Systems Engineering Effectiveness Survey*, Software Engineering Institute, 2012. Honour, E., *Systems Engineering Return on Investment*, Ph.D. thesis, University of South Australia, 2013.

[47] Weinberg, J., *Quality Software Management, Volume 4: Anticipating Change*, Dorset House, 1997.

[48] Feiler, P., et al., *Four Pillars for Improving the Quality of Safety-Critical Software-Reliant Systems*, SEI, 2013.

[49] Boehm, B., *Software Engineering Economics*, Englewood Cliffs, NJ: Prentice Hall, 1981.

[50] Davis, A. M., *Software Requirements: Objects, Functions and States*, Englewood Cliffs, NJ: Prentice Hall, 1993.

[51] Further detail on these models can be found in the following:

Blanchard, B., and W. Fabrycky, *Systems Engineering and Analysis*, Upper Saddle River, NJ: Prentice-Hall, 2011.

Boehm, B., "A Spiral Model of Software Development and Enhancement," *Computer*, Vol. 21, No. 5, May 1988.

Davis, A. M., E. H. Bersoff, and E. R. Comer, "A Strategy for Comparing Alternative Software Development Life cycle Models," *IEEE Transactions on Software Engineering*, Vol. 14, No. 10, October 1988, pp. 1453–1461.

Manifesto for Agile Software Development, 2001, http://agilemanifesto.org.

Martin, J., *Systems Engineering Guidebook: A Process for Developing Systems and Products*, Boca Raton, FL: CRC Press, 1997.

Project Management Institute, *Agile Practice Guide*, 2017.

Royce, W. W., "Managing the Development of Large Software Systems: Concepts and Techniques," *WESCON Technical Papers*, Vol. 14, August 1970, pp. A/1-1–A/1-9.

Young, R., *The Requirements Engineering Handbook*, Norwood, MA, Artech House, 2004, pp. 73–78.

[52] Forsberg, K., and H. Mooz, *Systems Engineering Overview*, Center for Systems Management, Cupertino, CA, 1996.

[53] Forsberg, K., and H. Mooz, "Application of the 'VEE' to Incremental and Evolutionary Development," *Proceedings of the INCOSE Conference*, St. Louis, MO, July 1995.

[54] Wheatcraft, L. S., M. J. Ryan, and T. E. Katz, *Needs and Requirements Manual: Needs, Requirements, Verification, Validation Across the Life Cycle*, Hoboken, NJ: John Wiley & Sons, 2024.

[55] Boehm, B. W., "Verifying and Validating Software Requirements and Design Specifications," *IEEE Software*, Vol. 1, No. 1, January, 1984, pp. 75–88.

2

Requirements Practice Framework

2.1 Introduction

As discussed in Chapter 1, systems engineering (particularly during the early processes of conceptual design and preliminary design) has a focus on gathering requirements in a formal systematic way, which is referred to generically as requirements engineering or business analysis, and, here, collectively called requirements practice. Before considering each of the major systems engineering processes in some detail in subsequent chapters, this chapter provides an overview of the extant requirements-practice body of knowledge, because it is an essential aspect of system development. For more detail, readers may consult many excellent references [1].

Section 2.2 makes the distinction between concepts, needs, and requirements. However, throughout this chapter and the remainder of the book, we will often use the single term "requirements" to mean both needs and requirements so as to avoid some tortuous combinations. For example, when we use the traditional terms requirements practice, requirements engineering, requirements analysis, and requirements elicitation, we mean for those terms to apply to both needs and requirements.

2.2 Concepts, Needs, and Requirements

When describing a system throughout its life cycle, we make the distinction between concepts, needs, and requirements. Concepts are typically narrative descriptions of ways in which the organization and its entities expect to manage, acquire, develop, deploy, operate, support, and retire the capability offered by the system. Needs are formal representations of expectations for an entity stated in the language and perspective of stakeholders. Needs are transformed into requirements, which are formal representations (model elements or textual statements) that communicate what an entity must do, and how well it must do it, to realize the intent of the needs. There may be more than one need for any given concept and more than one requirement defined for any need.

As illustrated in Figure 2.1, concepts, needs, and requirements exist at a number of levels corresponding to four major views: an enterprise view, in which enterprise leadership sets the enterprise strategies and concepts of operations; a strategic view, in which business management derives business concepts and needs and formalizes their requirements; an operational view, in which stakeholders in the business units define their concepts, needs, and requirements; and a systems view, in which system designers define the system, first in functional and then physical terms.

Because terms such as system and subsystem (system element) are level-specific (recall that we are following a systems approach using a systems paradigm as discussed in Chapter 1), we use the generic term entity which is, ". . . a single thing to which a need or requirement refers: an enterprise, business unit, project, system, or system element (which could be a product, process, human, or organization)" [2].

As shown in Figure 2.1 (based on [3]), the enterprise, strategic, and operational views are in the problem domain; the system view is in the solution domain. Recall that the problem domain is the responsibility of those who have ownership of the problem to be solved, so the descriptions of the system are predominantly in the language of the customer, focusing on what they need the system to be able to do, how well it should be done, and why. As discussed in Section 1.2.3, these descriptions are called functional (or often logical) descriptions. On the other hand, the solution domain is the responsibility of the developer (the performing activity) implementing the solution, most commonly a contractor, so the physical descriptions of the system in that domain are predominantly in engineering and physical terms, focusing on how the problem is to be solved—that is, what it will look like and how it will operate once it has been implemented. In many large organizations, there is a soft boundary between the problem domain and the solution domain in that

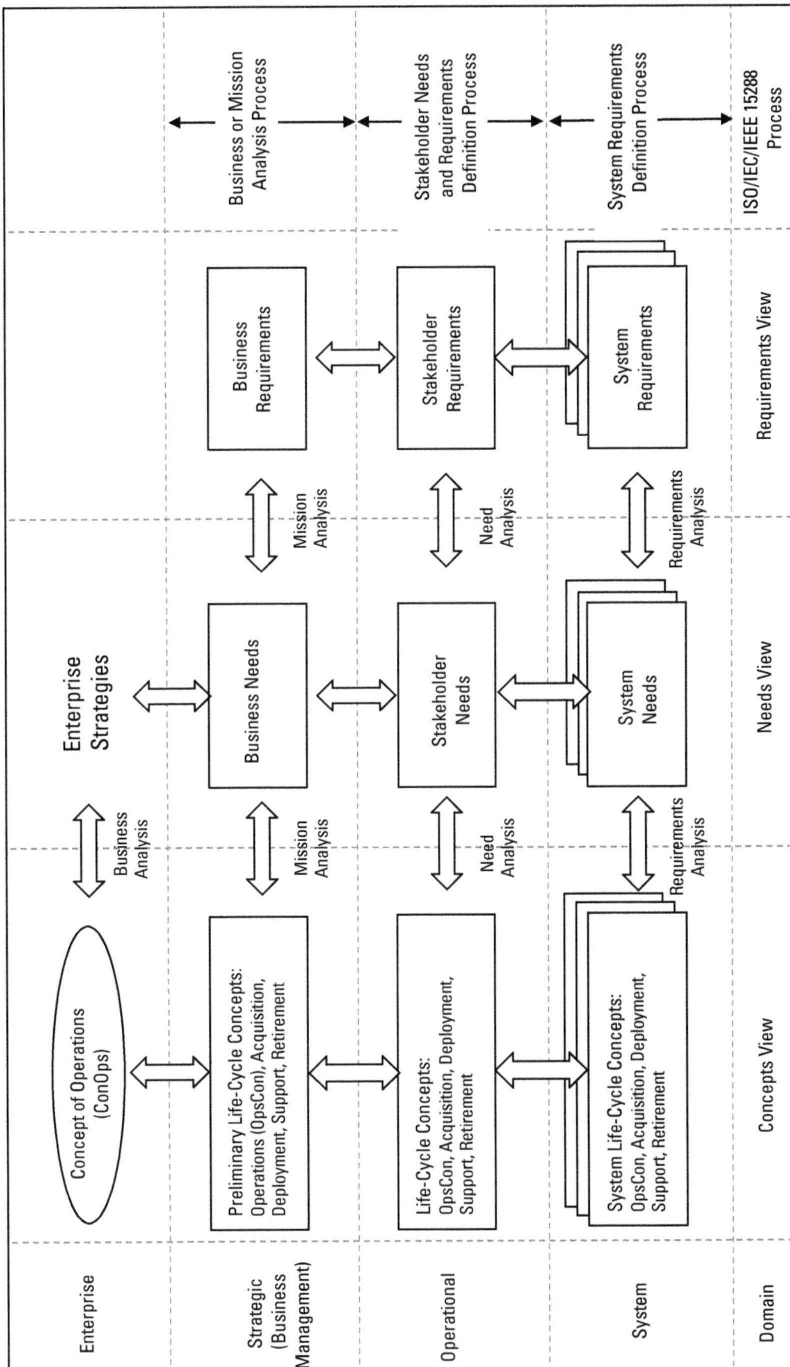

Figure 2.1 Domains, views, and the ISO/IEC/IEEE 15288 processes associated with the definition of concepts, needs, and requirements.

the customer (the acquirer or tasking activity) prepares a draft SyRS, which is taken to tender, responded to be tenderers, negotiated with the successful tenderer (the contractor), and then revised to become the SyRS that represents the selected solution of that contractor.

Although they coexist at each level of consideration, concepts, needs and requirements differ in the following ways [4, 5]:

- A concept is a written or graphic representation that concisely expresses how an entity will satisfy the problem it was defined to address. A number of concepts can apply to each entity, so it is useful to develop a minimal set of concepts that define the expectations of all entities. It is also useful to provide some structure to the set of concepts. As illustrated in Figure 2.1, life cycle concepts include operations concepts, acquisition concepts, deployment concepts, support concepts, and retirement (disposal) concepts.
- A need is the result of a formal transformation of one or more concepts into an agreed-to expectation for an entity to perform some function or possess some quality (within specified constraints with acceptable risk). Needs are generally capabilities stated as business needs in the language of business management, as stakeholder needs in the language of stakeholders at the business operations level, or as needs of the system. There can be more than one need defined for any one concept, but a need cannot exist without a parent concept.
- A requirement is the result of a formal transformation of one or more needs or parent requirements into an agreed-to obligation for an entity to perform some function or possess some quality (within specified constraints with acceptable risk). Requirements are formal statements that are structured and can be validated. Requirements are elaborated from needs through a process of requirements analysis, which is also called business analysis or mission analysis [6] at the highest level of application. There can be more than one requirement defined for any one of the needs, but a requirement cannot exist without a parent need.

Life-cycle concepts, needs, and requirements should be developed for entities at all levels of the organization. The activities associated with the development and transformations shown in Figure 2.2 [7] illustrate the relationships among concepts, needs, and requirements based on the INCOSE *Needs and Requirements Manual* (NRM) [8] and the INCOSE *Guide to Needs and Requirements* (GtWR) [9], both of which contain more detail.

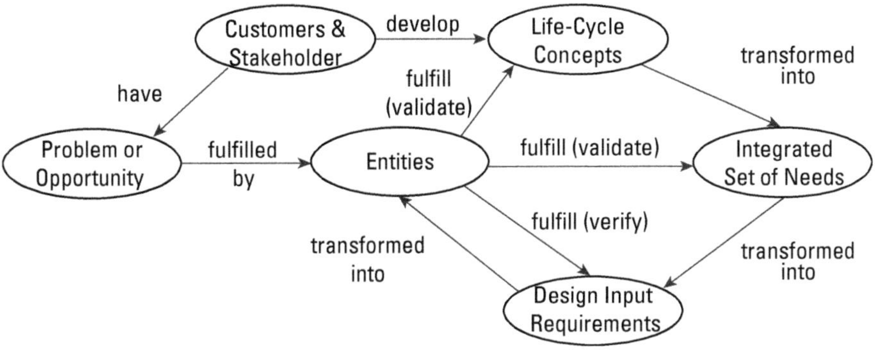

Figure 2.2 Entity-relationship diagram for customers, concepts, entities, needs, and requirements terms.

At the highest level, there is an enterprise view in which enterprise leadership sets business vision and strategic goals and objectives in the form of a concept of operations (ConOps), which is also called a strategic business plan (SBP). At the next two levels, there are the dual views: At the business management level there are business needs [10] and business requirements [11]; and at the business operations level there are stakeholder needs [12] and stakeholder requirements [13]. The transformation of needs into requirements occurs at each level. Here we are interested in the three major upper-level processes, described as follows:

- *BNR definition process:* At the strategic level, business management begins the process with the guidance contained in the ConOps, from which they define business needs, largely in the form of preliminary life cycle concepts (PLCs), which capture the preliminary acquisition concept, preliminary operational concept (OpsCon) [14], preliminary deployment concept, preliminary support concept, and preliminary retirement concept. Business needs are elaborated and formalized into business requirements, which are recorded in the business requirements specification (BRS) [15], which is also called the business requirement document [16].
- *SNR definition process:* Using the ConOps, the preliminary OpsCon, and the other PLCs as guidance, stakeholders from the business operations level develop stakeholder needs—in the form of a refined OpsCon and other life-cycle concepts (LCs)—and transform them into a formal set of stakeholder requirements, which are recorded in the stakeholder

requirement specification (StRS) [17], also called the integrated set of needs in the *NRM*.

- *System requirements definition process:* The requirements in the StRS are then transformed by requirements engineers into system requirements, which are contained in the SyRS [18]. This artifact is also referred to as the solution requirement specification [19], or simply the system specification [20]. A number of SyRS may be derived from the StRS—that is, perhaps, an SyRS is produced for each element of the capability described by the StRS (see Figure 2.3). Some organizations may also prepare individual LCs for each of a number of elements of capability that are developed to meet the business needs. The process continues for the system elements.

It should be noted that, in the past, almost every standard, textbook, and academic paper would have a slightly different spin on systems engineering terms, language, and processes. However, with the revision of ISO/IEC/IEEE 15288 in 2023 and the subsequent revision of the INCOSE *Systems Engineering Handbook* and the associated INCOSE manuals and guides, the language and approaches are gradually beginning to converge. The description above is consistent with ISO/IEC/IEEE 15288, the INCOSE *Systems Engineering Handbook*, and ISO/IEC/IEEE 29148 [21] with the exception that the business or mission analysis process in ISO/IEC/IEEE 15288 is called the BNR definition process in this text (the subordinate activities in the process are consistent).

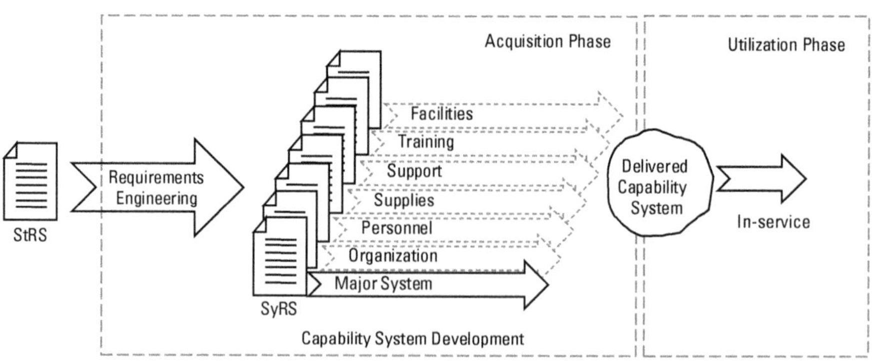

Figure 2.3 An example in which one SyRS is developed for each of the elements of the capability system described by the StRS.

Additionally, for practical implementation, the INCOSE *NRM* [22] combines what ISO/IEC/IEEE 15288 calls the BNR and SNR into an integrated set of needs, which is then transformed into design-input requirements (which ISO/IEC/IEEE 29148 calls the SyRS). Despite a preference for the structure of the *NRM*, we conform in this text to ISO/IEC/IEEE 15288 and ISO/IEC/IEEE 29148 and the INCOSE *Systems Engineering Handbook*, because systems engineers are certified against those documents, and many organizations use them to provide the frameworks for their projects (especially those projects that are already in progress).

In any case, it is of little consequence because the reader can readily make the translation; ISO/IEC/IEEE 15288's BNR and SNR are combined in the *NRM*'s integrated set of needs, and ISO/IEC/IEEE 29148's SyRS is called the design-input specification in the *NRM*. To assist throughout the remaining chapters, we occasionally use both sets of terms to reinforce the duality.

To help visualize the correlation, Figure 2.4 illustrates the relationship between the artifacts called for by ISO/IEC/IEEE 15288 and ISO/IEC/IEEE 29148, the relevant processes from ISO/IEC/IEEE 15288, and the artifacts from INCOSE's *NRM*.

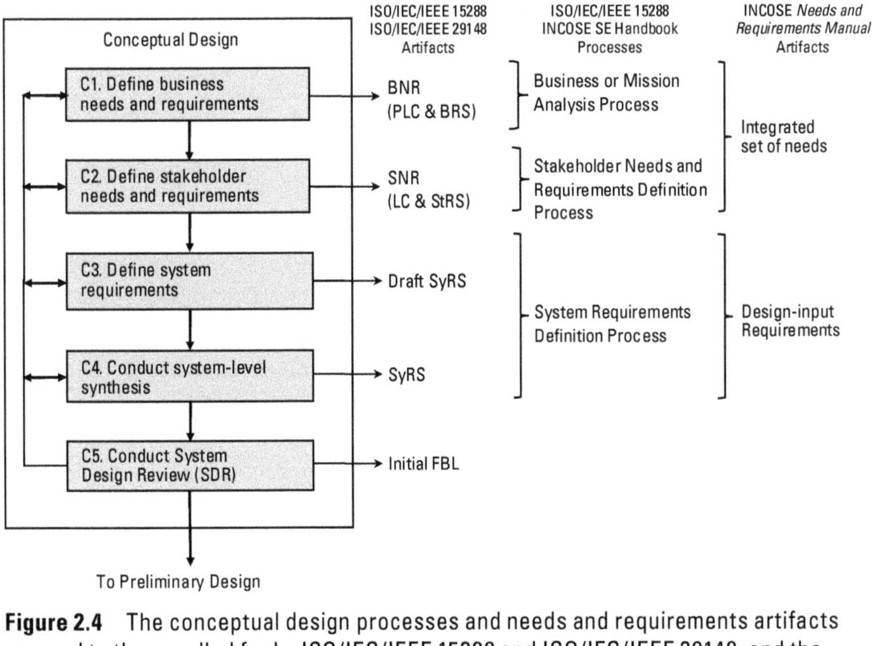

Figure 2.4 The conceptual design processes and needs and requirements artifacts mapped to those called for by ISO/IEC/IEEE 15288 and ISO/IEC/IEEE 29148, and the artifacts from INCOSE's *NRM*.

ANSI/AIAA G-043A-2012 [23] identifies that the terms "concept of operations" and "operational concept" are often used interchangeably but notes that an important distinction exists, because each has a separate purpose and is used to meet different ends. It is essential that these terms are used so that they are consistent with ANSI/AIAA G-043A-2012 and ISO/IEC/IEEE 29148 [24], as well as the way in which the terms are used by the US DoD and many other defense departments.

ISO/IEC/IEEE 29148 describes the ConOps as follows:

> The ConOps, at the organization level, addresses the leadership's intended way of operating the organization. It may refer to the use of one or more systems, as black boxes, to forward the organization's goals and objectives. The ConOps document describes the organization's assumptions or intent in regard to an overall operation or series of operations of the business with using the system to be developed, existing systems, and possible future systems. This document is frequently embodied in long-range strategic plans and annual operational plans. The ConOps document serves as a basis for the organization to direct the overall characteristics of the future business and systems, for the project to understand its background, and for the users of this International Standard to implement the stakeholder requirements elicitation.

ISO/IEC/IEEE 29148 describes the OpsCon as follows:

> A system OpsCon document describes what the system will do (not how it will do it) and why (rationale). An OpsCon is a user-oriented document that describes system characteristics of the to-be-delivered system from the user's viewpoint. The OpsCon document is used to communicate overall quantitative and qualitative system characteristics to the acquirer, user, supplier, and other organizational elements.

The ConOps is developed at the enterprise level and provides the context for the OpsCon, which is prepared by business management at the strategic level. Business management begins with the preparation of the preliminary OpsCon, which summarizes the needs of the system within its business context—the business needs. The preliminary OpsCon is then elaborated and refined into the OpsCon by engagement with stakeholders at the business operations level during the stakeholder needs and requirements definition process—the OpsCon therefore contains business needs and stakeholder needs.

The operational concept in the OpsCon, however, is just one of the concepts required to address the stakeholder needs across the system life cycle. Four more are required. They are described as follows:

- The acquisition concept describes the way the system will be acquired including aspects such as stakeholder engagement, requirements definition, solicitation and contracting issues, design, production, and verification.
- The deployment concept describes the way the system will be validated, delivered, and introduced into operations.
- The support concept describes the desired support infrastructure and manpower considerations for supporting the system after it is deployed. A support concept addresses operating support, engineering support, maintenance support, supply support, and training support.
- The retirement concept describes the way the system will be removed from operation and retired, including the disposal of any hazardous materials used in or resulting from the process.

PLCs are prepared by business management and contain the business needs defined as part of the business needs and requirements definition process. PLCs are elaborated and refined in much more detail in the LCs by stakeholders at the business operations level in the stakeholder needs and requirements definition process.

2.2.1 What Is a Requirement?

We should begin our consideration of requirements practice with a number of definitions of a requirement from the major related standards, listed as follows:

- MIL-STD-499B [25]: "Characteristics that identify the accomplishment levels needed to achieve specific objectives for a given set of conditions."
- IEEE-STD-1220 [26]: "A statement that identifies a product or process operational, logical, or design characteristic or constraint, which is unambiguous, testable or measurable, and necessary for product or process acceptability (by consumers or internal quality assurance guidelines)."

- EIA-632 [27]: "Something that governs what, how well, and under what conditions a product will achieve a given purpose."
- IEEE-610.12: "A condition or capability that must be met or possessed by a system or system component to satisfy a contract, standard, specification, or other formally imposed documents."
- *A Guide to the Business Analysis Body of Knowledge*® (*BABOK*® *Guide*) [28]: "A condition or capability that must be met or possessed by a solution or solution component to satisfy a contract, standard, specification, or other formally imposed documents."
- ISO/IEC/IEEE 29148 [29]: "A statement which translates or expresses a need and its associated constraints and conditions.'"

Perhaps more usefully, for the purposes of this book, a requirement can be defined as a statement of the following:

- The services and functions that the system should provide, the things it should do, or some action it should take (often called functional requirements);
- The qualities, properties, or attributes that the system must possess (often called nonfunctional requirements);
- Any restrictions or bounds under which the system should operate, or on the way in which the system is to be developed (also called constraints and sometimes also included in the category of nonfunctional requirements).

Each requirement statement may be supported by the following:

- Performance, verification statements supporting each requirement;
- Attributes supporting each requirement to facilitate its management;
- Definitions of other systems with which the system must integrate and to which it must interface;
- Information about the application domain within which the system must operate.

The requirements in the BRS and the StRS (the integrated set of needs) are not as formal as those in the SyRS (the design input requirements), but they have the same structure and follow the same rules as presented in the remainder of this chapter and in Chapter 9.

As well as these broad categories, an adjective is often included to describe the nature of the requirement. So, reference is often made to operational requirements, stakeholder requirements, environmental requirements, interface requirements, system requirements, regulatory requirements, safety requirements, design requirements, and many other categories of requirements.

In this text we simply consider requirements, since the selection of categories is relatively artificial. All other groupings are used loosely to describe various categories of requirement that are not part of a structured approach to writing a requirements document but are often convenient perspectives to take of the various subsets of requirements.

2.2.2 Requirements Characteristics and Attributes

As it is gathered or elaborated (regardless of how that is done), each requirement is to be examined to ensure that it has a number of important characteristics. Further, the single statement of each requirement is not sufficient on its own. To support analysis and management, each requirement should be recorded with associated attributes; the combination of the requirement statement; the associated performance, verification and rationale statements; and the associated attributes results in a requirement expression. This section examines the characteristics of a good requirement statement and the attributes that should be associated with each requirement and included in the requirement expression.

2.2.2.1 Requirements Characteristics

Each requirement statement should exhibit the following characteristics (based on [30, 31]):

- *Necessary:* The requirement must be necessary in that the system could not function in the desired way without this requirement. It also follows that a requirement cannot be necessary if it cannot be traced back to a higher-level requirement (and the associated need).
- *Singular:* A requirement cannot be a combination of two or more requirements—that is, the requirements set must be normalized so that requirements do not overlap. This also implies that the requirement is concise and clear.
- *Correct:* That the requirement must be correct is axiomatic—an incorrect requirement will result in a system that does not meet the real needs.
- *Unambiguous:* All readers of the requirement should come to the same understanding of what is meant by the requirement—there can be only one interpretation. This is difficult when the requirement is written in

natural language, particularly English, where a single statement can have more than one meaning.

- *Feasible:* A requirement must be achievable using extant design and engineering processes, extant technologies, and extant manufacturing procedures. A requirement should also be achievable within organizational constraints such as budget and schedule unless we take care to be unambiguous.
- *Appropriate-to-level:* Each requirement must be appropriate to the level at which it is stated. For example, a system-level requirement must be independent of the method of implementation (process) or physical solution—that is, it should say what is required, rather than how it should be implemented.
- *Complete:* The requirement must stand alone and not need any further explanation.
- *Conforming:* Each requirement must conform to a standard formal structure as defined by the organization, or perhaps in accordance with an external standard.
- *Verifiable:* Each requirement must be verifiable, in that it can be proved that the system meets or possesses the requirement. Verification may be by one or more means.

ISO/IEC/IEEE 29148 [32] and the INCOSE *Guide to Writing Requirements* [33] provide more detail on the characteristics of individual requirements.

2.2.2.2 Requirements Attributes

So far, we have been discussing requirement statements. However, a requirement is more than just a requirement statement. The full expression of a requirement includes associated attributes that aid in the management of the requirement. (The attributes may not be visible with the requirement statement but may be appended to it in the requirement document or in an accompanying or hosting database.) So, we can define a requirement expression: "A requirement expression comprises a requirement statement and a set of associated attributes" [34].

An attribute is ". . . additional information included with a requirement statement, which is used to aid in the management of that requirement" [35]. A number of sources define attributes [36, 37]. The most recent and most authoritative is the INCOSE *NRM*, which defines some 46 attributes, organized within four broad categories [38]:

- *Attributes to help define the requirement and its intent.* Examples include: rationale, trace to parent, trace to source, states and modes, and allocation/budgeting.
- *Attributes associated with the system verification or system validation.* Examples include: system verification or system validation success criteria, system verification or system validation strategy, system verification or system validation method, system verification or system validation responsible organization, system verification or system validation level, system or system validation verification phase, condition of use, system verification or system validation results, and system verification or system validation status.
- *Attributes to help manage the needs and requirements.* Examples include: unique identifier, unique name, originator/author, date requirement entered, owner, stakeholders, configuration management authority, change proposed, version number, approval date, date of last change, stability/volatility, responsible person, need or requirement verification status, need or requirement validation status, status, status of implementation, trace to interface definition, trace to dependent peer requirements, priority, criticality or essentiality, risk of implementation, risk mitigation, key driving need or key driving requirement, additional comments, and type/category.
- *Attributes to show applicability and allow reuse.* Examples include: applicability, region, country, state/province, application, market segment, business unit, and business line.

There is not sufficient space to discuss fully here a taxonomy for attributes nor for how such attributes could be defined and used. Here, we simply note the following 18 mandatory attributes from the INCOSE *NRM* [39]:

- *A1 Rationale:* As each requirement is elicited/elaborated, a rationale must be attached to record the reason for the requirement's existence. Rationale is essential to guide subsequent design.
- *A2 Trace to Parent:* Requirements at one level are implemented at the next level of the system architecture via allocation. A child requirement is one that is an elaboration, derivation, or decomposition of an allocated parent. The achievement of each of the child requirements is necessary to the achievement of the parent requirement.

- *A3 Trace to Source:* There are likely to be a number of sources of requirements. The originator (person, organization, document, or process) must be recorded to ensure that all requirements can be justified. Additionally, sources may need to be consulted when negotiation and compromises are required during requirements analysis.
- *A5 Allocation/Budgeting:* Requirements at one level are allocated or budgeted to parts of the architecture for implementation. There are two types of allocation, described as follows.
  - Allocation (budgeting) of performance, resources, quantity, or quality from one level to the next level (such as performance, power, mass, reliability, accuracy, precision, and quality);
  - Allocation of responsibility, where requirements at one level are assigned to subsystems and system elements at the next level for implementation.
- *A6 System Verification or System Validation Success Criteria:* The criteria that must be met to provide objective evidence that the system has successfully met the need or requirement.
- *A7 System Verification or System Validation Strategy:* A high-level description of the approach to be used to verify that the system meets a design input requirement or to validate the system meets a need.
- *A8 System Verification or System Validation Method:* The method to be used (test, demonstration, inspection, or analysis) to provide objective evidence.
- *A9 System Verification or System Validation Responsible Organization:* The organization that is responsible for system verification and system validation for a need or requirement or sets of needs and requirements for the system.
- *A15 Unique Identifier:* Each requirement must be able to be identified as a unique statement. Consequently, when a requirement is recorded, it must be allocated a unique identifier that remains with the requirement throughout the design process. This is essential for requirements management and for requirements traceability, in particular. Although managing the allocation of unique numbers is difficult in a paper-based system, the process is straightforward when conducted within the database underlying an automated requirements management tool. Each requirement also has performance and verification statements associated with it. While these require separate identifiers, it is useful to use a system whereby the association between the three types of requirements is immediately obvious. One recommended system uses the same stem number for all three types of requirements and

appends the letters P and V to denote the performance and verification statements associated with the requirement statement. See Chapters 8 and 9 for more details.

- *A18 Originator/Author:* The person defining and entering a need or requirement expressions into the needs/requirements database.
- *A19 Owner:* The person or organizational element that maintains the need or requirement, has the right to say something about this need or requirement, approves changes to the need or requirement, and reports the status of the need or requirement.
- *A28 Need or Requirement Verification Status:* An indication that a need or requirement has been verified.
- *A29 Need or Requirement Validation Status:* An indication that the need or requirement has been validated.
- *A32 Trace to Interface Definition:* This attribute provides a link tracing the interface requirements to where the interaction referenced in the requirements are defined.
- *A33 Trace to Dependent Peer Requirements:* This attribute links requirements that are related to, or dependent on, each other (other than parent-child) at the same level. The relationship with other requirements must be identified to assist in forward and backward traceability.
- *A34 Priority:* Requirements are often stated in terms of their importance to individual stakeholders. The priority of each requirement must therefore be captured with the requirement so that requirements engineers (and designers subsequently) have some design space within which to work during requirements analysis and negotiation. This priority value is allocated from a business management or stakeholder perspective and indicates how important the achievement of each requirement is in the context of the overall system. Stakeholders are tempted to state all requirements as having the highest priority (mostly for the fear that the lower priority requirements will be subsequently discarded), but they should be encouraged to allow some flexibility within which design can be conducted. The most important (mandatory) requirements are often stated as shall; the next most important are stated as shall, where practicable; the next most important can be stated as preferred or should; and the next most important can be stated as may. These levels are often simplified to a three-level (mandatory, highly desirable, desirable) or a two-level (mandatory, optional) scheme. When initially collecting requirements, it may be sufficient to ask stakeholders to rate the requirement's priority as high, medium, or low. A more detailed system could ask them to rate priority on a scale of

1 to 10, which would allow some numerical analysis of priority during subsequent design, if that is desirable. In baseline specifications, priority can therefore be allocated to the requirement in two ways—either in the text of the statement using shall, should, and may; or all statements could be stated using a term such as "shall" and an indication of priority could be appended as an attribute in a numerical form or in the form of high, medium, or low. The latter method is preferred as it avoids errors that may creep in during retyping the requirement when the priority is changed.

- *A35 Criticality or Essentiality:* It is also useful to ask the business management and stakeholders to identify those requirements that are critical in terms of their contribution to system performance. While seemingly related to priority, this category identifies show-stoppers—those requirements that are essential to the success of the overall system. In other words, a requirement that is deemed critical must be met or possessed by the system or else the system cannot be introduced into service. In some cases, it may be useful to have a rating system, but it is most often sufficient to identify each requirement in a binary fashion as critical or noncritical.
- *A36 Risk of Implementation:* Each requirement contributes to overall system risk in its own way and the management of risk is a critical component of systems engineering and project management. It is useful to consider the risk associated with each requirement as it is collected, either by allocating a risk rating (1–10, for example) or by writing an associated risk statement.
- *A37 Risk Mitigation:* This attribute indicates whether a need or requirement is part of a risk mitigation action.

2.2.3 What Not How

In general, business, stakeholder, and system requirements should be statements of what a system should do, rather than how it should do it. However, for a number of reasons, this is often too simplistic in practice. First, it may be essential that the system performs a function in a particular way (such as for interoperability, or to meet extant standards). We could also specify a solution to meet broader concepts such as support and commonality with existing solutions. Additionally, a specific statement is often less confusing than an abstract statement of the problem. A specific statement is therefore the best guidance that a conceptual designer can give a lower level designer. Further, for good business reasons, business management may well dictate not only what is required but also how it is to be implemented. Finally, the people who specify the system are often the domain experts—they are therefore best placed to state how the system should be developed and how it should operate. Still, unless one of those reasons prevails, we should ensure that the BRS, StRS, and SyRS provide logical specifications (because they are in the problem domain), not physical specifications (which belong in the solution domain).

2.2.4 Emergent Properties

Not all properties of a system are exhibited separately by its constituent elements. That is, the system may have properties that only exist when the system elements are integrated; called the emergent properties of a system, they are described as follows:

- Those that are possessed by the system as a whole;
- Those that only emerge after the individual system elements have been integrated;
- Those that cannot be exhibited by elements of the system in isolation;
- Those that depend on interactions among system elements, including their environment.

Examples of emergent properties are the "-ilities," such as maintainability, reliability, availability, and usability. Other examples include safety, security, comfort, and ease of use. Emergent properties must therefore be defined from the top down—they cannot be obtained by specifying requirements for individual elements and then hoping that the desired properties will exist on integration.

Perhaps the bicycle is one of the best examples of a system that possesses emergent properties, since it can only perform its principal function when all system elements, including the rider, have been integrated. If any one of the major elements is removed, the system cannot function.

We also must consider the possibility of the completed system exhibiting emergent properties that were not considered as part of its design and that, in some cases, may be undesirable in the end result.

2.3 Requirements Elicitation and Elaboration

Requirements elicitation and elaboration involve working with business managers and stakeholders to investigate the problem to be solved and to identify their needs and requirements. At the beginning of this chapter, we considered the following three levels of requirement statement:

- Business needs and business requirements in the BRS;
- Stakeholder needs and stakeholder requirements in the StRS;
- System requirements in the SyRS.

We then discussed requirements practice by referring to "requirements" in general terms. Before discussing in subsequent chapters how these requirements are part of systems engineering, it is useful here to discuss how the design moves through those three levels of requirement statement. In general, requirements in the BRS, StRS, and SyRS, outlined as follows, are described hierarchically through the processes of elicitation and elaboration [40].

- *Elicited requirements:* Elicited requirements can be attributed to the source (business manager or stakeholder), either directly or through negotiation. These requirements are normally gathered via interview or a structured workshop. The source of an elicited requirement will be a relevant stakeholder need.
- *Elaborated requirements:* Elaboration is used to move between the levels in the requirements hierarchy—the process is also often referred to as requirements flowdown. The source of an elaborated requirement will be a parent need or parent requirement. Elaborated requirements are of the following two types:
  - *Decomposed requirements:* Elicited requirements tend to be at a high level because they were obtained from those directly involved in business management and business operations supported by the system of interest. Decomposition entails breaking a higher level requirement into those lower level requirements that are explicitly required by it. For example, if an aircraft needs to be able to land on a certain class of airfield, then it must be able to land within the appropriate runway length, adhere to strict mass-related constraints, be able to interface to the appropriate air traffic control systems, and so on. In decomposition, the need for the child requirements is obvious in the parent. An elicited requirement can be decomposed into two or more subordinate requirements whose total meaning is equivalent to that of the original requirement. It should be noted that the original requirement normally becomes redundant and is replaced by the lower level requirements. The essence of the upper level requirement often becomes a heading or is noted

in the rationale for the decomposed requirements, although it is good practice to retain the hierarchical structure of the RBS by abstracting the original requirement to a heading in the structure.

- *Derived requirements:* In addition to being high-level, elicited requirements also tend to be from the business management and operations perspectives and will not necessarily address all aspects of a systems design. Derivation entails requirements engineers drawing some inference from the higher level requirements to obtain lower level statements. That is, business management or the stakeholders did not state the requirement directly, but the derived requirement is a necessary part of the system design if one or more of the directly stated requirements are to be met. Like decomposed requirements, derived requirements replace the original parent requirement.

The use of elaboration (decomposition and derivation) is discussed in more detail in Chapter 5, which includes examples of each activity.

Business, stakeholder, and system requirements are not collated randomly; requirements analysis is necessary for, among other things, elaborating a statement of business and stakeholder need through a systematic exposition of what the system must do in order to meet that need [41]. Gathering, analyzing, and elaborating requirements is an essential step to ensure that the system, once built, meets business needs and stakeholder needs [42]. High-level requirements are elaborated further into a number of lower level requirements, which are then grouped, along with other elaborated requirements, through the requirements analysis process [43] into the business management–level descriptions in the BRS, the business operations–level descriptions in the StRS, and the system-level description in the SyRS.

The requirements in the SyRS are then further elaborated into lower level requirements, which are then rearranged from the logical groupings of the SyRS into physical groupings of requirements (the CIs). The CIs are documented in a number of development (subsystem) specifications, which then define "how" the system will be implemented. The mapping of the logical groupings of requirements in the SyRS into the physical groupings in the CIs in the development (subsystem) specifications informs the definition of project deliverables.

For well-behaved systems, the requirements in each of the BRS, StRS, SyRS, and system-element specifications are normally arranged in a hierarchical fashion (albeit grouped logically in the first three and physically in the last). This hierarchy is useful because it allows the system to be decomposed into statements of a complexity that can be more easily be understood and communicated, as well as managed within the limitations of humans and our organizations.

A hierarchical description of a system is also useful in that it supports requirements traceability, which is an essential element of effective systems engineering as well as project management. Traceability is assisted by a hierarchical numbering of requirements. As illustrated in Figure 2.5, Requirements 1.1.1, 1.1.2, and 1.1.3 are child requirements of parent Requirement 1.1, which is a child of Requirement 1. For Requirements 1.1.1, 1.1.2, and 1.1.3 to be children of parent Requirement 1.1, they must be derived or decomposed from it—that is, Requirement 1.1 is a superset of Requirements 1.1.1, 1.1.2, and 1.1.3; and Requirement 1 is a superset of Requirements 1.1 and Requirements 1.2.

Elicitation involves engagement with the stakeholders in order to establish their requirements for the new system. There are a number of techniques for this engagement, such as facilitated structured workshops, brainstorming, interviews, surveys, questionnaires, process models, use cases and user scenarios, simulations, prototypes, observation of work studies (time and motion studies), participation in work activities, observation of the system's organizational and political environment, technical artifact review, benchmarking, competitive system assessment, reverse engineering, and market analysis. (See Section 2.4.3 for more detail.)

2.3.1 Elicitation Dimensions

Requirements elicitation involves engagement with the stakeholders in order to establish their requirements for the new system. There are a number of techniques for this engagement, such as facilitated structured workshops,

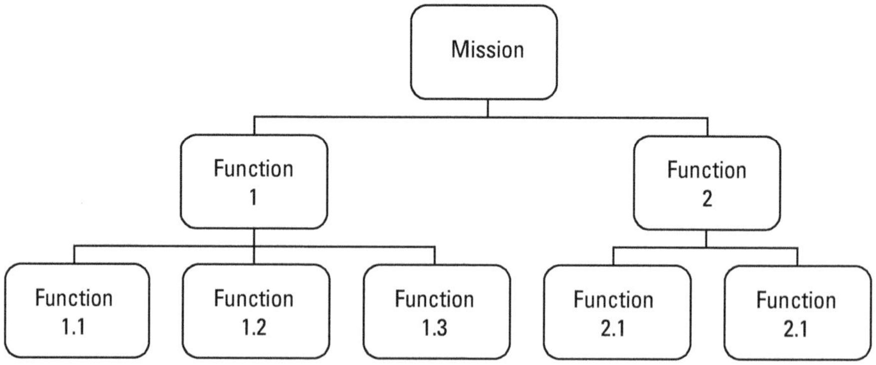

Figure 2.5 The hierarchical elaboration of requirements.

brainstorming, interviews, surveys, questionnaires, process models, use cases and user scenarios, simulations, prototypes, observation of work studies (time and motion studies), participation in work activities, observation of the system's organizational and political environment, technical documentation review, benchmarking, competitive system assessment, reverse engineering, and market analysis. (See Section 2.5 for more detail.)

Requirements elicitation is not at all straightforward. It is not just a matter of "collecting" requirements—that is, simply writing down the requirements as they are gathered from stakeholders. Requirements engineers must be more than simple scribes in this process—they must be able to add value to what the stakeholders say. There are, therefore, seven dimensions to requirements elicitation [44], described as follows:

- *Understanding the business:* Requirement engineers must understand how the new system contributes to the business or organization. Doing so requires them to understand the needs of business management and the nature of the business and those aspects of it that will have an impact on the new system.
- *Understanding the application domain:* Requirement engineers must also understand the general application domain within the business in which the new system is to be implemented.
- *Understanding the specific problem:* Understanding of the general application domain must be focused further, so requirements engineers must also understand the specific problem addressed by the system.
- *Understanding the needs and constraints of system stakeholders:* The specific needs and constraints of stakeholders must be understood by requirements engineers, as there may be several different sets of criteria by which the final system is judged. Additionally, the specific context within which stakeholders describe their requirements must also be understood, as must the language they use, their predilections, and their biases. Further, the need to trade off various requirements relies on an understanding of the needs of stakeholders.
- *Understanding acquisition and project management:* Requirements engineers are heavily involved in defining the system requirements, which are, in effect, the system scope upon which project managers develop the project WBS (and the consequent schedule and budget). Additionally, the SyRS developed by requirements engineers becomes the centerpiece of the statement of work underpinning the contract signed with the supplier or developer. Requirements engineers must therefore understand how systems are acquired and projects are managed.

- *Understanding requirements engineering and systems engineering:* Requirements engineers stand on the boundary between the business and the project. As well as understanding the business environment in which the system will be deployed, the requirements engineer must also be able to translate that environment into the systems engineering and engineering domains. It is very difficult to contribute to a process that is not understood. Those undertaking requirements elicitation must therefore understand, and preferably have significant experience with, systems engineering and requirements engineering.
- *Understanding the technologies and engineering involved:* Although it is common to assert that a good manager can manage in any field, even at a high level, a manager is not performing to full potential if the managed processes and contributing technologies are not understood. This is certainly true at the level of requirements elicitation, where the requirements engineer must understand the technologies involved as well as be familiar with their engineering.

For further detail, Young [45] provides an excellent coverage of the skills required, the characteristics, and the job description, of requirements analysts.

It should be noted that the understanding outlined here is very rare and is not naturally occurring—requirements engineers must be trained. There are two broad approaches: Engineers can be trained to become systems engineers and requirements engineers and then be provided with exposure to business and business practices; or business managers can be educated in technology, engineering, systems engineering, and then requirements engineering. While the latter is not impossible, the former is the most probable.

2.3.2 Difficulties in Eliciting and Elaborating Requirements

Unfortunately, needs and requirements elicitation is not a simple matter of transferring knowledge from stakeholders to conceptual designers, which means that the effective development of needs and requirements can be difficult and challenging. Stakeholders will probably not have given the new system much consideration and will not have had a chance to develop a clear view of their requirements—they may even be hearing about the system for the first time when contacted by conceptual designers requesting their input to the design. Even if they have an informed view, different stakeholders have conflicting requirements, they are not normally aware of the requirements-engineering process and of the acquisition process in general, they are often not willing to compromise their expectations to meet any budget, and they are rarely aware of the limitations that technology has on their expectations. This section briefly discusses these and other difficulties.

Instead of focusing on the logical architecture of the system, stakeholders sometimes identify particular physical solutions that appear to meet their requirements. That is, instead of concentrating on what the system is required to do, there is an ever-present urge to describe how the system should be designed or built. While this is dangerous because it can preclude effective solutions from consideration, requirements engineers can use the predilection to some advantage, because the stakeholder can at least visualize (and therefore even describe) something that meets their perceived needs.

Further, stakeholders sometimes become confused between a desire and a need. It is vital that system-level requirements are based on the true needs of the system and not merely a wish list of additional functionality. This reinforces the need for forward and backward traceability between the endorsed business needs and requirements and the SyRS. Additional (unendorsed) functionality may add complexity and expense onto the system without enhancing the system's ability to meet the true system-level requirements. This gold-plating of the requirements must be avoided.

Another difficulty is that stakeholders themselves do not always contribute positively to the requirements-engineering process. Some of the reasons for this are described as follows:

- Stakeholders usually do not really know what they need, do not know what questions need to be asked or answered, and have almost never considered the problem in terms of the detail that must be specified [46].
- Stakeholder expectations are often unrealistic.
- Stakeholders often have no desire to compromise their unrealistic demands when faced with cost constraints.
- Stakeholders do not understand the requirements-engineering process and often do not participate in reviews when required, or with the regularity required.
- Stakeholders will often not commit to a set of written requirements because they fear that they will then be committed to a particular solution; this is particularly evident when they either have a poor understanding of the problem or have had insufficient time to devote significant effort to its solution.
- Stakeholders often insist on modifying or adding new requirements after the requirements set has been endorsed and the cost and schedule have been fixed.

- Stakeholders are not normally willing to devote the time to understand properly the intricacies of the products and the technologies involved; nor are they normally interested in understanding the engineering, systems-engineering, and requirements-engineering processes involved in developing the requirements set.
- Different stakeholders often have different perspectives resulting in conflicting requirements, or at least similar requirements expressed differently. Even a single stakeholder may have different perspectives over time. Additionally, as key positions change, the new incumbent may have a significantly different view of requirements to that of the previous stakeholder representative.

Information on the application domain is not always collected in one place and may involve specialist terminology. It is highly unlikely in the development of large systems that any one stakeholder (or any single group of stakeholders, for that matter) is able to articulate all system-level requirements. Additionally, the language employed by stakeholders in most complex systems tends to be arcane to outsiders and even to other specialist stakeholders of the same system. Some considerable time may have to be invested by requirements engineers to be able to comprehend the domain knowledge of all stakeholders.

Requirements are often poorly stated because the real requirements are not known or understood. It cannot be assumed that the potential users of the system know what they need from the system—which is only to be expected since they may have never seen it or operated it. Requirements engineers must recognize this and try a number of different communication and elicitation techniques. As a corollary, stakeholders sometimes firmly believe that they know exactly what they need until they are presented with a system that has their specified functionality whereupon they realize that they do not require some of the features included and bemoan the lack of others. To overcome this, stakeholders must be presented with prototypes, models, and simulations as soon as possible to ensure that there is a strong correlation between the stakeholders' perceived requirements and the appropriate system requirement statements encapsulated in the SyRS. As mentioned earlier, the stakeholders' predilection for stating solutions in physical terms can be used to advantage by requirements engineers to confirm requirements through analogy by comparing stakeholders' stated needs with existing systems or prototypes that perform similar functions.

Conflicting requirements often result when two or more requirements are mutually exclusive such that satisfaction of one requirement will result in failing to meet another requirement. Conflicts must be resolved early in the requirements-engineering process, as they can be a source of ambiguity. Conflicts can also occur where the same requirement appears twice, is worded in slightly different ways, or requires different levels of performance for the same function. Every requirement must be unique and stated only once to avoid confusion and conflict.

A number of organizational and political factors, described as follows, also influence requirements elicitation and elaboration:

- Changes in economic and business environment are a constant factor in the work lives of most stakeholders, and they may initially show some understandable animosity to the suggestion that they should devote valuable time to contributing to yet another change in the way they conduct business.
- Even without considering the constant need to change, the stakeholders who best understand the problem to be solved are invariably not available to help specify requirements for the new system. Stakeholders in modern "right-sized" organizations are probably busier than they have ever been at any other point in time. The pressures of their daily workplace may preclude them from being sufficiently available to contribute to the requirements-engineering process.
- Organizations and systems involve people, and the human element is a strong influence on requirements engineering. Political, organizational, or personal bias is therefore often evident in early statements of requirement. Requirements engineering is not concerned with removing these factors, which will arguably always exist whenever humans are involved in the process. Bias does, however, need to be identified as early as possible to ensure that subsequent decisions are made on an informed basis, and the bias is formally recorded as a constraint.
- Additionally, it cannot be assumed that every stakeholder is a supporter of the new system. Some may prefer that available funds are expended on another project that is more important to them. Others may not agree that the system is necessary at all and consequently fail to contribute adequately to the requirements-engineering process.

The final challenge is to know when enough is enough, as it is difficult to judge when the requirements analysis effort is complete. Although experienced personnel are best placed to make this judgment, no two system developments are likely to be the same, and the stopping criteria will require careful consideration for each system design.

2.3.3 Techniques for Eliciting Requirements

Requirements engineers select an appropriate requirements elicitation methodology to suit the system being developed. Most requirements-engineering literature (see, for example, [47–54]) suggests techniques such as facilitated structured workshops, brainstorming, interviews, surveys, questionnaires, process models, use cases and user scenarios, simulations, prototypes, observation of work studies (time and motion studies), participation in work activities, observation of the system's organizational and political environment, technical documentation review, benchmarking, competitive system assessment, reverse engineering, and market analysis. The selection of technique is critical since inappropriate use can doom a project to failure [55]. Frameworks (see, for example, [56–57]) have been proposed to assist in selecting an appropriate requirements-engineering technique for a particular project.

Sections 2.3.3.1 to 2.3.3.12 describe each of the major requirements elicitation/elaboration techniques in more detail.

2.3.3.1 Facilitated Structured Workshops

Perhaps the fastest (and most efficient) way to elicit requirements is through structured requirement workshops. It must be remembered that workshops are part of the iterative cycle of requirements development—the top-down approach greatly assists here in allowing participants to focus on appropriate levels of abstraction (level of detail). The iterative cycle does not start with a blank sheet of paper, however. A workshop should always start with a draft set of requirements and other early conceptual design artifacts, particularly the visual artifacts we describe at the end of this chapter. These visual representations (and other drawings and diagrams) are very useful as discussion and communication tools as well as requirements analysis tools. If possible, the set of draft upper level artifacts should be sent out to participants before the workshop to allow them to prepare. It is therefore essential that a first iteration of requirements engineering is conducted before the workshop is held. Conducting a workshop that is not informed by some well-focused preliminary documents normally leaves stakeholders with a poor opinion of the new system even before conceptual design is well under way. Another way to ensure that stakeholders make valuable contributions is to conduct a period of briefing/training in requirements elicitation before beginning the workshop.

Significant thought must be given to the administration and the mechanics of the conduct of the workshop. As mentioned earlier, it is essential that the workshop is not only structured but also facilitated to avoid serious digression. The location, resources, and other logistical considerations must be taken into account to ensure that the workshop runs smoothly; the inability to run a simple workshop without problem is not likely to give stakeholders confidence that their new system is in good hands. A popular way of avoiding workplace distractions is to hold the events at a convenient off-site location where stakeholders can devote their attention to articulating their needs. There must also be a scribe nominated to record the design as it is iterated.

The main disadvantage of workshops is the additional travel costs that may be incurred in gathering stakeholders in the one location and the difficulty associated with getting many busy people in the same place at the same time. Technologies that facilitate remote access (such as videoconferencing) may assist in reducing costs.

2.3.3.2 Brainstorming Sessions

Brainstorming sessions have a similar format to structured workshops but are considerably less formal. It may well be worth conducting a brainstorming session with a small number of selected key stakeholders and designers before running a structured workshop as a means of obtaining the first iteration of the requirements engineering products. Brainstorming is particularly sensitive to the nature of the people involved as well as their number—the value of the exercise tends to diminish as the number of participants increases. As with a structured workshop, brainstorming sessions should be facilitated and can usefully be held off site to allow participants to focus on the outcomes without the distractions of daily work.

2.3.3.3 Interviews, Surveys, and Questionnaires

Interviews, surveys, and questionnaires offer a formal process through which requirements may be garnered from individual stakeholders. Their use should be considered carefully, however, in that useful information is only provided in response to focused, well-balanced questions, which implies that a measure of requirements engineering has been conducted before the interviews are held or before the surveys and questionnaires are sent out. All three processes can also be very time-consuming, which again requires that careful thought should be given to the questions to ensure that responses add to an understanding of the system's requirements, rather than add more noise to the requirements-definition process.

In some respects, interviews should be one-on-one structured workshops. Similar preparation is required—the interview should never be approached cold by either the requirements engineer or the stakeholder being interviewed. Interviews can be effective but are very time-consuming, and a structured workshop may be preferred if there is a large number of stakeholders. As an exercise in good stakeholder management, however, it is generally advisable to interview at least the key stakeholders.

While there is no satisfactory replacement for the personal contact possible in an interview, there are occasions when surveys and questionnaires are useful requirements elicitation tools. Their utility is generally limited, however, to activities such as the collection of statistical data to reinforce or validate requirements collected by interview or workshop.

2.3.3.4 Use Cases and Operational Scenarios

One of the difficulties with asking stakeholders about their requirements for the new system is that stakeholders will probably have been trained on an existing system and may have difficulty in setting aside current operating procedures when considering requirements for the new system. Use cases and operational (or user) scenarios (sometimes also generically called storyboarding or user stories in Agile) are a simple but powerful way of assisting in providing stakeholders with a new frame of reference, particularly at the beginning of the requirements-elicitation process where they may have difficulty in breaking free from existing practices.

Additionally, stakeholders will find it difficult to understand requirements when they have been written as a series of simple "shall" statements. Use cases and scenarios are given in a narrative (and often visual) style to illustrate how the system will operate and how users will interact with it. Use cases and scenarios are, therefore, quite useful, because users often find it easiest to describe their view of the system in terms of examples of how it might be used.

For that reason, use cases provide a good framework within which to elicit requirements, as well as serving throughout the project as upper level descriptions of the required functionality. Use cases and scenarios also perform the important role of defining the purpose of the system, which is useful during acceptance testing to ensure that the delivered system is "fit-for-purpose"—the use cases developed at the start of development to assist in eliciting requirements then effectively become test cases at the end of the development to assist in verification/validation of the system.

To ensure a complete description of a system, it is important that the use cases and scenarios capture the exception conditions as well as the normal operational conditions. That is, users are often keen to describe the way in which the system works in terms of the positive aspects of its implementation, rather than describe what to do when the normal process is prevented from occurring due to events such as system failure, actions by a competitor or adversary, or simply operator error.

Use cases and use-case modeling, which are popular in software requirements engineering, are described in a number of excellent references [58].

2.3.3.5 Process Models

Process models provide descriptions of the major activities involved in a particular process as well as their sequencing. Process models are normally provided at a number of different levels and from a number of different perspectives to allow for an understanding at different levels of abstraction from coarse to fine resolution and to ensure that all views of the process are considered.

Process models are particularly well-suited to systems such as transaction-based systems where more formal methodologies such as entity-relation models can be used to define formal database structures. For requirements engineering during conceptual design of most other types of systems, however, process models are too low-level to be useful, and use cases are preferred.

2.3.3.6 Prototypes

Requirements elicitation can be enhanced by the ability of requirements engineers to mock up a system, or part of a system, that meets the collected requirements and then present that system to stakeholders for their consideration. Prototypes and simulations can be very complex, or they can be as simple as a series of role-playing exercises in which users are asked to describe how they would imagine the system operating under particular circumstances. At the highest level, use cases and operational scenarios could also be considered to be a form of paper-based prototyping (a form of virtual walkthrough) in which stakeholders describe their requirements in a range of typical operations for the system.

Simulations and prototypes can also be used for requirements validation, because they are a quite useful (visual) way of showing stakeholders how the designers have captured and interpreted their requirements.

Prototypes have two main forms, described as follows:

- Throwaway prototypes that are intended to elicit or validate requirements but not expected to form part of the developed system;
- Evolutionary prototypes that are intended to deliver a working system to the customer in incremental steps.

It is important to identify the purpose of the prototype. Although both forms are used to assist in eliciting and elaborating requirements, the nature of the prototype must be explained to the stakeholders when they are viewing it. If it is an evolutionary prototype, they can expect to see it, or elements of it, again. Care must be taken to explain that a throwaway prototype is just that, and they cannot expect to see it again. Dangers with throwaway prototypes include the following:

- Stakeholders who are detail-oriented often pay more attention to the structure of the prototype than to its operation. This is particularly true of software prototypes where stakeholders are sometimes blinded by the (often rudimentary) graphical user interface and cannot see past it to the functionality being demonstrated.
- Having seen the prototype in software, stakeholders sometimes find it difficult to understand why it will take so long to develop a system that has the functionality they have just witnessed. They struggle to see the difference between a mocked-up prototype version and a properly tested, robust, operational version. Again, this is particularly true of software prototypes—while a pilot would never expect to take away and fly in a cardboard cut-out prototype used to illustrate cockpit layout, stakeholders often see the functionality in a software prototype and somehow extrapolate it into a working system.
- Developers of code for the prototype are reluctant to throw it away and feel compelled to use it in the final system, despite the fact that, again, the code is not well-designed or properly tested.

2.3.3.7 Modeling and Simulation

Mathematical modeling and simulation are widely used and are becoming increasingly popular as computers and software become more capable. As the name suggests, mathematical models provide a mathematical representation of a physical system or design. Simulation then uses the models to determine both static and dynamic characteristics of the system under varying conditions. An example of a static model would be one that represents the structural characteristics of a bridge. Loads and stresses can be changed in magnitude, direction, and position to ensure that the structure will withstand the loads associated with its intended purpose. An example of a dynamic system is an aircraft's autopilot system, which is a classic example of a control system that can be used to determine characteristics such as stability and response times.

Models and simulations are excellent tools for performing what-if analyses and determining the limits of the design. Variables can be changed and conditions easily altered to allow designers to investigate the likely behavior of the design in varying conditions. Simulation is widely used for a variety of different synthesis tasks. Environmental design and human-machine interface

(HMI) design are two examples of synthesis tasks that can benefit from simulation. The fidelity of the simulation (and the models) must be defined and confirmed prior to their use as synthesis tools.

To build and use a mathematical model, designers need to define the problem and formulate the model to replicate the problem. The model can be investigated with pen and paper if desired, but the more common approach is to convert the model into a computer program for ease of use and speed. Once this is done, perhaps the most crucial step must be performed: The performance of the model must be validated to ensure that it is an accurate representation of the problem. It is pointless to collect data from a model if the model is incorrect (that is, it is not representative of the problem). Once the model has been built and validated, designers can experiment with it via simulation and collect important design data.

2.3.3.8 Analysis of Similar Systems

In a similar vein to the use of prototypes, it may be useful to assist stakeholders to investigate their requirements by asking them to describe what they like and don't like about an existing system that is similar in functionality (even if only in part) to the system under consideration. At the very least, requirements engineers will be able to develop an awareness of the stakeholders' level of understanding of the new system. Even if that is poor, requirements engineers should still gain an understanding of user interface issues and perhaps even of some required functionality in the new system. Additionally, requirements engineers may gain some insight into the new system by scanning problem reports from similar existing systems.

2.3.3.9 Observation of the Workplace

Although time-and-motion studies may not always be a suitable way of eliciting requirements (particularly at the systems level) an understanding of the work environment and the workplace context is essential to the development of stakeholder requirements. By observing stakeholders in their natural environment, requirements engineers may often identify requirements that the stakeholder had not stated directly through more formal requirements elicitation techniques such as interviews or workshops. When describing their workflow, stakeholders often assume skills, processes, and tools and do not describe them explicitly. They are also not always consciously aware of all their work practices that are relevant to the new or modified system.

If the workplace cannot be observed directly, key stakeholders can be selected to be permanent members of the requirements engineering team if it is considered that stakeholder input is continuously required. In some cases, stakeholder involvement may continue past the requirements stage and a stakeholder representative may be attached to the teams performing the lower level design to ensure that the workplace context is always taken into account.

2.3.3.10 Documentation Review

In addition to observing the workplace, requirements engineers may review relevant business and technical documents. Although few organizations have very many information-rich documents, many organizations have standard operating procedures (SOPs) that may detail how particular operations are conducted. Those documents may, therefore, be useful as alternate sources of information about the workplace. Care must be taken, however, because the documents may be good descriptions of the old system but not be appropriate for the new one.

2.3.3.11 Market Analysis

Many systems do not have a readily available, homogeneous set of stakeholders. Consumer products such as motor vehicles, for example, have a wide range of stakeholders whose requirements are many and varied, often with some conflict. Market analysis techniques may be useful in identifying a common subset of achievable requirements that maximize the potential market for the system.

2.3.3.12 Competitive System Assessment and Trials

As with prototypes and simulations, competitive system assessment and trials are aimed at providing stakeholders with some tangible manifestation of early requirements to assist in requirements elicitation and validation. One simple, quick way of providing depth to early requirements is to identify a number of existing systems that seem suitable and ask stakeholders to evaluate them in the context of their needs.

2.4 Requirements Validation

Requirements must be validated to ensure that individual requirements are appropriate and that the requirements set is an adequate description of the desired system. This process is intended to detect any problems and gaps in the requirements before they are used as a basis for system development. In particular, as illustrated in Figure 2.6, the SyRS must be validated to show that it meets the stakeholders' original requirements in the StRS, which must be validated to show that it meets the business requirements in the BRS.

Requirements validation is very difficult—somehow, we have to ensure that we have a complete set of requirements that, in their aggregation, represents

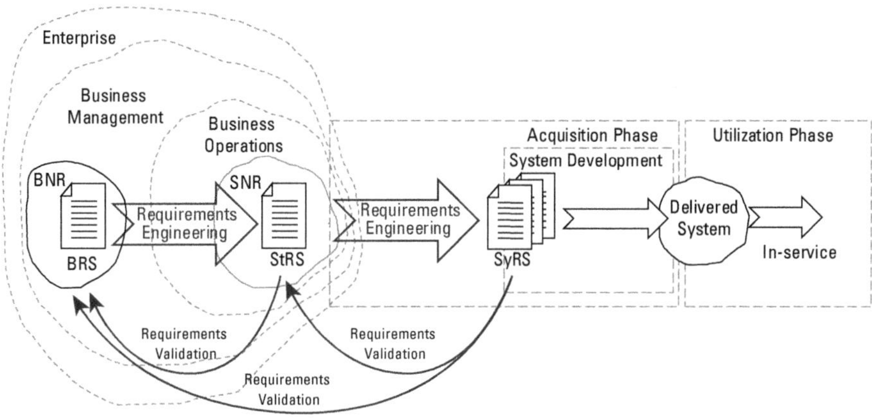

Figure 2.6 Validation of the requirements in the SyRS.

the system required by the stakeholders and meets the business needs. This is difficult because there is no significant baseline we can refer to for the validation. Later in the life cycle, we will look back to the previous baseline; for example in preliminary design, we look back to the SyRS; in detailed design and development, we look back to the development specifications; and in construction and production we look back to the product specifications. In conceptual design, however, we only have the high-level business descriptions as the start point. It is therefore not a straightforward matter to validate that the StRS and then the SyRS represent the complete set of requirements associated with meeting those needs. Validation is normally conducted by a formal requirements review, or a series of reviews.

We must also accept that we can only guarantee zero errors in our requirements-engineering process by spending an infinite amount of time; at some point we must draw a line and say that we are sufficiently prepared to be able to continue the systems engineering process.

Broadly, there are two aspects to requirements validation, listed as follows and discussed in Sections 2.4.1 and 2.4.2:

- Each requirement must be valid.
- The complete set of requirements must be valid.

2.4.1 Validating Individual Requirements

We need to be able to convince ourselves that each requirement meets our conditions of being necessary, singular, correct, unambiguous, feasible, appropriate to level, complete, conforming, and verifiable. Since our requirements engineers undertook this task when writing the requirements, validation is probably best achieved by an independent reviewer.

We are also interested in making sure that we are complete in describing all aspects associated with each requirement. That is, we need to ensure that, in addition to the requirements statement, we have articulated performance and verification requirements (and added an appropriate rationale). We also need to ensure that we have addressed associated constraints and exception conditions.

In some senses, validating individual requirements is relatively straightforward. Validating the set of requirements is much more difficult, however—as soon as we are faced with more than a couple of dimensions our human intuition quickly deserts us.

2.4.2 Validating the Requirement Set

Once we have validated each requirement and the associated statements, we must convince ourselves that the set of requirements meets the overall need. In short, the set needs to meet the following criteria:

- That it is complete (all requirements included, irrelevant ones excluded);
- That it is consistent (internally in that requirements do not conflict, as well as with external interfaces);
- That it is comprehensible (a reader can "visualize" the system from it);
- That it is able to be validated (can be shown to meet the needs);
- That it is feasible (that it is achievable and affordable, for example).

These sorts of characteristics were relatively easy to check in individual requirements but are much harder tasks when considering the set as a whole.

2.4.2.1 Characteristics of a Good Requirement Set

This section examines each characteristic in a little more detail and then discusses some ways of validating the requirements set.

2.4.2.1.1 Complete

A good requirements set is complete—that is, it contains all the requirements necessary to describe the system. Being able to assess completeness, therefore, requires that we can determine when we have finished the elicitation and elaboration, negotiation, and allocation processes. Despite ensuring that we make the best attempt at validating requirements, we must accept that our requirements-engineering process can only guarantee that we have gathered all requirements with zero errors by spending an infinite amount of time—somewhere we must draw a line and say that we are sufficiently prepared to continue the systems engineering process.

When is that point? Well, that is largely a matter of judgement, which is one of the reasons that the most capable of our systems engineers are allocated to requirements engineering. We can, however, come to some understanding of what our requirements would look like at the point we are happy to continue. We are aided by the fact that requirements engineering is an iterative process that will continue throughout the life cycle—so we only have to be confident at this stage that we have iterated sufficiently to have captured (and adequately organized) the bulk of the requirements so that any further iteration will only produce minor variations.

2.4.2.1.2 Correct

The set of requirements must be an accurate representation of the needs, sources, or higher level requirements from which it was transformed.

2.4.2.1.3 Consistent

A consistent set of requirements ensures that individual requirements are unique and do not conflict internally or with external interfaces. It is possible in some formal specification techniques to define a formal property of consistency and prove it using formal methods. These formal techniques are really only useful (cost-effective) in limited applications such as when the problem domain is small and well-bounded, or (as is the case in safety-critical systems) we are striving to guarantee consistency.

There are, however, some things that can be done to ensure consistency. While we can't feed the requirements through some automated tool and have a consistency factor returned and all inconsistent requirements flagged, we can use some tools to assist. For example, a requirements-management tool can be used to search for any requirement that does not have a parent requirement, or that does not spawn at least one child requirement.

Tools can also assist in presenting portions of the requirements set (ordered by type, for example) to assist in looking for potential conflicts between requirements. While the tool assistance is useful, consistency checks require considerable intuition and are therefore best conducted by experienced requirements engineers.

2.4.2.1.4 Comprehensible

Like consistency, comprehension is also difficult to describe formally. If natural specification language has been employed, we can look for structure and syntax within each requirement, but we don't have the same sort of assistance in ensuring that the set of requirements will be able to be comprehended by a broad range of stakeholders, requirements engineers, and lower-level designers. Visualization techniques (such as the use of prototypes) therefore greatly aid the process of ensuring that the stakeholders understand what the collection of "shall" statements is describing. Inclusion in the SyRS of figures and drawings as well as the visual artifacts of requirements engineering also greatly assists comprehension of the requirements set.

2.4.2.1.5 Able to Be Validated

The transformation must be formal and able to be validated, not just for individual requirements, but also for the set. It must be able to be shown that achievement of the set of requirements will result in meeting the set of needs from which it was transformed.

2.4.2.1.6 Feasible

As with consistency, it is relatively simple to check that an individual requirement is realistic. For example, even a nonexpert would easily be able to identify an unrealistic requirement such as a mobile phone with a battery life of 50 days—even at first glance, it does not seem to be feasible and, even if it is, it doesn't seem like it will be practicable or affordable.

The combination of feasible individual requirements does not necessarily sum to a feasible set of requirements, however. For example, would our set of requirements be feasible for the following seemingly realistic individual requirements for a portable, external hard drive for a computer?

- To weigh less than 1 kg;
- To have a storage capacity of 1 TB;
- To be directly accessible on an Ethernet network;
- To be able to be dropped from 1m without damage;
- To survive in temperatures of ±50°C;
- To retail for less than $100.

While each of these requirements seems perfectly feasible, even at first glance, we cannot be so readily sure that the set is feasible (that is, all requirements can be met simultaneously). Again, as soon as we go past a couple of dimensions our human intuition quickly deserts us.

We need experienced requirements engineers who have seen such systems defined before. Even then, they can probably only determine that the set is feasible by continuing the requirements/design process to the point where the design space is sufficiently well-bounded and understood.

Once more, then, we have a case for not worrying too much about getting to too low a level as we conduct the design; we should continue to the point where we are happy with the content and then "write it up" at the (higher) level appropriate to the baseline document.

2.4.2.2 Validation Methods

Although there are few available formal techniques, there are a number of ways to assist in validation of the requirement set.

First, of course, a good top-down requirements-engineering process already has a significant measure of validation incorporated into it. As discussed in Chapter 5, when we develop the mission statement, we ask ourselves whether this single sentence is necessary, singular, correct, unambiguous, feasible, appropriate-to-level, complete, conforming, and verifiable. We then develop half a dozen goals and ask ourselves whether the achievement of those goals would result in the achievement of the mission and then whether each goal is necessary, singular, correct, unambiguous, feasible, appropriate-to-level, complete, conforming, and verifiable. Subsequently, we analyze objectives to ensure that they would result in the achievement of the relevant goals and to ensure that the objectives themselves are also necessary, singular, correct, unambiguous, feasible, appropriate-to-level, complete, conforming, and verifiable. Furthermore, the requirements-engineering artifacts have been developed top down allowing us to validate requirements at each level.

When we were developing use cases and operational scenarios, we should also have developed a series of test cases (these are system validation test cases). Although our design only exists in paper form, we can "do" some of these tests. Walk-throughs (inspections) involve experienced systems engineers and designers "walking through" the requirements using scenarios and test cases to ensure that the system has accommodated such a condition and is likely to produce the required response. Walk-throughs are therefore a form of "mental" prototyping.

Again, the design artifacts discussed at the end of the chapter assist here as the stakeholder can be walked through any scenario at any chosen level of detail, using a visual format rather than reading centimeters of document containing "shall" statements. Other forms of soft and hard simulation, prototyping, and modeling can be used to put the virtual system through its paces to determine whether the requirements stand up to a virtual "trial."

It is not always possible to produce simulations and models, but any forms of visualization greatly assist during validation. Visualization is particularly important for stakeholders who are not familiar with imagining what systems would look like from paper-based descriptions (even two-dimensional drawings).

2.4.2.3 Validation Plan

Since there are a number of ways (no single one of which is guaranteed of success) to validate individual requirements and the requirements set, a suitable validation plan must be developed early on in the process. The plan should articulate the validation methods; who is going to be involved; and who adjudicates conflicts, timing, major events such as review workshops, locations, and so on.

Since requirements engineering is a complex process and the requirements documents are necessarily complex, reviewing the requirements set will also be complex and can quickly become unmanageable unless considerable thought has been given to the process and all participants have had sufficient time to prepare.

2.4.2.4 Review Meetings

As with all formal meetings, the major work is conducted before the meeting—reviews are costly in time, and care must be taken to minimize the impact on the schedules of the reviewers. Additionally, the reviewers should be chosen carefully to ensure a balance across the stakeholders.

Before the meeting, documents are distributed early to give attendees an opportunity to identify and document any conflicts, omissions, or ambiguities. These responses are returned and an agenda is created. The meeting is then held to discuss any problems with the individual requirements or the set of requirements. Action items are identified and allocated to appropriate individuals. The meeting chair follows up after the meeting to ensure that all action items have been completed.

2.5 Requirements Management

Requirements management provides the processes by which the requirements set and any changes to that set are managed throughout the system life cycle. Requirements management involves establishing and ensuring compliance with a formal procedure for the elicitation/elaboration, verification, and traceability of requirements. The many detailed requirements in lower level specifications have been derived from a relatively simple statement made by a stakeholder; these lower level requirements must be managed as they are developed to ensure that each of the lower level requirements can be justified. Additionally, a formal procedure is required for the control of changes to requirements.

2.5.1 Requirements Change Management

Throughout the entire life cycle of the system, we need to manage the configuration of the system elements. This configuration management activity begins as soon as configuration items are defined at the end of preliminary design (in the ABL). Before that, however, we must manage the configuration of the requirements set—this process involves a formal change-management process to ensure that changes to the requirements set occur in a controlled way.

2.5.1.1 Change-Management Policies

Change-management policies are developed to define the procedures, processes, and standards that are to be used to manage changes to requirements during the requirements-engineering process. Change-management policies should cover the following:

- The process by which a change is proposed;
- Definition of the information required to support each change request;
- The process used to analyze the reason for the change and to assess the impact of the change;
- The change-control authority (person or board) that considers and approves the change;
- The process by which the change-control authority receives, considers, and approves the change;
- Tool support for the change-management process.

2.5.1.2 Change-Management Process

The requirements change-management process has four main activities, described as follows:

- *Problem identification and analysis:* A problem with one or more requirements could arise during requirements analysis or allocation, through a revised stakeholder need, from operational problems, or as a result of a change in technology. Once identified, the problem is analyzed and a change is proposed. Change proposals (change request forms) would normally contain the following:
  - A statement outlining the change;
  - Reasons for the change;

    - Available alternatives;
    - Preferred alternative detail;
    - Impact of the change, especially interface impacts;
    - Draft design documentation reflecting the change.
- *Change analysis:* The proposed change is analyzed to determine which requirements are affected by the change and to analyze the impact (on cost and time, among other effects) of the change.
- *Change approval:* The change is submitted to the change authority for approval. The change is described in a change proposal outlining why the change is required, what options are available, and which option is preferred. The change proposal may be rejected if the change is not considered to be necessary (or has already been proposed), the change is unacceptable to the stakeholders, or there are insufficient funds to support the change.
- *Change implementation:* Once the change has been approved, the documentation set is amended accordingly. Any amendments are validated (in accordance with the current validation status).

2.5.2 Change Management: Traceability

Requirements cannot be managed effectively without requirements traceability, which allows us to identify where each requirement came from, what requirements are related to it, and which requirement or requirements stemmed from it (that is, which is its child or are its children). If a requirement must change for some reason, good traceability allows us to identify all requirements affected by the change. It has been estimated that some 35% of requirements for the average project will have been added after the requirements-engineering process is deemed to have ended [59]. Consequently, traceability is essential to support change management and reinforces the project management function of project scope management.

Broadly, there are two main sorts of traceability, described as follows:

- *Forward traceability:* Forward traceability gives us confidence that each requirement has been addressed in the design. Through forward traceability, design decisions can be traced from any given high-level requirement (a parent requirement) down to a detailed design decision (a child requirement). For each requirement, we must be able to find at least one child in the subordinate design documents. If there is more than one child, we must be able to identify all of them.

- *Backward traceability:* We must be able to trace each requirement back to at least one requirement in the parent requirement specification. This backward traceability ensures that additional requirements (not formally endorsed by the customer) have not slipped into—through requirements creep—the design. Any aspect of the design that cannot be traced back to a higher level requirement is likely to represent unnecessary work for which the customer is most probably paying a premium. That is, an orphan requirement must, by definition in such a process, be out-of-scope.

Backward and forward traceability can require that an enormous amount of information be recorded. Consequently, one of the key decisions to be made early in the requirements-engineering process is the degree to which traceability is implemented and the point at which traceability and change management start. If change management starts too early (with the mission, goals, and objectives, for example) the large number of changes in the early iterations are not important, and tracking them leads to a huge amount of relatively useless information. It is generally better to allow the design to mature to a reasonable degree, than it is to insert traceability information as soon as the requirements are entered into a database.

2.5.3 Requirements Management Tools

Due to the large number of requirements elicited/elaborated during the development of the SyRS, a requirements management tool is generally necessary to assist in the management of requirements. In particular, it is almost impossible to have requirements traceability without implementing the requirements in some automated context.

A simple internet search reveals that there are approximately two dozen requirements management tools commercially available [60]. Rather than attempting to review a subset of the tools, this section describes some of the most important features that a requirements management tool should have to assist during the system-development process. In particular, a requirements management tool needs to do the following:

- *Support requirements elicitation:* The selected requirements management tools must support the initial capture of requirements, which may involve importing requirements statements from existing requirements documentation or extracting individual requirements from within a larger, more general source document. Once the requirements have been captured, the management tool must support the editing and addition of requirements attributes. Because the requirements management tool will trace all changes and there is a very large rate of change of requirements during elicitation, there is some advantage in not populating a requirements management tool until the rate of requirements churn has reached an acceptably low level.

- *Allow readers to browse:* It is important for stakeholders to be able to relate to the requirements database in a natural way, so requirements management tools must allow the stakeholders to browse through the requirements in any order.
- *Allow for the retrieval of specific requirements:* The tool must permit a specific requirement to be retrieved directly by its unique identifier, short title, or any part of its text.
- *Support forward and backward traceability:* Traceability is critical to the systems engineering and requirements-engineering processes. Requirements management tools must support and maintain traceability between the different levels of requirements (from business to stakeholder to system to subsystem to assembly to component, as well as in the reverse direction), and between logical design and physical architecture.
- *Support the generation of good requirements:* Some requirements management tools assist with the aim of producing good requirements by providing automated functions to check for inconsistencies and ambiguities. These functions are valuable, but stakeholders should establish a degree of confidence in the process initially and always cross-check that the requirements are well-written.
- *Generate reports:* A good requirements management tool can generate reports of requirements based on selected criteria (a selected attribute, selected combinations of attributes, or some textual string) to assist in validation of the requirements set.
- *Allow import and export of requirements in various formats:* It is also essential that the tool permits import and export of requirements in various database, word-processing, and spread-sheet formats, and that it interfaces with other requirements management tools. This ensures that the overall requirements management process can be maintained even when business managers, stakeholders, contractors, and subcontractors use different tools.
- *Support change control and change impact assessment:* Requirements change over time. Performance levels may change, new requirements

may be added, and old requirements may be deleted. The requirements management tool must support this configuration management function. Other configuration management functions such as impact analysis, change proposals and approvals, and change history should also be available. For example, support for some form of impact analysis is necessary to enable stakeholders to investigate the potential impact on the system of adding, deleting, or modifying requirements. As well as enabling changes, the tool should support the ability to restrict rights of users (to read-only, for example) in order to prevent unauthorized changes and to keep track of which changes were made by whom.

- *Support allocation and logical-to-physical translation:* During preliminary design, a physical architecture is established and logical requirements allocated to it. Some requirements management tools allow the physical architecture to be documented within the tool and support the process of requirements allocation. This feature, combined with an ability to perform impact analysis, becomes extremely valuable as the system development passes through the more detailed design stages.
- *Allows the use of the full array of requirements-engineering processes:* It is essential that the requirements engineers do not have to modify their preferred methodologies and processes, because they have adopted a tool that instantiates a different set of processes. Although most tools are delivered with built-in specification templates based on popular specification standards, requirements engineers should also be able to format the information to their specifications.

2.6 Requirements-Engineering Tools

Many tools exist to assist in requirements engineering, including the context diagram, functional flow block diagrams (FFBDs), RBS, and N2 diagrams. Other tools include structured analysis, data flow diagrams (DFDs), control flow diagrams (CFDs), IDEF diagrams, behavior diagrams, action diagrams, state/mode diagrams, process flow diagrams, function hierarchy diagrams, state transition diagrams, entity relationship diagrams (ERDs), structured analysis and design, object-oriented analysis (OOA), unified modeling language (UML), structured systems analysis and design methodology (SSADM), and quality function deployment (QFD). Each of these techniques focuses on gathering requirements in a formal systematic way.

This text focuses on the philosophy of requirements engineering using the upper level tools of the context diagram, the FFBD, and the RBS, leaving the reader to consult the many excellent references for the more detailed tools [61], particularly MBSE [62].

This section provides a brief introduction to the RBS and FFBD, which are used in the remaining chapters to illustrate the processes, activities, and steps necessary to implement the methodology presented. Chapter 6 details the use of a context diagram.

2.6.1 RBS

The requirements-engineering process can be assisted by the use of a requirements framework around which the logical description of the system can be based. Here we call the requirements framework the RBS, because, as illustrated in Figure 2.7, the requirements are shown in a tree-structured elaboration of the mission. The words are deliberately chosen to differentiate this structure from the well-known project management document called the WBS [63]. The RBS is grouped logically (functionally), and the WBS is structured by physical work packages (including physical system elements) and contains other project-related work; the logical groupings of the RBS are then allocated to the physical groupings of the physical system elements in the WBS. The RBS is also called a functional hierarchy.

In addition to being a useful requirements analysis tool, the RBS can be used to capture the business requirements in the BRS, which can be further refined to reflect stakeholder requirements in the StRS, and then the system requirements in the SyRS. When completely populated, the RBS for a system provides an outline of the SyRS. In the first instance, the system is described

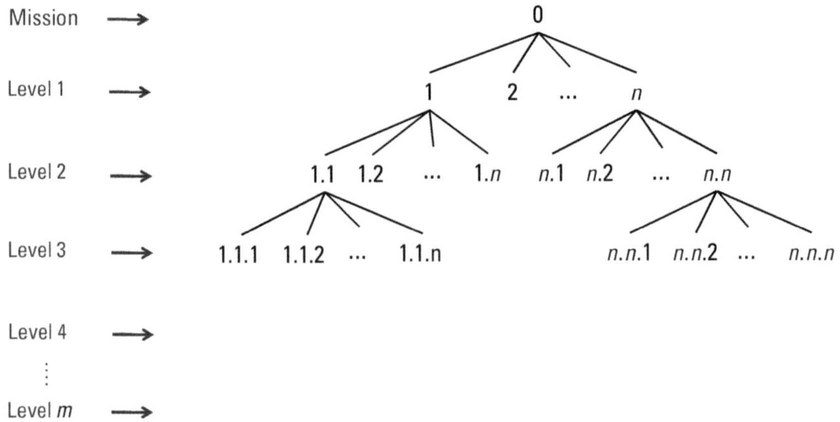

Figure 2.7 A simple RBS showing the hierarchical elaboration of requirements from the mission statement down to *m* levels.

by the words or short phrases of the RBS; in the SyRS, these short titles are elaborated into well-formed requirement statements articulating functional, performance, and verification requirements.

There are a number of advantages associated with the establishment of a suitable framework such as the RBS early in requirements analysis. First, the framework will act as a reference source during requirements analysis to ensure that all aspects of the system requirements are addressed and that important areas are not omitted. The framework allows multiple people to work on the analysis simultaneously as it facilitates the effective allocation to individuals of responsibility for sections of requirements. The framework assists in avoiding duplication of requirements in different sections, which is undesirable as it raises the probability of conflict between requirements and often leads to ambiguity and confusion.

It should be noted, of course, that not all complex systems can be represented adequately in a hierarchical decomposition, and we must remember that we might need to extend the representation to describe fully the system to reflect its possible emergent properties. Still, almost all workable human-made systems are hierarchical, so the RBS is an adequate representation [64].

When developing the RBS, the following general rules should be heeded:

- The RBS is to be displayed hierarchically. The level of the RBS illustrated should be such that the portion displayed is visible on a single A4 sheet of paper, at no less than 10-point font, with no lines leaving or entering the page.
- Naming of RBS elements should be consistent so that the key word phrases are either verb-based or noun-based. The discipline of using three words is useful, however, as it forces the author to stay focused on the function, particularly the verb (which is the function that has to be performed).
- The length of the key word phrases should be consistent (three or four words are normally sufficient).
- The RBS elements are to be logical (functional) not physical.
- Each key word phrase must be unique (since it must be fleshed out into a unique requirement).
- As illustrated in Figures 2.5 and 2.7, the RBS elements should be numbered with a hierarchical numbering system so that parent requirements can be traced to their children and vice versa.
- There must be more than one child at each level—having only one child indicates that the child is redundant (simply a copy of the parent) or additional children must be elaborated.

2.6.2 FFBDs

The hierarchical representation of requirements in the RBS can be supported by FFBDs [65], a simple example of which is shown in Figure 2.8. While the RBS is useful to show a hierarchical decomposition of the requirements, FFBDs are useful in showing the interrelationship of the functions to be performed by the system in accomplishing its mission. Some functions must be performed sequentially; some may be performed in parallel; sometimes alternative paths may be taken based on certain conditions—the FFBDs can show these relationships much more easily than the RBS.

As with the RBS, FFBDs are developed top-down through hierarchical decomposition. Each top-level block can be expanded (decomposed) as necessary to ensure that the functionality of the system is adequately defined, as illustrated in Figure 2.8.

When developing FFBDs, the following general rules should be followed:

- Each function is to be presented in a single box enclosed in a single solid line.
- As with the RBS, naming of the FFBD elements should be consistent so that the key word phrases are unique, either verb-based or noun-based, and are no more than three or four words long. The same names should be used for the function as it is displayed in both the FFBD and the RBS.
- As shown in Figure 2.8, the FFBDs should be displayed as a hierarchical elaboration.
- The level of the FFBD illustrated should be such that the portion displayed is visible on a single A4 sheet of paper, at no less than 10-point font.
- As with the RBS, the FFBDs also make use of a hierarchical numbering system. Where possible, the same numbering system is to be used in the RBS and the FFBDs.
- As with the RBS, naming of the FFDB elements should be consistent so that the key-word phrases are unique, either verb-based or noun-based, and are no more than three or four words long. The same names should be used for the function as it is displayed in both the FFBD and the RBS.
- The lines that connect the blocks have arrows to show the flow of functions, not the elapsed time between them. Consistency in drawing is required—conventionally, the flow is from left to right of the FFBD.

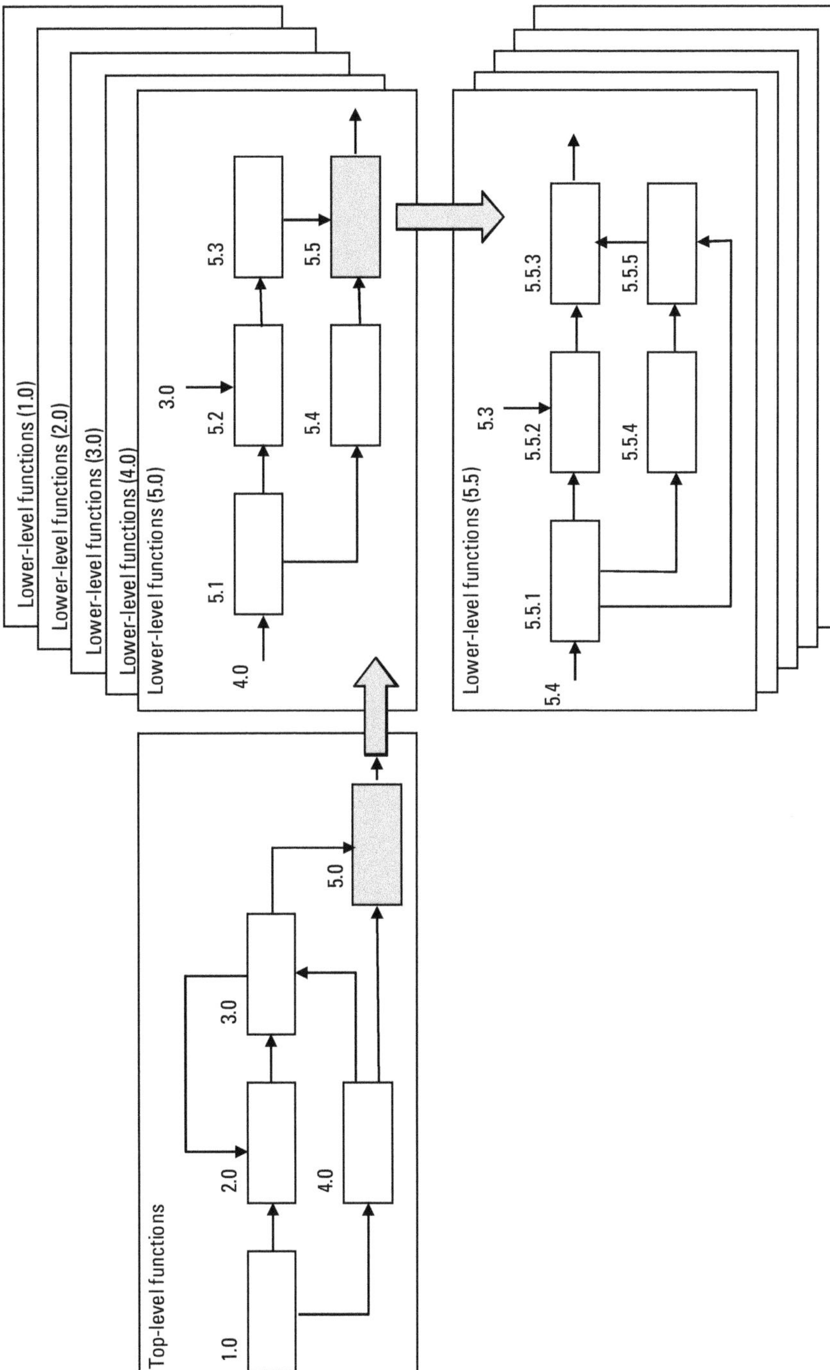

Figure 2.8 Example hierarchical decomposition of FFBDs.

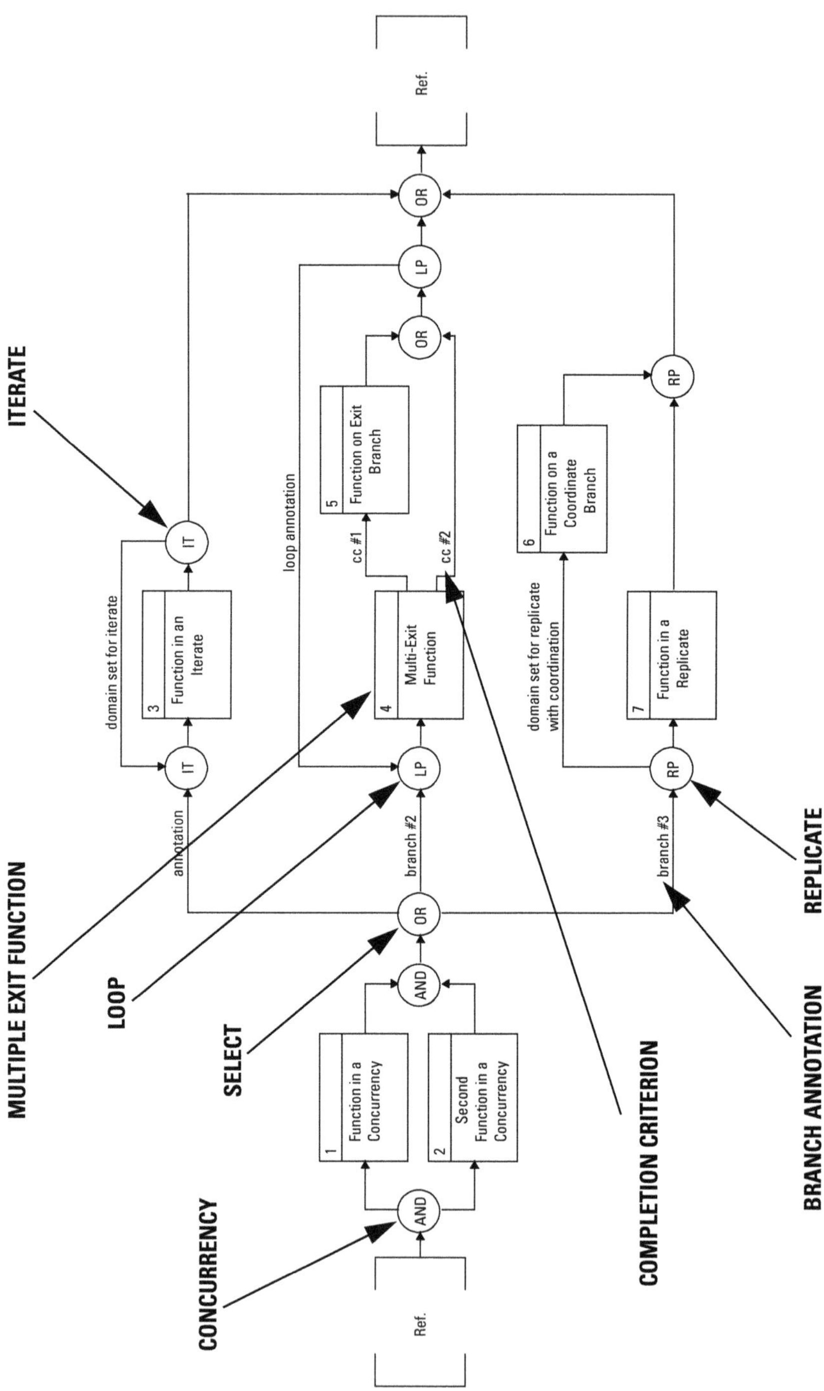

Figure 2.9 Example FFBD showing additional control constructs (reproduced with permission of Vitech Corporation).

If useful, additional blocks may be used to show such additional control constructs as decision points, repetition, iteration, concurrent flow (logical AND), and alternative flow (logical OR); Figure 2.9 shows an example [66].

Review Questions

1. Define what is meant by a need and a requirement.
2. Briefly describe the business needs and requirements definition process, including the roles of the PLC and the BRS.
3. Briefly describe the stakeholder needs and requirements definition process, including the roles of the LCs and the StRS.
4. Briefly describe the System requirements definition process, including the roles of the SyRS.
5. Briefly describe the role and content of the ConOps and the OpsCon.
6. Briefly describe what is meant by a functional requirement, a nonfunctional requirement, and a constraint.
7. Briefly describe the nine characteristics of a good requirement.
8. Briefly describe four main categories of attributes that may be associated with each requirement.
9. Briefly explain five of the 18 main attributes that may be associated with each requirement.
10. Briefly explain why system-level requirements should describe what the system should do rather than how it should be done. Outline why this rule can sometimes be broken.
11. Briefly describe what is meant by the term emergent properties.
12. Briefly describe what is meant by the terms elicitation, elaboration, decomposition, and derivation, in the context of requirements engineering.
13. Briefly describe the seven dimensions to requirements elicitation.
14. Briefly describe the major difficulties in eliciting and elaborating requirements.
15. Briefly explain the major techniques for eliciting requirements.
16. Briefly describe the requirements-engineering activity of requirements validation. Describe the difference between validating each requirement and validating the complete set of requirements.
17. Briefly describe the meanings of the characteristics of a good set of requirements (that it is complete, correct, consistent, comprehensible, able to be validated, and feasible).
18. Briefly describe what is involved in requirements management.

19. Briefly describe requirements change management, including the principal activities.
20. Briefly describe requirements traceability (including forward traceability and backward traceability).
21. List eight of the 10 desirable features of a requirements management tool.
22. Briefly describe the RBS and its role as a requirements framework.
23. Briefly describe FFBDs and their role in defining functional requirements.

Endnotes

[1] In addition to information contained in the general systems engineering references listed at the end of Chapter 1, specific detail on requirements engineering can be found in the following:

Adzic, G., *Specification by Example*, Shelter Island, NY: Manning Publications, 2011.

Alexander, I. F., and L. Beus-Dukic, *Discovering Requirements: How to Specify Products and Services*, Chichester, UK: John Wiley & Sons, 2007.

Alexander, I. F., and R. Stevens, *Writing Better Requirements*, London: Addison Wesley, 2002.

Andriole, S., *Managing Systems Requirements: Methods, Tools and Cases*, New York: McGraw-Hill, 1996.

Aurum, A., and C. Wohlin (eds.), *Engineering and Managing Software Requirements*, Springer, 2005.

Berenbach, B., et al., *Software and Systems Requirements Engineering in Practice*, New York: McGraw-Hill, 2009.

Bray, I. K., *An Introduction to Requirements Engineering*, Boston, MA: Addison Wesley, 2002.

Bron, J.-Y., *System Requirements Engineering: A SysML Supported Requirements Engineering Method*, Wiley-ISTE, 2020.

Davis, A. M., *Software Requirements: Analysis and Specification*, Englewood Cliffs, NJ: Prentice Hall International, 1990.

Davis, A. M., *Software Requirements: Objects, Functions and States*, Englewood Cliffs, NJ: Prentice Hall International, 1993.

Davis, A. M., *Just Enough Requirements Management: Where Software Development Meets Marketing*, New York: Dorset House Publishing, 2005.

Dorfman, M., and R. H. Thayer, *System and Software Requirements Engineering*, IEEE Computer Society Press, 1990.

Ferdinandi, P. L., *A Requirements Pattern: Succeeding in the Internet Economy*, Boston, MA: Addison-Wesley, 2001.

Fernandes, J., and R. Machado, *Requirements in Engineering Projects*, Switzerland: Springer, 2016.

Fitchett, P., and J. Haslam, *Writing Engineering Specifications*, London: Spon Press, 1988.

Gause, D., and G. Weinberg, *Exploring Requirements: Quality Before Design*, New York: Dorset House Publishing Company, Incorporated, 1989.

Gilb, T., *Competitive Engineering: A Handbook For Systems Engineering, Requirements Engineering, and Software Engineering Using Planguage*, Butterworth-Heinemann, 2005.

Goldsmith, R. F., *Discovering Real Business Requirements for Software Project Success*, Norwood, MA: Artech House, 2004.

Gottesdiener, E., *Requirements by Collaboration: Workshops for Defining Needs*, Reading, MA: Addison Wesley, 2002.

Grady, J., *Systems Requirements Analysis*, New York, NY: McGraw-Hill, 2006.

Hass, K. B., D. J. Wessels, and K. Brennan, *Getting it Right: Business Requirement Analysis Tools and Techniques*, Vienna, VA: Management Concepts, 2008.

Hatley, D. J., P. Hruschka, and I. A. Pirbhai, *Process for System Architecture and Requirements Engineering*, New York, NY: Dorset House Publishing Company, 2000.

Hay, D. C., *Requirements Analysis: From Business Views to Architecture*, Upper Saddle River, NJ: Prentice Hall, 2003.

Hood, C., et al., *Requirements Management: The Interface Between Requirements Development and All other Systems Engineering Processes*, Oberhaching, Germany: Springer, 2008.

Hooks, I. F., and K. A. Farry, *Customer-Centered Products: Creating Successful Products Through Smart Requirements Management*, New York, NY: AMACOM, 2001.

Hossenlopp, R., and K. B. Hass, *Unearthing Business Requirements: Elicitation Tools and Techniques*, Vienna, VA: Management Concepts, 2008.

Hull, E., K. Jackson, and J. Dick, *Requirements Engineering*, London: Springer, 2011.

International Institute for Business Analysis, *A Guide to the Business Analysis Body of Knowledge® (BABOK® Guide)*, Version 3, 2015.

ISO/IEC/IEEE 29148 *Systems and Software Engineering—Lifecycle Processes—Requirements Engineering*, 2018.

Jackson, M., *Software Requirements and Specifications: A Lexicon of Software Practice, Principles and Prejudices*, Reading, MA: Addison Wesley, 1995.

Jonasson, H., *Determining Project Requirements, Second Edition: Mastering the BABOK® and the CBAP® Exam*, Auerbach Publications, 2012.

Kossmann, M., *Requirements Management: How to Ensure You Achieve What You Need from Your Projects*, Gower, 2013.

Kotonya, G., and I. Sommerville, *Requirements Engineering: Processes and Techniques*, West Sussex, England: John Wiley & Sons, 2000.

Kulak, D., and E. Guiney, *Use Cases: Requirements in Context*, Boston, MA: Addison-Wesley, 2004.

Laplante, P. A., *Requirements Engineering for Software and Systems*, Boca Raton, FL: CRC Press, 2014.

Larson, E., and R. Larson, *Practitioner's Guide to Requirements Engineering—Part 1: Requirements Planning*, Minneapolis, MN: Watermark Learning Publications, 2009.

Leffingwell, D., and D. Widrig, *Managing Software Requirements: A Unified Approach*, Reading, MA: Addison-Wesley, 2000.

Leffingwell, D., *Agile Software Requirements: Lean Requirements Practices for Teams, Programs, and the Enterprise*, Reading, MA: Addison-Wesley, 2011.

Loucopoulos, P., and V. Karakostas, *Systems Reauirements Engineering*, New York, NY: McGraw-Hill, 1995.

Lutowski, R., *Software Requirements: Encapsulation, Quality, and Reuse*, Boca Raton, FL: Taylor and Francis, 2005.

Macaulay, L., *Requirements Engineering*, London: Springer, 1996.

Mate J. L., and A. Silva (eds.), *Requirements Engineering for Sociotechnical Systems*, Hershey, PA: Information Science Publishing, 2005.

Pohl, K., *Requirements Engineering: Fundamentals, Principles, and Techniques*, Heidelberg, Germany: Springer-Verlag, 2010.

Pohl, K., *Requirements Engineering Fundamentals*, Santa Barbara, CA: Rocky Nook, 2011.

Pohl, K., and C. Rupp, *Requirements Engineering: Fundamentals, Principles, and Techniques*, Heidelberg, Germany: Springer-Verlag, 2011.

Purdy, D., *A Guide to Writing Successful Engineering Specifications*, New York: McGraw-Hill, 1991.

Rinzler, B., *Telling Stories: A Short Path to Writing Better Software Requirements*, John Wiley & Sons, 2009.

Robertson, S., and J. Robertson, *Requirements-Led Project Management*, Boston, MA: Addison-Wesley, 2005.

Robertson, S., and J. Robertson, *Mastering the Requirements Process*, Harlow, England: Addison-Wesley, 2013.

Sommerville, I., and P. Sawyer, *Requirements Engineering*, Chichester: UK, John Wiley & Sons, 1997.

Thayer, R. H., and M. Dorfman (eds.), *Software Requirements Engineering*, Los Alamitos, CA: IEEE Computer Society Press, 1997.

van Lamsweerde, A., *Requirements Engineering: From System Goals to UML Models to Software Specifications*, Chichester, UK: John Wiley & Sons, 2009.

Wheatcraft, L. S., M. J. Ryan, and T. E. Katz, *INCOSE Needs and Requirements Manual: Needs, Requirements, Verification, Validation Across the Lifecycle*, John Wiley & Sons, 2024.

Wiegers, K. E., and J. Beatty, *Software Requirements*, Redmond, WA: Microsoft Press, 2013.

Wieringa, R., *Requirements Engineering: Frameworks for Understanding*, John Wiley & Sons, 1996.

Withall, S., *Software Requirements Patterns (Best Practices)*, Redmond, WA: Microsoft Press, 2007.

Young, R., *Effective Requirements Practices*, Boston, MA: Addison-Wesley, 2001.

Young, R., *The Requirements Engineering Handbook*, Norwood, MA: Artech House, 2004.

Young, R., *Project Requirements: A Guide to Best Practices*, Vienna, VA: Management Concepts, 2006.

[2] Ryan, M. J., and L. S. Wheatcraft, "On a Cohesive Set of Requirements Engineering Terms," *Systems Engineering*, Vol. 20, No. 2, 2017, pp. 118–130.

[3] Ryan, M. J., "An Improved Taxonomy for Major Needs and Requirements Artifacts," *INCOSE International Symposium IS2013*, June 2013.

[4] Ryan, M. J., and L. S. Wheatcraft, "On a Cohesive Set of Requirements Engineering Terms," *Systems Engineering*, 2017.

[5] Wheatcraft, L. S., M. J. Ryan, and T. E. Katz, *Needs and Requirements Manual: Needs, Requirements, Verification, Validation Across the Lifecycle*, Hoboken, NJ: John Wiley & Sons, 2024.

[6] International Institute for Business Analysis, *A Guide to the Business Analysis Body of Knowledge® (BABOK® Guide)*, Version 3, 2015.

[7] Ryan, M. J., and L. S. Wheatcraft, "On a Cohesive Set of Requirements Engineering Terms," *Systems Engineering*, 2017.

[8] Wheatcraft, L. S., M.J. Ryan, and T. E. Katz, *Needs and Requirements Manual: Needs, Requirements, Verification, Validation Across the Lifecycle*, Hoboken, NJ: John Wiley & Sons, 2024.

[9] International Council on Systems Engineering (INCOSE), INCOSE-TP-2010-004-04, *Guide to Writing Requirements*, 2024.

[10] ISO/IEC/IEEE 15288, *Systems and Software Engineering—System Lifecycle Processes*, 2023.

[11] ISO/IEC/IEEE 29148, *Systems and Software Engineering—Lifecycle Processes—Requirements Engineering*, 2018.

[12] ISO/IEC/IEEE 15288, *Systems and Software Engineering—System Lifecycle Processes*, 2023.

[13] ISO/IEC/IEEE 29148, *Systems and Software Engineering—Lifecycle Processes—Requirements Engineering*, 2018.

[14] ANSI/AIAA G-043B-2012, *Guide for the Preparation of Operational Concept Descriptions*, American National Standards Institute, American Institute of Aeronautics and Astronautics (sponsor), 2012.

[15] ISO/IEC/IEEE 29148, *Systems and Software Engineering—Lifecycle Processes—Requirements Engineering*, 2018.

[16] International Institute for Business Analysis, *A Guide to the Business Analysis Body of Knowledge® (BABOK® Guide)*, Version 3, 2015.

[17] ISO/IEC/IEEE 29148, *Systems and Software Engineering—Lifecycle Processes—Requirements Engineering*, 2018.

[18] ISO/IEC/IEEE 29148, *Systems and Software Engineering—Lifecycle Processes—Requirements Engineering*, 2018.

[19] International Institute for Business Analysis, *A Guide to the Business Analysis Body of Knowledge® (BABOK® Guide)*, Version 3, 2015.

[20] MIL-STD-499B, *Military Standard—Systems Engineering—Draft*, Washington D.C.: US Department of Defense, 1994.

[21] Wheatcraft, L. S., M. J. Ryan, and T. E. Katz, *Needs and Requirements Manual: Needs, Requirements, Verification, Validation Across the Lifecycle*, Hoboken, NJ: John Wiley & Sons, 2024.

[22] Wheatcraft, L. S., M. J. Ryan, and T. E. Katz, *Needs and Requirements Manual: Needs, Requirements, Verification, Validation Across the Lifecycle*, Hoboken, NJ: John Wiley & Sons, 2024.

[23] ANSI/AIAA G-043B-2012, *Guide for the Preparation of Operational Concept Descriptions*, American National Standards Institute, American Institute of Aeronautics and Astronautics (sponsor), 2012.

[24] ISO/IEC/IEEE 29148, *Systems and Software Engineering—Lifecycle Processes—Requirements Engineering*, 2018.

[25] MIL-STD-499B, *Military Standard—Systems Engineering—Draft*, Washington DC: US Department of Defense, 1994.

[26] IEEE-STD-1220, *IEEE Standard for Application and Management of the Systems Engineering Process*, New York: IEEE Computer Society, 2005.

[27] EIA-632, *Processes for Engineering a System*, Washington, D.C.: Electronic Industries Association (EIA), 2003.

[28] International Institute for Business Analysis, *A Guide to the Business Analysis Body of Knowledge® (BABOK® Guide)*, Version 3, 2015.

[29] ISO/IEC/IEEE 29148, *Systems and Software Engineering—Lifecycle Processes—Requirements Engineering*, 2018.

[30] ISO/IEC/IEEE 29148, *Systems and Software Engineering—Lifecycle Processes—Requirements Engineering*, 2018.

[31] International Council on Systems Engineering (INCOSE), INCOSE-TP-2010-004-04, *Guide to Writing Requirements*, 2024.

[32] ISO/IEC/IEEE 29148, *Systems and Software Engineering—Lifecycle Processes—Requirements Engineering*, 2018.

[33] International Council on Systems Engineering (INCOSE), INCOSE-TP-2010-004-04, *Guide to Writing Requirements*, 2024.

[34] Ryan, M. J., and L. S. Wheatcraft, "On a Cohesive Set of Requirements Engineering Terms," *Systems Engineering*, 2017.

[35] Ryan, M. J., and L. S. Wheatcraft, "On a Cohesive Set of Requirements Engineering Terms," *Systems Engineering*, 2017.

[36] Wheatcraft, L. S., M. J. Ryan, and J. Dick, "On the Use of Attributes to Manage Requirements," *Systems Engineering*, Vol. 19, No. 5, September 2016, pp. 448–458.

[37] IEEE-STD-1233, *IEEE Guide for Developing System Requirements Specifications*, New York, NY: The Institute of Electrical and Electronic Engineers (IEEE), 1998. Note that a similar list is in IEEE 29148.

[38] Wheatcraft, L. S., M. J. Ryan, and T. E. Katz, *Needs and Requirements Manual: Needs, Requirements, Verification, Validation Across the Lifecycle*, John Hoboken, NJ: Wiley & Sons, 2024.

[39] Wheatcraft, L. S., M. J. Ryan, and T. E. Katz, *Needs and Requirements Manual: Needs, Requirements, Verification, Validation Across the Lifecycle*, Hoboken, NJ: John Wiley & Sons, 2024.

[40] Hitchins, D., *Systems Engineering: A 21st Century Systems Methodology*, Chichester, England: John Wiley & Sons, 2007.

[41] Grady, J. O., *System Requirements Analysis*, Burlington MA: Academic Press, 2006.

[42] Daniels, J., and T. Bahill, "The Hybrid Process That Combines Traditional Requirements and Use Cases," *Systems Engineering*, Vol. 7, No. 4, 2004.

[43] IEEE-STD-1220, *IEEE Standard for Application and Management of the Systems Engineering Process*, New York: IEEE Computer Society, 2005.

[44] See also a similar taxonomy in: Kotonya, G., and I. Sommerville, *Requirements Engineering: Processes and Techniques*, Chichester, UK: John Wiley & Sons, 2000.

[45] Young, R., *The Requirements Engineering Handbook*, Norwood, MA, Artech House, 2004.

[46] Brooks, F., *The Mythical Man Month: Essays on Software Engineering*, Chapel Hill, NC: Addison-Wesley, 1995.

[47] Alexander, I. F., and R. Stevens, *Writing Better Requirements*, Addison-Wesley, London, 2002.

[48] Goguen, J., and C. Linde, "Techniques for Requirements Elicitation", *International Symposium on Requirements Engineering*, San Diego, California: IEEE Computer Society Press, 1993, pp. 152–164.

[49] Hull, E., K. Jackson, and J. Dick, *Requirements Engineering*, Springer, London, 2011.

[50] Katonya, G., and I. Sommerville, *Requirements Engineering: Processes and Techniques*, Chichester, England: John Wiley and Sons, 2000.

[51] Sommerville, I. and P. Sawyer, *Requirements Engineering: A Good Practice Guide*, Chichester, England: John Wiley and Sons, 1997.

[52] Young, R. R., *Effective Requirements Practices*, London: Addison-Wesley, 2001.

[53] Young, R. R., *The Requirements Engineering Handbook*, Norwood, MA: Artech House, 2004.

[54] IEEE-STD-1233, *IEEE Guide for Developing System Requirements Specifications*, New York: The Institute of Electrical and Electronic Engineers (IEEE), 1998.

[55] Verner, J., et al., "Requirements Engineering and Software Project Success: an Industrial Survey in Australia and the U.S.," *Australasian Journal of Information Systems*, Vol. 13, No. 1, September 2005.

[56] Coulin, C., D. Zowghi, and A. E. K. Sahraoui, "A Situational Method Engineering Approach to Requirements Elicitation Workshops in the Software Development Process," *Software Process Improvement and Practice 11*, 2006, pp. 451–464.

[57] Tsumaki, T., and T. Tamai, "Framework for Matching Requirements Elicitation Techniques to Project Characteristics," *Software Process Improvement and Practice*, Vol. 11, 2006, pp. 505–519.

[58] Specific detail on use-case modeling can be found in the following:

Bittner, K., and I. Spence, *Use Case Modelling*, Boston, MA: Addison-Wesley, 2003.

Cockburn, A., *Writing Effective Use Cases*, Boston, MA: Addison-Wesley, 2001.

Kulak, D., and E. Guiney, *Use Cases: Requirements in Context*, Boston, MA: Addison-Wesley, 2004.

Rinzler, B., *Telling Stories: A Short Path to Writing Better Software Requirements*, John Wiley & Sons, 2009.

[59] Jones, C., *Software Quality—Analysis and Guidelines for Success*, International Thomson, 1997.

[60] See also: Andriole, S., *Managing Systems Requirements: Methods, Tools and Cases*, New York: McGraw-Hill, 1996.

[61] Specific detail on analysis tools can be found in the following:

Blanchard, B., and W. Fabrycky, *Systems Engineering and Analysis*, Upper Saddle River, NJ: Prentice Hall, 2011.

Coad, P., and E. Yourdon, *Object Oriented Analysis*, Upper Saddle River, NJ: Prentice Hall, 1990

Defense Systems Management College (DSMC), *Systems Engineering Management Guide*, Washington, DC: US Government Printing Office, 1990.

Eisner, H., *Essentials of Project and Systems Engineering Management*, New York: John Wiley & Sons, 2008.

Gause, D., and G. Weinberg, *Exploring Requirements: Quality Before Design*, New York: Dorset House Publishing Company, 1989.

Goodland, M., and C. Slater, *SSADM: A Practical Approach*, London: McGraw-Hill Publishing Company, 1995.

Grady, J. O., *Systems Requirements Analysis*, Burlington, MA: Academic Press, 2006.

Heap, G., J. Stanway, and A. Windsor, *A Structured Approach to Systems Development*, London: McGraw Hill Book Company, 1992.

Hitchins, D., *Putting Systems to Work*, Chichester, UK: John Wiley & Sons, 1992.

Kotonya, G., and I. Sommerville, *Requirements Engineering: Processes and Techniques*, West Sussex, England: John Wiley & Sons, 2000.

Lacy, J., *Systems Engineering Management: Achieving Total Quality*, New York: McGraw Hill, 1992.

Leffingwell, D., and D. Widrig, *Managing Software Requirements: A Unified Approach*, Reading, MA: Addison-Wesley, 2000.

Macaulay, L., *Requirements Engineering*, London: Springer, 1996.

Maier, M., and E. Rechtin, *The Art of Systems Architecting*, Boca Raton, FL: CRC Press, 2009.

Martin, J., *Principles of Object-Oriented Analysis and Design*, Upper Saddle River, NJ: Prentice Hall, 1993.

Martin, J., *Systems Engineering Guidebook: A Process for Developing Systems and Products*, Boca Raton, FL: CRC Press, 1997.

Martin, J., and J. Odell, *Object-Oriented Analysis and Design*, Upper Saddle River, NJ: Prentice Hall, 1992.

Rumbaugh, J., et al., *Object-Oriented Modeling and Design*, Upper Saddle River, NJ: Prentice Hall, 1991.

Sage, A., and J. Armstrong, *Introduction to Systems Engineering*, New York, NY: John Wiley & Sons, 2000.

Shlaer, S., and S. Mellor, *Object-Oriented Systems Analysis*, Upper Saddle River, NJ: Prentice Hall, 1988.

Stevens, R., et al., *Systems Engineering: Coping with Complexity*, Hertfordshire, England: Prentice Hall Europe, 1998.

Young, R., *Effective Requirements Practices*, Boston, MA: Addison-Wesley, 2001.

Westerman, H., *Systems Engineering Principles and Practice*, Norwood, MA: Artech House, 2001.

[62] Excellent coverage of MBSE and its application can be found in the following texts:

Buede, D. M., and W. D. Miller, *The Engineering Design of Systems: Models and Methods*, John Wiley & Sons, 2006.

Delligatti, L., *SysML Distilled: A Brief Guide to the Systems Modeling Language*, Pearson Education, 2010.

Dori, D., *Model-Based Systems Engineering with OPM and SysML*, Springer, 2016.

Friedenhal, S., A. Moore, and P. Stein, *A Practical Guide to SysML: The Systems Modelling Language*, The MK/OMG Press, 2015.

Hetherington, D., *Simple SySML for Beginners: Using Sparx Enterprise Architect*, Asatte Press, 2019.

Holt, J., *Systems Engineering Demystified: A Practitioner's Handbook for Developing Complex Systems Using a Model-Based Approach*, Birmingham, UK: Packt, 2021.

Holt, J., and S. Perry, *SysML for Systems Engineering: A Model-Based Approach*, Institution of Engineering and Technology, 2013.

Micouin, P., *Model-Based Systems Engineering: Fundamentals and Methods*, John Wiley & Sons, 2014.

Vitech Corporation, *A Primer for Model-based Systems Engineering*, Second Edition, 2011.

[63] Project Management Institute (PMI), *Guide to the Project Management Body of Knowledge*, Newtown Square, PA: Project Management Institute (PMI), Version 6, 2017.

[64] Simon, H., *The New Science of Management Decision*, Englewood Cliffs, NJ: Prentice-Hall, 1977.

[65] Blanchard, B., and W. Fabrycky, *Systems Engineering and Analysis*, Upper Saddle River, NJ: Prentice-Hall, 2011. (Note that these generic FFBDs can be prepared by any one of the graphical methods including Integrated DEFinition (IDEF) modeling method, the behavioral diagram method, and the N^2 charting method, which are compared and discussed in: Blanchard, B., D. Verna, and E. Peterson, *Maintainability: A Key to Effective Serviceability and Maintenance Management*, New York: John Wiley & Sons, 1995. The DSMC management guide (Defense Systems Management College (DSMC), *Systems Engineering Management Guide*, Washington, DC: US Government Printing Office, 1990, pp. 6.3–6.5) explains how FFBDs can be used during function identification.)

[66] Long, J., *Relationship Between Common Graphical Representations Used in Systems Engineering*, National Council on Systems Engineering (NCOSE) Symposium, 1995.

3

Introduction to the Conceptual Design Methodology

3.1 Introduction

Conceptual design (also called *concept design*, *logical design*, or *functional design*) is the most critical of all of the life cycle activities because it is responsible for the expansion of the system definition from a short statement of business need into a logical architecture that may be tens or hundreds of pages long. While other organizations may be involved, the conceptual design of the system is in the problem domain and therefore is normally the responsibility of the customer who must determine what the system needs to do, how well it needs to do it, and how the system-level functionality is to be validated—after all, it is their business from which the needs for the system have arisen.

This chapter begins by examining the guidance available for conceptual design in extant standards, handbooks, manuals and guides and then presents a conceptual design methodology (CDM) that provides a framework for the conduct of conceptual design, which is then elaborated in some detail in Chapters 4 to 8.

3.2 Guidance for the Conduct of Conceptual Design

Despite the considerable effort expended in systems engineering and requirements engineering standards, there are surprisingly few methodologies available for the conduct of conceptual design. At one level, this is not so surprising because standards such as EIA/IS-632 [1] (MIL-STD-499 [2]), IEEE-STD-1220 [3], EIA-632 [4], and ISO/IEC/IEEE 15288 [5] have a true systems engineering background—they are therefore focused on *what* must be done and do not prescribe *how* it should be done. Similarly, although the INCOSE *Systems Engineering Handbook* [6] provides more detail on what must be done, it does not provide a methodology for how to do conceptual design. All these standards and handbooks are useful, but an organization cannot "do" them because they (deliberately) provide little or no guidance on "how" to perform any aspect. The INCOSE *NRM* [7] and its supporting guides provide more detail within the framework of the INCOSE *Systems Engineering Handbook*, and they do provide some guidance on how to conduct conceptual design, but more is needed.

3.2.1 EIA/IS-632 (MIL-STD-499B): Systems Engineering

EIA/IS-632 (a slightly revamped form of MIL-STD-499) and IEEE-STD-1220 are two of the most commonly used systems engineering standards—both are regularly specified in contracts to be used as the framework within which logical (functional) design is conducted. This section introduces those two standards and illustrates that, while they do describe what must be done regarding conceptual design in more detail than EIA-632 and ISO/IEC/IEEE 15288, they are still not very useful in describing how to do it.

Figure 3.1 illustrates the process that is the crux of EIA/IS-632 (MIL-STD-499) [8]. Each activity is elaborated by the standard with statements such as the following:

- *Requirements analysis:* "The performing activity shall analyze customer needs, objectives, and requirements in the context of customer missions, utilization environments, and identified system characteristics to determine functional and performance requirements for each primary system function. Prior analyses shall be reviewed and updated, refining mission and environment definitions to support system definition. Requirements analysis shall be conducted iteratively with functional analysis . . ."
- *Functional analysis/allocation:* "The performing activity shall define and integrate a functional [logical] architecture to the depth needed to support synthesis of solutions for people, products, and processes and management of risk. Functional analysis/allocation shall be conducted iteratively."

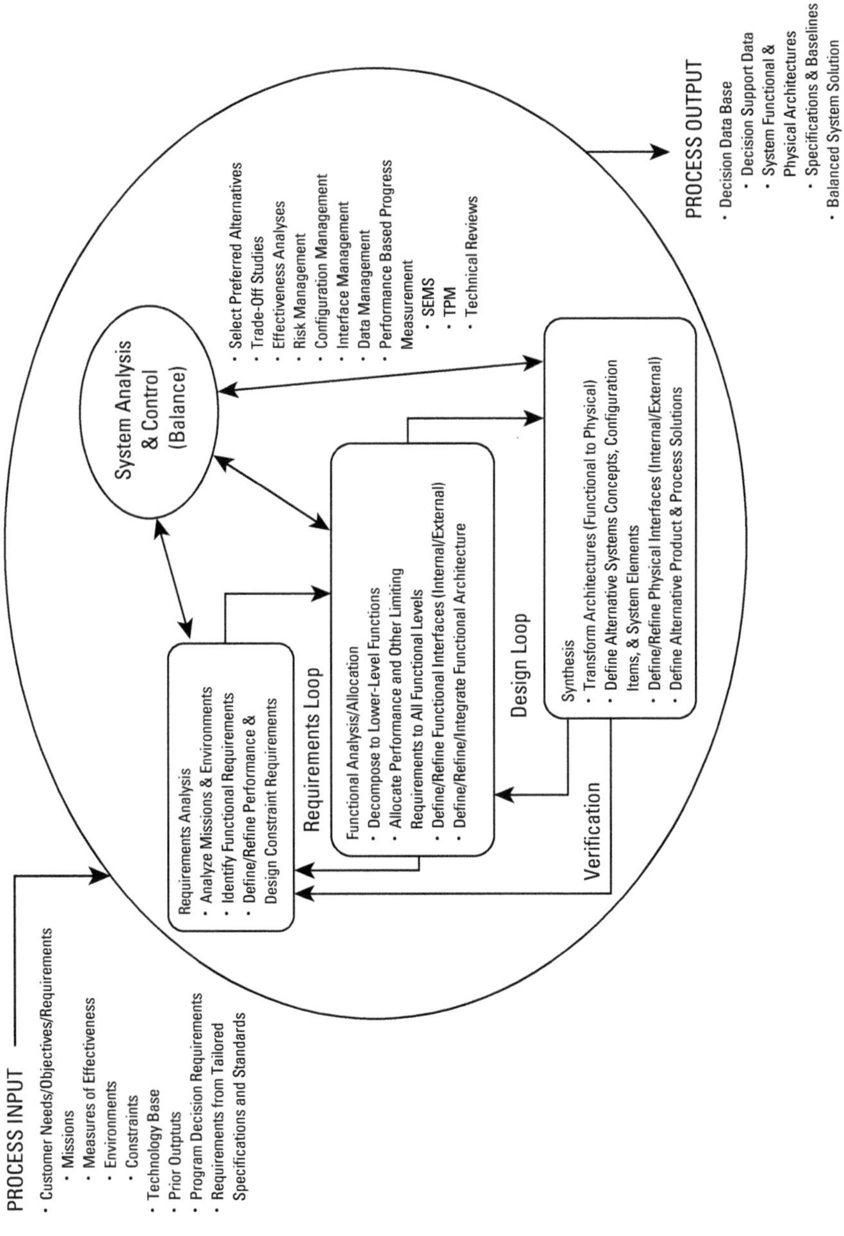

Figure 3.1 The EIA/IS-632 (MIL-STD-499) systems engineering process.

- *Synthesis:* "The performing activity shall define and design solutions for each logical set of functional and performance requirements in the functional [logical] architecture and integrate them as a physical architecture. The performing activity shall conduct synthesis iteratively with functional analysis/allocation to define a complete set of functional and performance requirements necessary for the level of design output required. Requirements analysis shall be used to verify that solution outputs can satisfy customer input requirements."

The standard therefore mandates what shall be done, not how; the performing activity is then required to develop methodologies, processes, and procedures to describe how it is to be done.

3.2.2 IEEE-STD-1220: IEEE Standard for Application and Management of the Systems Engineering Process

IEEE-STD-1220 is a systems engineering standard that is of a similar vintage and considerably more prescriptive than MIL-STD-499 and, therefore, much more helpful at the working level of systems/requirements engineering. Figure 3.2 illustrates the IEEE-STD-1220 [9] top-level systems engineering process (SEP), which, despite being shown in a different form, is largely the same as that of MIL-STD-499—that is, it has the major steps of requirements analysis, functional analysis, and synthesis, with verification and iteration. More usefully than MIL-STD-499, however, IEEE-STD-1220 provides lower level processes to support the SEP. For example, Figure 3.3 illustrates the steps associated with requirements analysis. The standard then provides a paragraph or two suggesting in outline what each of the subordinate activities might entail.

So, IEEE-STD-1220, like MIL-STD-499, mandates what shall be done to perform logical design (albeit with much better lower level guidance than MIL-STD-499). However, the performing activity is still required to develop methodologies, processes, and procedures to describe how it is to be conducted. When invoked in a contract, systems engineering standards are therefore very useful in mandating what shall be done with regard to requirements engineering as part of the contract, but they must be supported by a methodology that describes how to conduct logical design.

3.2.3 EIA-632: Processes for Engineering a System

EIA-632 represents a significant change from the interim standard and focuses on the enterprise and its policies and procedures rather than focusing narrowly

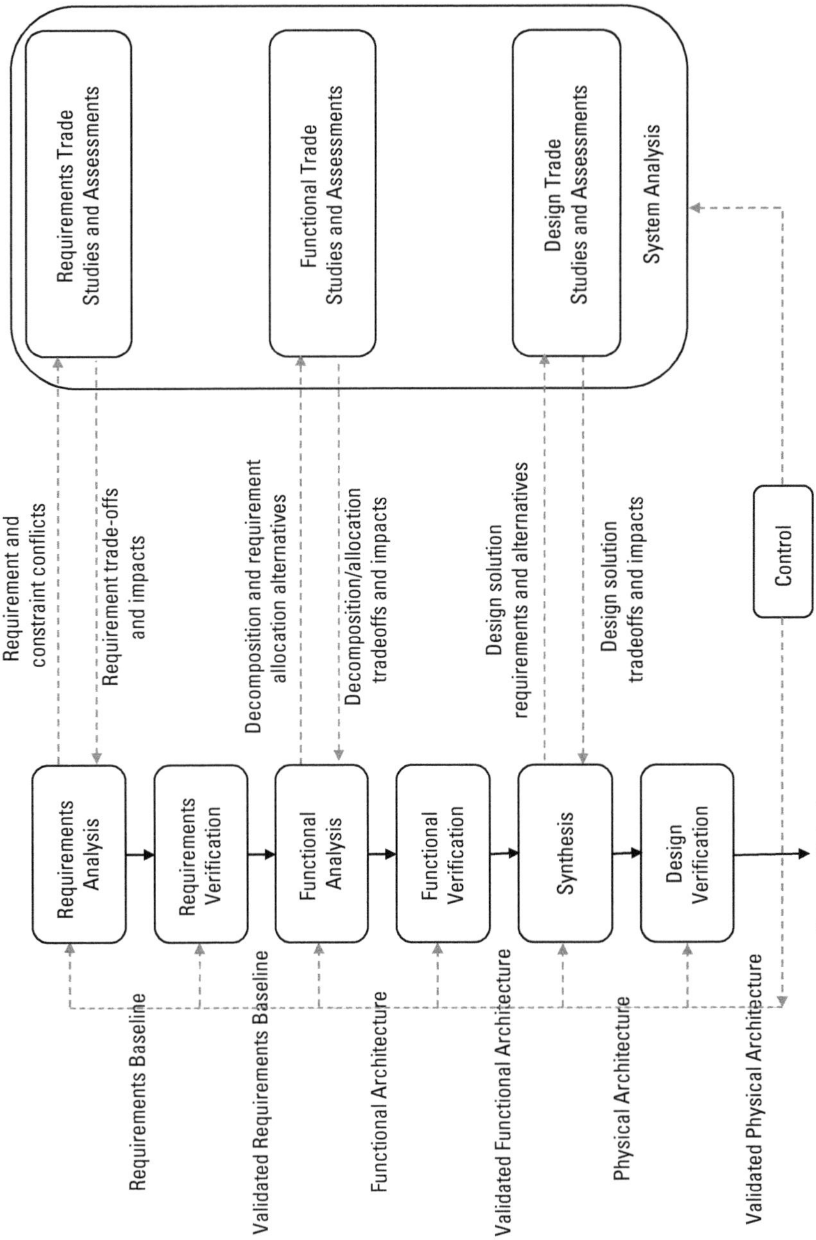

Figure 3.2 The IEEE-STD-1220 systems engineering process.

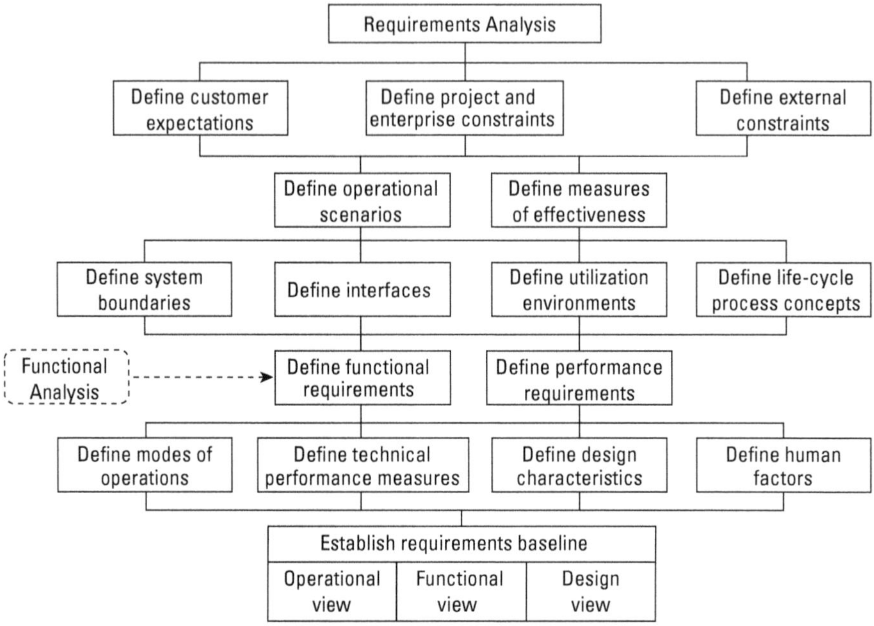

Figure 3.3 The requirements analysis process of IEEE-STD-1220.

on a particular system development being undertaken by that enterprise. By operating at the enterprise level, EIA-632 avoids being a prescriptive standard and concentrates on what must be done to engineer a system successfully and completely. Figure 3.4 [10] illustrates the application of EIA-632. The standard has been designed to help enterprises develop policies and procedures relevant to their role in engineering systems within their industries. Projects being conducted by the enterprise are then responsible for further development of company policies and procedures into project-specific plans, schedules, and so on.

EIA-632 introduces 13 processes required to successfully engineer a system. Within these 13 processes, there are 33 requirements set out in the standard. Not all of the 33 requirements will be relevant or required in every situation; therefore, EIA-632 is expected to be tailored to individual development projects. Figure 3.5 [11] shows the 13 processes contained in EIA-632 grouped into the five broad categories described in the standard. The categories and processes are arranged into a flow diagram that effectively describes the EIA-632 logic. Note the conceptual design activities exist only as a subset of the overall EIA-632 process.

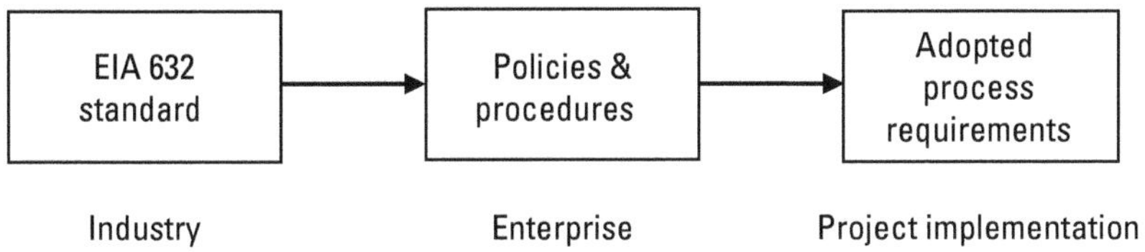

Figure 3.4 Application of EIA-632.

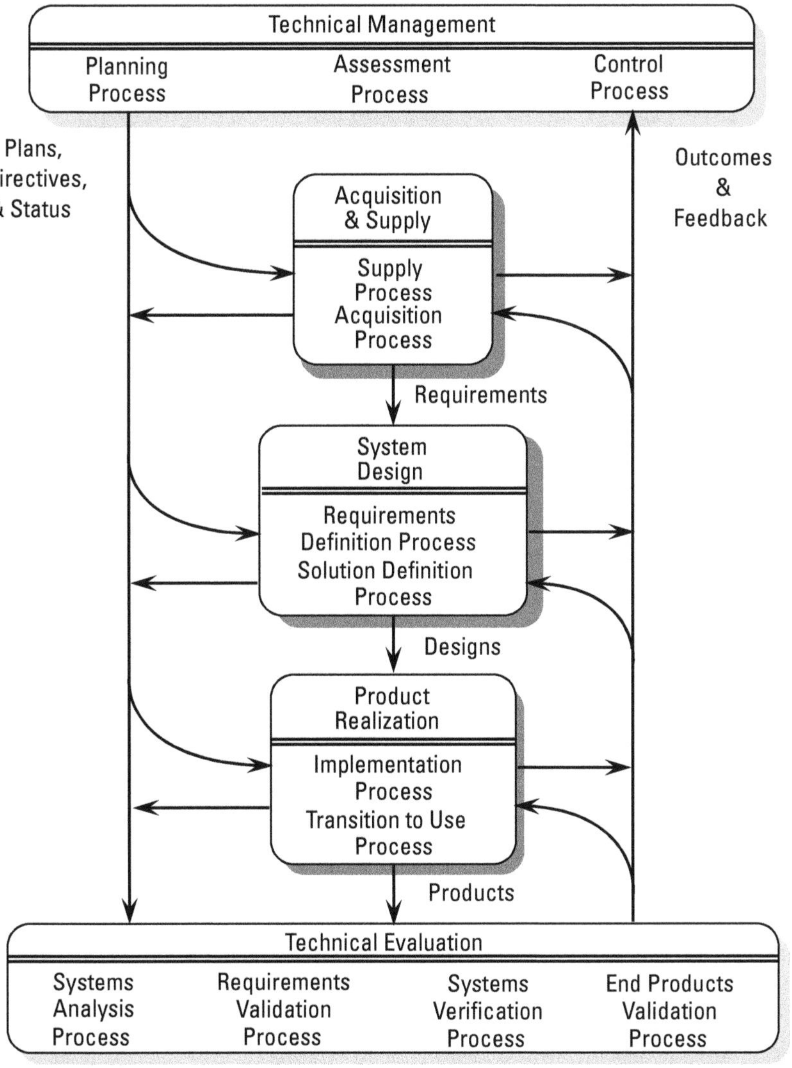

Figure 3.5 EIA-632 processes.

EIA-632 focuses on process rather than method. The methods for achieving the requirements are expected to come from enterprise policy and procedures. Each of the process areas shown in Figure 3.5 is dealt with separately by considering the inputs and outputs of the process and the requirements within the process.

Of the 13 processes, the Requirements Definition Process is most concerned with conceptual design (although the Solution Definition Process has a to some extent) but the process description is focused on what to do rather than how to do it.

3.2.4 ISO/IEC/IEEE 15288: System and Software Engineering—System Life-Cycle Processes

Although the standards considered so far are still in use, the premier systems engineering standard in the modern world is ISO/IEC/IEEE 15288. Like EIA-632, it is focused on industry and business levels, providing guidance on what must be done, deliberately avoiding anything about how it should be done. ISO/IEC/IEEE 15288 is useful to us here, however, because it proposes a common process framework covering the life cycle of human-made/modified systems in order to improve communication and cooperation among the parties that create, utilize, and manage modern systems. ISO/IEC/IEEE 15288 identifies four process groups, each of which includes a number of processes:

- Agreement Processes;
- Organizational Project-Enabling Processes;
- Technical Management Processes;
- Technical Processes;

Figure 3.6 illustrates the 30 ISO/IEC/IEEE 15288 [12] processes organized within the four process groups in a manner that reflects the flow of processes in a generic project (recognizing that ISO/IEC/IEEE 15288 expects processes to be iterative and recursive, not sequential as shown here). The processes that directly relate to conceptual design (highlighted in gray in Figure 3.6), are listed as follows:

- Business or Mission Analysis Process;
- Stakeholder Needs and Requirements Definition Process;
- System Requirements Definition Process.

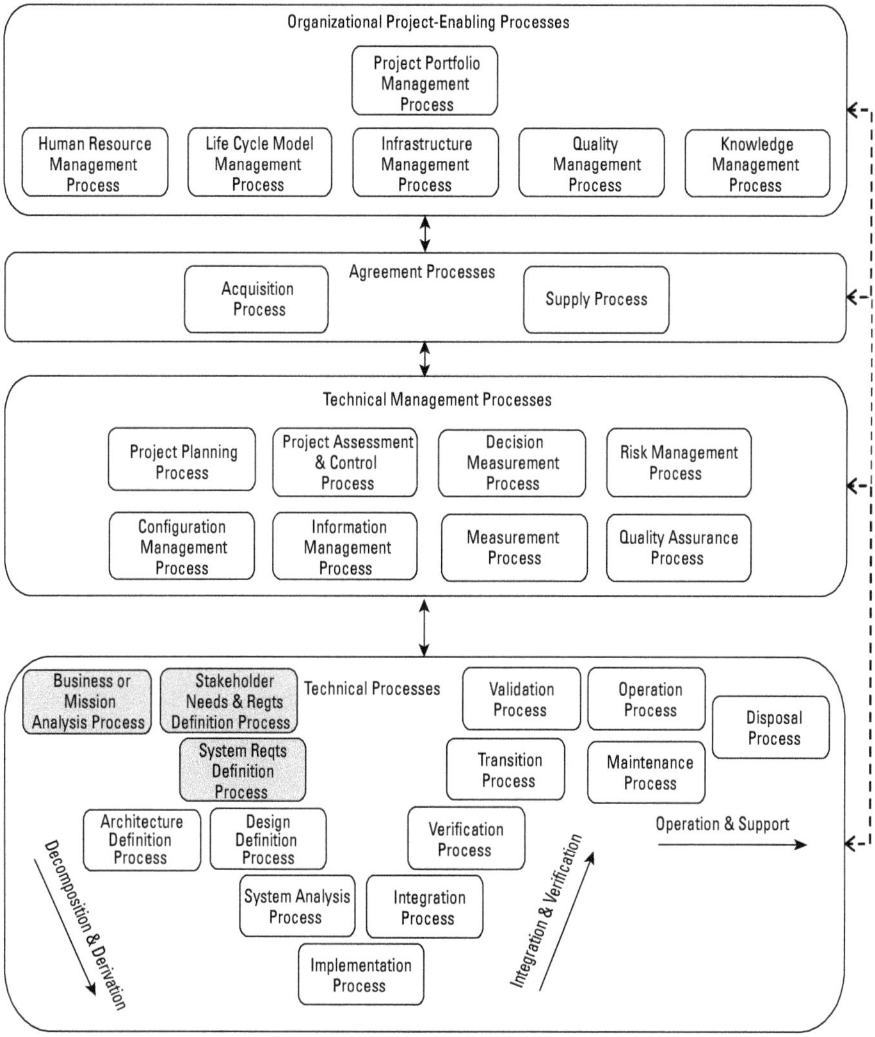

Figure 3.6 The ISO/IEC/IEEE 15288 processes organized in a generic implementation. The Business or Mission Analysis Process, the Stakeholder Needs and Requirements Definition Process, and the System Requirements Definition Process are highlighted as the three processes of most interest in conceptual design.

3.2.5 INCOSE *NRM:* Needs, Requirements, Verification, and Validation across the Life Cycle

In support of the INCOSE *Systems Engineering Handbook* [13], the INCOSE *NRM* [14] provides considerably more detail on the development of concepts, needs, and requirements. The *NRM* is intended for those whose role is to perform needs, requirements, verification, and validation activities throughout the system life cycle and is addressed to practitioners of all levels of experience. Those new to project management and system development should find the detailed guidance useful. Those more experienced should be able to find new insights across all phases of the life cycle, particularly in terms of a data-centric perspective to which they may not have been exposed. The *NRM* is supported by the following INCOSE guides:

- *Guide to Needs and Requirements* [15];
- *Guide to Writing Requirements* [16];
- *Guide to Verification and Validation* [17];
- *Guide to Security Needs and Requirements* [18].

The broad scope of the *NRM* fills in much of the detail that needs to be elaborated to support the ISO/IEC/IEEE 15288 and the INCOSE *Systems Engineering Handbook*. It is a valuable resource, particularly as it focuses on not only the processes and artifacts across the life cycle, but also on the associated verification and validation activities.

Figure 3.7 summarizes the *NRM* approach to the activities associated with conceptual design, resulting in the principal artifacts of the integrated set

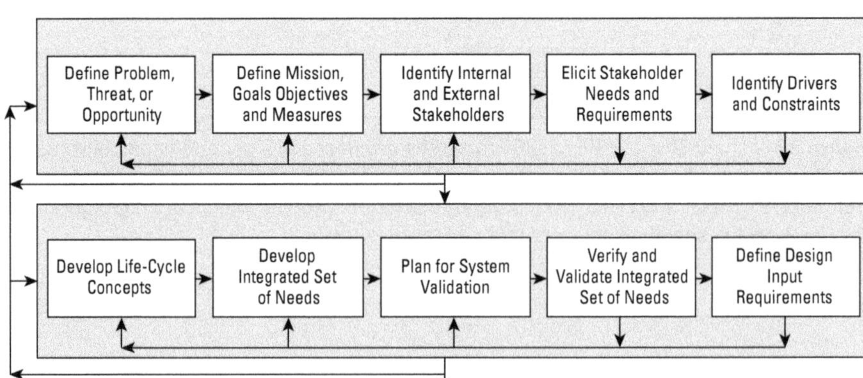

Figure 3.7 The INCOSE *NRM* processes associated with conceptual design.

of needs and the design input requirements. In addition to providing valuable guidance, the *NRM* uses a helpful example throughout to illustrate the various aspects of the life cycle activities and artifacts.

Because of its wide mandate, the *NRM* is focused mostly on what and why. Although it provides some considerable guidance on how to complete the life cycle activities (with an excellent example to illustrate) it does not have the space for much detail. It does, however, provide an excellent framework that is much more detailed than ISO/IEC/IEEE 15288 and the INCOSE *Systems Engineering Handbook.*

3.3 Introduction to the CDM

This section describes a CDM that was developed to provide a step-by-step approach to conceptual design, culminating in the development of the SyRS and the conduct of the SDR. CDM is in lockstep with the INCOSE *NRM* and its supporting guides and therefore conforms to it. In order to allow designers to demonstrate conformance to IEEE-STD-1220 or to ISO/IEC/IEEE 15288, if required by their contract, CDM has been deliberately designed so that its activities can be readily shown to map onto the processes outlines in either of the standards.

This section outlines the key aspects that underpin CDM along with the upper level processes. Next it details the major processes, activities, steps, and tasks, which are detailed and illustrated in the remaining chapters.

3.3.1 Key Aspects of CDM

The world is inherently complicated, and complicated problems do not usually have simple solutions. Yet, we cannot specify, develop, manage, and operate complicated systems by using processes that are themselves complicated beyond our overall capacity to understand and manage. There are, therefore, a number of challenges to our judgement and intuition in articulating requirements for complicated systems.

First, we must use a top-down approach in which the system is viewed as a whole—this facilitates the subsequent analysis of the system and its interfaces and allows for the design to accommodate desired attributes such as quality, reliability, maintainability, survivability, and security, which may not be attained by taking a bottom-up approach. Yet, humans are most comfortable with solving problems using bottom-up methodologies that best fit the application of our limited intuition—our system design process must be able to accommodate this predilection. Additionally, our understanding of the system cannot be expected to be complete until we have developed the complete design, and even then, we will most likely still be learning about it, particularly its emergent properties; we therefore require an iterative design approach. Finally, traceability ensures that design decisions can be traced from any system-level requirement down to at least one lower level requirement and be traced from any lower level requirement back to a higher level requirement.

3.3.1.1 Top Down

The principal aims of conceptual design of a system are to articulate the business and stakeholder needs and to analyze and document the requirements flowing from those needs. Focus during conceptual design is on the development of a logical architecture of the system (the "whats," captured in the FBL, the centerpiece of which is the SyRS. Once the logical architecture has been validated, focus then shifts to the development of a physical architecture (the "hows") of the system captured in the ABL. It is possible to attempt logical design by simply identifying as many lower level "shall" statements as possible and grouping them together to describe the functions of the system; unfortunately, however, complicated problems cannot be solved that way, nor can complicated systems be developed adequately using such a bottom-up approach. A top-down focus is essential because systems have emergent properties that may not be discovered by taking a bottom-up approach.

In the top-down process, before any work can commence on developing the system, the fundamental needs and requirements must be clearly and completely articulated. Because business managers and stakeholders have most probably stated their needs in a fairly general way, every project should begin with a concise statement of the system's mission, which is then elaborated by statements of the system-level goals and objectives. The hierarchical elaboration of the mission statement into goals, and then of each of the goals into objectives, is an essential component of the top-down design of a system.

There is another very important reason for taking a top-down approach. The problem must first be addressed by enterprise management and business management and a number of decisions made before passing the subsequently constrained solution space in the BNR on to the business operations stakeholders, who will further consider and constrain the solution space in the SNR before passing on to systems designers to develop the SyRS.

By way of illustrating the need for the top-down approach, the following example (also included by the author in the INCOSE *NRM*) [19] shows why it is essential that enterprise and business management appropriately constrain the solution space.

Consider ACME Timber, an early 1900s wood-felling company that employs approximately 200 axemen and whose principal output is lumber from two sawmills. ACME's management has just returned from a 1930s trade show where they witnessed a demonstration of the chainsaw, which has the potential to revolutionize wood felling.

Imagine in the first instance that the company simply nominates one of its operations managers to define a set of requirements for the procurement of chainsaws and the conversion of ACME's workforce from axemen to chainsaw operators. That is, they approach the opportunity bottom-up.

The first issue is that stakeholders at the operational level are not really in a position to describe what they want from the introduction of the new chainsaw capability. They certainly are experienced in managing axemen and felling trees with axes, but neither operations managers nor axemen will be able to describe in any manner how the new device is to be operated, since they have no familiarity with the operating procedures, training, safety, or the maintenance necessary for the new device. (Of course, they may also be reluctant to participate since they are going to be retrained at best or, at worst, let go once the chainsaw is in service.) Similarly, the logistics and procurement staff at the operational level will know, in considerable detail, logistics information to support of current operations but will know nothing of the support for a device that will need a different maintenance methodology; significantly different support materials such as fuel and lubrication; different storage; and many more parts of much greater variety.

Nonetheless, if the nominated operations manager identifies the mission to use the chainsaw to produce more lumber, they will then go on to identify the needs and requirements by taking a systems approach to the new tool. They will examine the chainsaw as well as the other elements of capability and begin to identify that the employees may change from the historically all-male crews to mixed crews, along with the needs to upgrade personal protection (goggles, hearing protection, gloves, and helmet), to develop safety procedures, and to adjust individual and collective training. Additional support will be required in terms of fuel supply, spare parts, and maintenance staff. Fuel will need to be delivered by a newly acquired fuel truck for each work site, and the maintainer will need limited field facilities at each site. Axemen likely own their axes and bring them to the site each day; the chainsaw, on the other hand, will need to be secured in a truck and brought to and from the site each day. Additionally, while axemen will be responsible for sharpening their own axes, the chainsaw will need the chain tightened and the teeth sharpened by the maintainer. Once the wood-felling team can cut down more wood, more wood-clearing vehicles will be required to take the wood to more sawmills (or at least more saw lines at each of the existing sawmills). More trucks will be required to clear the lumber from sawmills to the marketing outlets, more of which may be required. Unfortunately, however, delivering more wood into a fixed market will invariably cause the price of wood to drop, resulting in lower profits, which was not the intention.

In other words, unless business management clearly articulates the mission behind the acquisition of a chainsaw capability, the project may fail because even though the lower level stakeholders were earnest in their systems approach, the business intent was not communicated properly to them. Before passing on the task, business management must understand first how this new technology will transform the business, how the existing workforce will be affected, and how the new tool will be phased into operations to ensure business continuity (i.e., new devices deployed across all operators at once, or team by team, or region by region). Business management must draft the initial versions of the acquisition concept, deployment concept, operating concept, maintenance concept, and retirement concept.

It is most likely that the reason to procure chainsaws is to cut down the same amount of wood more efficiently (with fewer operators, for example) resulting in lower production costs, thus making ACME more profitable. This is quite a different project to the one based on cutting down more trees. In both cases, all the lower level issues of protection, safety, training, fuel, maintenance, and so on will still apply. If, however, the same amount of wood is to be cut down more efficiently, there will be no need for additional wood-clearing vehicles and the current sawmills and marketing outlets will suffice.

This example illustrates the need for a top-down approach; and it is easy to understand why the lack of attention at the business level is a principal reason for project failure.

3.3.1.2 Use of Intuition and Bottom-Up Analysis Within a Top-Down Construct

There is a natural upper bound to the number of concepts—the magical number seven, plus or minus two—that any human can retain in conscious immediate memory (see Miller [20]). We cope with the complications of life around us by combining and organizing information into coherent patterns that Miller called chunks. This natural limitation and our consequent tendency to chunk information have two main implications for the garnering of requirements from stakeholders. First, stakeholders struggle to consider more than seven, plus or minus two, concepts at any one time; and second, we tend to aid our understanding of complicated issues by chunking in a bottom-up manner.

It is therefore very difficult to get individuals, particularly groups of individuals, to participate in a top-down decomposition process when their natural decision-making processes tend to be based on using their intuition to perform a bottom-up aggregation of those issues that are within their comprehension and experience. One useful solution to this dilemma is to break the top-down process into a series of small steps, each of which may be completed using an intuitive bottom-up approach.

For example, crafting a complete balanced mission statement in one (top-down) pass is very difficult for most stakeholders, particularly if the system is large and complicated, has a number of interfaces, and is a new system (that is, it contains elements with which they have no direct experience). It is therefore often productive to ask stakeholders to simply list (in a bottom-up manner) the five to seven key words or phrases that are candidates to be considered for inclusion in the mission statement. Once each of these candidate need elements has been assessed for suitability, a mission statement can easily be developed by combining the elements into a complete balanced sentence. Consequently, the stakeholders' natural tendency to think in a bottom-up manner can be supported, but the output is useful because it contributes to the design within a top-down construct. In CDM, all activities have been designed to allow bottom-up aggregation to occur within a top-down construct.

3.3.1.3 Use of Iteration

It is not reasonable to assume at the start of a system design that we understand everything that we need to know about the system to design it properly. We therefore require an iterative process to allow us to develop that understanding and to incorporate it into the design as we proceed—a sequential design process can only incorporate our understanding as it exists at the outset of the design. Iteration must be controlled, however, if we are to capture adequately the design artifacts that result, as our understanding matures.

Second, everyone will invariably start a design with a different mindset, background, and experience level. Simon [21] recognized that any two people might perceive the same situation differently in any attempt to correct the gap between "what is" and "what should be"—that is, to solve a complicated problem. Iteration in a collective design process allows us to enter a series of controlled divergent/convergent thinking phases before finally converging our collective understanding to a common agreed-on set of mission, goals, and objectives. Van Gundy [22] suggests how to use this divergence/convergence as part of creative problem solving (CPS) to address problems in a structured way.

Iteration is therefore an essential part of logical design. Even if it were not, however, as noted earlier in this chapter, systems engineering standards such as IEEE-STD-1220 incorporate iteration as part of their mandated processes. A supporting design methodology must therefore be able to support iteration during design.

3.3.1.4 Traceability

As shown in Chapter 2, traceability ensures that design decisions can be traced from any system-level requirement down to at least one lower level requirement (forward traceability) and that any lower level requirement can be justified by being able to be traced to a higher level requirement (backward traceability). Traceability is an essential component of the effective management of system development, since it allows stakeholders to be convinced that their upper level requirements have been translated adequately into lower level ones. Additionally, traceability allows designers to justify the presence of any lower level requirement by tracing it back to an approved upper level requirement—this is particularly necessary for those requirements that have been derived by the designers and not directly stated by the stakeholders.

Traceability therefore makes an essential contribution to the scope management being conducted by the project manager. It assures all those involved that upper level requirements have been accommodated by the design, guarantees that all lower level requirements can be justified, and ensures that no unnecessary or unauthorized requirements have been included in the design.

3.4 Summary of CDM Processes and Artifacts

As discussed in Chapter 2, the five major processes that are conducted during conceptual design and the major artifacts that are produced, are the following:

- Definition of the BNR, comprising the PLC and the BRS;
- Definition of SNR comprising the LC and StRS;
- Definition of system requirements in the draft SyRS;
- System synthesis to finalize the SyRS with the preferred solution in mind;
- The SDR, at which the initial FBL is established.

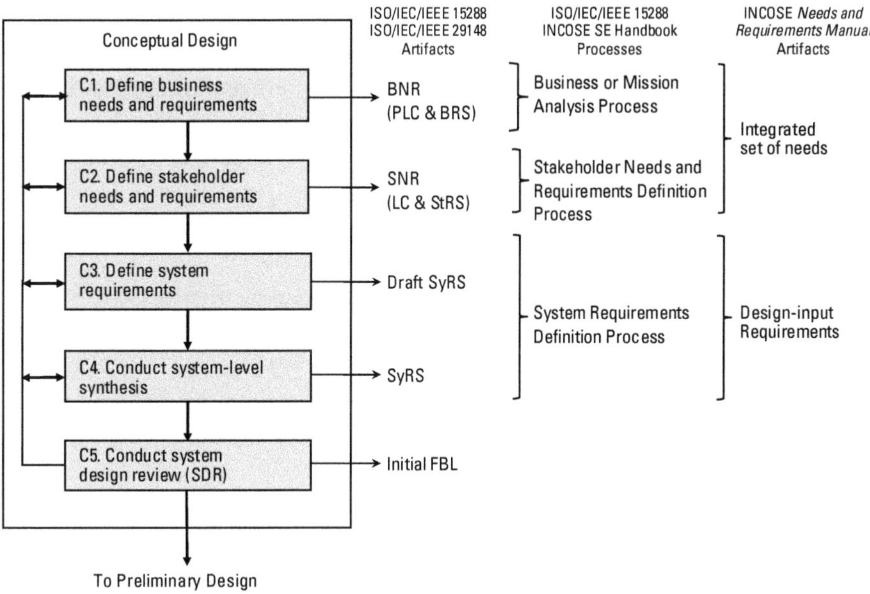

Figure 3.8 CDM processes and artifacts mapped to those called for by ISO/IEC/IEEE 15288 and ISO/IEC/IEEE 29148, and the artifacts from the INCOSE *NRM.*

To help visualize the relationships among those processes and artifacts, Figure 3.8 also shows the processes and artifacts called for by ISO/IEC/IEEE 15288 and ISO/IEC/IEEE 29148 and the artifacts from the INCOSE *NRM*.

Note that the CDM processes are iterative, even at the highest level of abstraction—at each stage the improved understanding of the system is used to revisit the previous process or processes if necessary. This section briefly outlines the five major CDM processes (and associated activities), which are described in more detail in the following five chapters.

3.4.1 C1—Define BNR

Before any work can commence on developing the system, business management needs to articulate, clearly and completely, the basic BNR. While this need appears obvious, a surprising number of system developments commence without a clear and complete understanding of the fundamental BNR. Not surprisingly most of these developments founder or, at the very least, are introduced into service with poor levels of user acceptance.

Figure 3.9 summarizes the activities and steps involved with the process definition of BNR resulting in the development of the PLC and the BRS. Note once again that the activities are iterative, allowing current understanding to improve the quality of previously generated artifacts. Sections 3.4.1.1–3.4.1.5 outline the activities of Figure 3.9, which are then described in much more detail in the following chapters.

3.4.1.1 C1.1—Identify Major Stakeholders and Constraints

Requirements engineering begins with business management's identification of the major stakeholders with whom we are to engage to elicit needs and requirements and to verify our design artifacts as we proceed. We also must identify the business, project, external, and design constraints, which provide important context within which to consider the system requirements—and are, in fact, the source of many requirements.

3.4.1.1.1 C1.1.1—Identify Major Stakeholders

The first step in developing the BNR (Figure 3.9, activity C1.1.1) requires the identification of the major stakeholders for all aspects of the new system including those that will be the source of any constraints the business wishes to place on the system and its development. These stakeholders are the individuals and groups from the business units at the organization's operational level, identified by business management as those who will be responsible for writing and agreeing to the BNR prior to their endorsement. These major stakeholders also have a significant role during later activities to assist with the requirements analysis effort—that is, they will most likely lead the development of the SNR.

As discussed in Chapter 4, the identification of stakeholders is not a matter of simply listing all those who may have an interest in the system. Since it is from these stakeholders that key aspects of the system will be elicited, they must be selected carefully. The difficult issues with stakeholder identification and management are described in considerable detail in Chapter 4.

3.4.1.1.2 C1.1.2—Identify Business and Project Constraints

Constraints are requirements that are imposed on the system by circumstance, force, or compulsion. They therefore limit absolutely the options open to a designer by imposing immovable boundaries and limits [23]. Before focusing on the detail of the desired system, it is therefore essential to identify the business and project constraints (Figure 3.9, step C1.1.2) that are relevant to the system and its acquisition—a system will fail if it cannot operate under

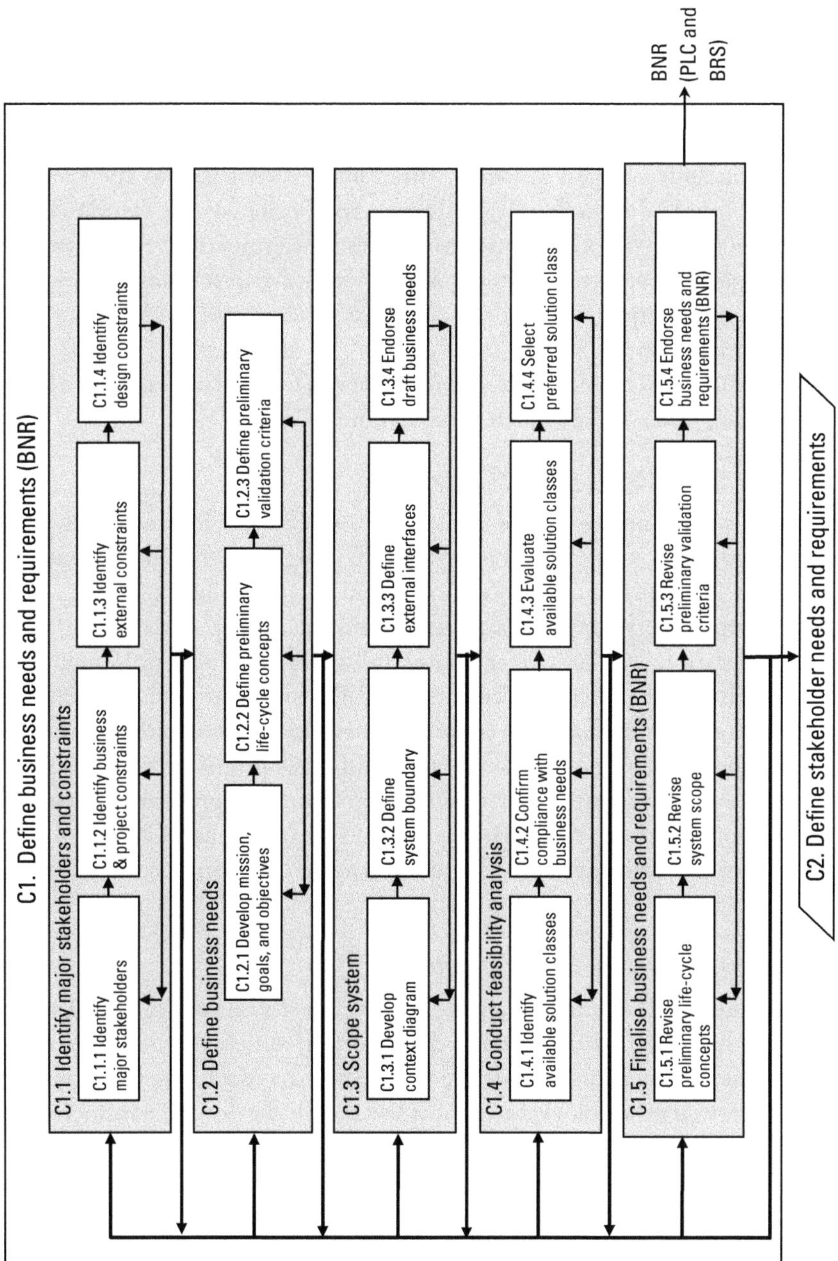

Figure 3.9 Summary of C1—Define business needs and requirements.

the constraints imposed upon it. This analysis provides essential information about the development environment for the system and begins the top-down approach to system development.

Business constraints include management guidance; organizational policies, procedures, standards, or guidelines that guide system development and procurement; human resource limitations, and budget restrictions. Project constraints include budget, schedule, the resources allocated to the project, externally imposed deliverables, timeframes, and business-wide standards for processes such as quality assurance and systems engineering. Additionally, projects might be constrained to conform to particular standards; toolsets, metrics, documentation sets and templates, technology use, and control and reporting mechanisms.

The issues associated with the identification of business and project constraints are discussed in much more detail in Chapter 4.

3.4.1.1.3 C1.1.3—Identify External Constraints

In addition to business-imposed constraints, there are wider external constraints (Figure 3.9, step C1.1.3) on system development that arise from the requirement for conformance to national and international laws and regulations, compliance with industry-wide standards, along with ethical and legal considerations. Other external constraints include the requirement for interoperability and the capabilities required for interfacing with other systems. Additionally, the capability of competitors and the availability of human resources, specific skill sets, technologies, and tools might provide external constraints. Again, an important aspect of top-down design is to understand these constraints before considering needs and requirements. Chapter 4 discusses the issues associated with the identification of external constraints in much more detail.

3.4.1.1.4 C1.1.4—Identify Design Constraints

Design constraints (Figure 3.9, step C1.1.4) include those factors that directly affect how the system is designed and the way in which the design is conducted. Typical constraints include the state-of-the-art of relevant technologies as well as extant methodologies and tools to assist in the design, development, construction, and production of the system. Such issues must be addressed by business management who will address factors such as the risk associated with embracing new technologies, the impact of upgrading construction facilities, and workforce reskilling issues. The issues associated with the identification of design constraints are discussed in much more detail in Chapter 4.

3.4.1.2 C1.2—Define Business Needs

Having established the context by identifying major stakeholders and constraints, the next activity is to define business needs. The major activities are the definition of the mission, goals and objectives; preliminary operational scenarios; preliminary validation criteria; and PLC. Chapter 5 discusses each of these in much more detail.

3.4.1.2.1 C1.2.1—Define Mission, Goals, and Objectives

It is axiomatic that the development of every system should begin with a concise statement of the mission of the system, elaborated by statements of the system-level goals and objectives (Figure 3.9, step C1.2.1). The mission statement should be quite short (stated in a single sentence) and may be expressed in only a few lines, although it must have a word or phrase for every important aspect of the system. While business management often finds it difficult to state the mission in a single, short sentence, the development is invariably doomed to failure if the owners of the mission cannot articulate its mission succinctly at the outset. Interestingly, it will often take a small number of participants several hours to craft an agreed-on mission statement—the process is useful therefore, not only to start the design process, but to serve as a mechanism for developing a common understanding of the system.

The discipline of limiting the mission statement to a single sentence is important because the inability to express the mission in a single sentence tends to translate into the inability to constrain the scope of the system sufficiently to enable subsequent requirements analysis. The urge to write two sentences in English is normally driven by two separate, or at least loosely coupled, thoughts (which probably relate to two separate, or at least loosely coupled, solutions). For similar reasons, conjunctions should not be used in the mission statement (except, of course, to connect adjectives and adjectival phrases). As well as being contained in a single sentence, the mission should be short (the sentence should be read in a single breath) because any convoluted concepts in the mission will invariably translate into a convoluted set of ambiguous requirements.

The mission statement is then expanded on and qualified by short declarative statements of the system goals and objectives. Goals are normally relatively broad statements, each of which spawns a number of more specific objectives (although in business planning these are sometimes treated in reverse order, and objectives are considered to lead to goals). Chapter 5 details how to develop the mission, goals, and objectives for a system.

3.4.1.2.2 C1.2.2—Define Preliminary Life-Cycle Concepts

Once the mission, goals, and objectives have been articulated, business management, in conjunction with stakeholders from business operations, articulates the operational concept (Figure 3.9, step C1.2.2) proposed for the system. First, the general operational environment for the system is described to identify all of the environmental factors that may influence the operation of the system. (These may or may not have been identified in the development of the mission, goals, and objectives.) Specific operational scenarios are then described to depict the full range of circumstances under which the system is required to operate. It is not necessary to describe every possible scenario, but all types of operation must be represented. Scenarios also need to represent the perspectives of all stakeholders.

Business management must also give some guidance on the life-cycle concepts related to the acquisition, deployment, operation, support, and retirement of the system. While the systems engineering procedures that follow will ensure a life cycle focus, it is important that there is an early focus on the major cost drivers that will affect all aspects of the system throughout its life. Additionally, there are a number of life cycle-related trade-offs at a business-case level.

Chapter 5 contains more detail on the definition and use of life-cycle concepts during conceptual design.

3.4.1.2.3 C1.2.3—Define Preliminary Validation Criteria

An important top-level activity is the identification and definition of preliminary validation criteria—Figure 3.9, step C1.2.3. Broadly, validation criteria encompass any mechanism by which business management will measure satisfaction with the capability system at the end of the acquisition phase. Key criteria include performance in each operational scenario, safety, reliability, supportability, maintainability, ease of use, and time and cost to train. Chapter 5 goes into more detail on measures and measure hierarchies.

3.4.1.3 C1.3—Scope System

The next phase in understanding business needs and requirements involves the development of an understanding of the scope of the system development effort (called scoping, in the vernacular). This scoping activity helps to establish a clear understanding of what the system is expected to do by defining the system context, system boundary, and external interfaces.

To assist with the scoping activity, a tool called a context diagram (Figure 3.9, step C1.3.1) may be used to illustrate the related systems, relevant regulatory environments, stakeholders, external systems, interfaces, and so on. Different systems may, of course, have significantly different context diagrams.

Definition of the system boundary (Figure 3.9, step C1.3.2) is essential early in the acquisition phase so that it is clear which elements are included in the system and which are outside. This activity is also particularly important to the project manager who is principally responsible for defining project scope—that is, what is to be included in the project as well as what is to be excluded.

Interfaces with external systems (Figure 3.9, step C1.3.3) must also be defined as these will place a considerable number of requirements on the system. While these external systems are not directly related to the project, the success of the fielded system will depend on its ability to interface to its external environment.

Finally, the draft business needs are endorsed by business management (Figure 3.9, step C1.3.4) before going on to transform those needs into formal requirements. This endorsement is most likely undertaken through a review conducted by the business owner in the presence of business management and the major stakeholders from business operations. Chapter 6 details scoping the system.

3.4.1.4 C1.4.1 to C1.4.4 Feasibility Analysis

Once business management have defined their needs—the next major activities are related to the transformation of those needs into formal requirements. At this stage of conceptual design, we have been careful to ensure that business management have stated their needs and requirements in logical terms—that is, in terms of the problem domain (the business and operational environment). For any problem domain, however, there is normally a range of potential solution spaces within the solution domain. Each alternative solution class may represent a completely different type of project; the feasibility analysis is therefore an essential step to narrowing the solution domain so that the subsequent project can be managed effectively.

If we continue without narrowing down the design into one of those domains, we will have great difficulty coping with the breadth of the design. Consequently, we next undertake a system-level feasibility analysis, the aim of which is to select a desired solution space (which is then to be elaborated further by the stakeholders at the business operations level).

Once the business needs have been identified and articulated satisfactorily, the feasible system-level alternatives available to meet the new needs can be developed. Feasible alternatives are considered in terms of available resources such as money, time, personnel, and materials. As illustrated in Figure 3.9, the following steps are completed during feasibility analysis:

- *Step C1.4.1.* Identify the alternative system-level solution classes capable of satisfying the Business Needs.
- *Step C1.4.2.* Confirm compliance of each solution class with the Business Needs and, if not completely compliant, note the likely level of achievement.
- *Step C1.4.3.* Evaluate the alternative solution classes in terms of feasibility, performance, effectiveness, technical and project risk, and other selected measures of performance.
- *Step C1.4.4.* Select the best of the alternative solution classes (ensuring that the options are narrowed down as much as is practicable at this stage).

The feasibility analysis and selection of the resultant desired solution class may require the draft business needs to be revisited in light of the more detailed investigations of the feasibility analysis. The hierarchical representation of business needs (mission, goals, and objectives) is now further elaborated (decomposed and derived) and formalized into a balanced set of business requirements, which are recorded in the BRS.

Chapter 6 further discusses feasibility analysis.

3.4.1.5 C1.5—Finalize BNR

At this stage, business management now understand the system much better than they did at the beginning when they first drafted the mission statement; not only has much more detail been elicited, but a solution class has been identified, providing a much more concrete view of the system and its operation. Consequently, the BNR can be finalized by revisiting earlier artifacts and revising them to incorporate the increased understanding. The preliminary life cycle scenarios are revised (Figure 3.9, step C1.5.1); the solution scope in terms of the mission, goals, objectives, context diagram, boundary, and external interfaces (Figure 3.9, step C1.5.2) is defined; and the preliminary validation criteria (Figure 3.9, step C1.5.3) are established. Finally, the BNR are endorsed (Figure 3.9, step C1.5.4) by business management, most likely at a major review. Finalize BNR is described in more detail in Chapter 6.

3.4.2 C2—Define Stakeholder Needs and Requirements

The endorsement of the BNR signifies that business management are confident that they have defined the problem domain in sufficient detail to communicate their needs and requirements to the stakeholders at the business operations level. These stakeholders then go on to develop their needs and requirements (the SNR) within the context of the BNR (the BRS and PLC). Figure 3.10 summarizes the activities involved with the process of the definition of SNR resulting in the development of the LCs and the StRS. Note once again that the process is iterative, allowing us to use our current understanding to improve the quality of previously generated artifacts.

The activities of Figure 3.10 are described briefly in the following sections and in more detail in Chapter 7.

3.4.2.1 C2.1—Define Stakeholder Needs

Once the BNR have been confirmed within the preferred solution class, the nominated stakeholders at the business operations level can begin to articulate their needs and requirements through the following steps:

- *C2.1.1—Identify stakeholders.* Although the major stakeholders were identified during the definition of BNR, the first act of the process of defining SNR is to refine the stakeholders by naming individuals within the major stakeholder groupings and adding lower level stakeholders at the business operations level.
- *C2.1.2—Refine mission, goals, objectives.* Within the context set by the BNR, stakeholders articulate their needs through the refinement of the mission, goals, and objectives.

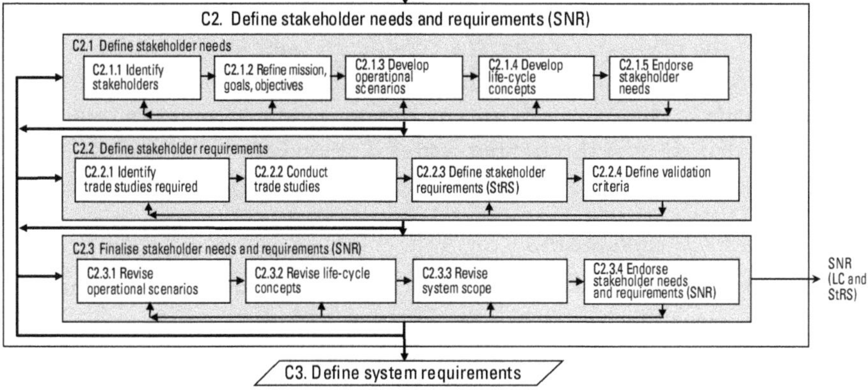

Figure 3.10 Summary of C2—Define stakeholder needs and requirements.

- *C2.1.3—Develop operational scenarios.* The preliminary operational scenarios defined by business management are now defined in more detail and other lower level scenarios (often called vignettes) are developed as the design progresses.
- *C2.1.4—Develop life-cycle concepts.* Similarly, the other PLCs are developed further and fleshed out with details from the various stakeholders to be more complete.
- *C2.1.5—Endorse stakeholder needs.* Stakeholder needs are endorsed first by the stakeholders at the business operations level and then by business management.

3.4.2.2 C2.2—Define Stakeholders Requirements

Once the stakeholder needs have been endorsed, the process of transforming their needs into formal stakeholder requirements is undertaken. As noted earlier, the RBS is a useful framework for refining the BRS into the detail of the StRS while also assisting in the maintenance of traceability.

3.4.2.2.1 C2.2.1—Identify Trade Studies Required

Similarly to how business management made use of a feasibility analysis to narrow down the alternative solution classes, stakeholders at the business operations level will have a number of alternative ways in which their needs can be met. The selection of the most suitable of the available options is best undertaken by some form of trade (trade-off) study. The first step, C2.2.1, lists those trade studies that are required, groups related trade studies for efficiency, prioritizes studies, and allocates appropriate resources.

3.4.2.2.2 C2.2.2—Conduct Trade Studies

As noted in the previous section, the conduct of trade studies did not begin here—the feasibility analysis conducted earlier can be considered as a system-level trade study—and will not end here as trade studies are an important design tool that are used throughout system development and the remainder of the life cycle. Chapter 7 provides further details on the conduct of trade studies.

3.4.2.2.3 C2.2.3—Define StRS

The development of the StRS, an essential first step toward successful system development, is given a number of titles, including simply stakeholder requirements or user requirements. A reader of the StRS should be able to understand completely the likely applications or missions for which the system is intended; the major operational characteristics to be exhibited by the system;

the operational constraints that limit the design and development of the system; the external systems and interfaces with which the system must operate; the operational and support environment within which the system must exist; and the support concept to be employed to underpin the system and enable it to continue performing in accordance with business and stakeholder needs.

3.4.2.2.4 C2.2.4—Define Validation Criteria

Each requirement in the StRS must be accompanied by a validation statement—that is, the stakeholders must not only say what it is that they require but also how it is they will know that the system can achieve that requirement.

3.4.2.3 C2.3—Finalize SNR

At this stage, the stakeholders now understand the system much better than they did at the beginning of this process—the broad guidance of the BNR has been significantly refined, and much more detail has been added as a result of the additional needs elicited from stakeholders and the decisions made as a result of a number of trade studies. Consequently, the SNR can be finalized by revisiting earlier artifacts and revising them to incorporate the enhanced understanding. The operational scenarios are revised (Figure 3.10, step C2.3.1); as are the LC (Figure 3.10, step C2.3.2); and the solution scope in terms of the mission, goals, objectives, context diagram, boundary, and external interfaces (Figure 3.10, step C2.3.3).

Finally, the SNR are endorsed (Figure 3.10, step C2.3.4) by stakeholders and by business management, most likely at a major review. Endorsement of the SNR is a critical step in developing the system in order to provide confidence to the system designers that all the fundamental stakeholder requirements have been captured. Endorsement of the StRS is not to be taken lightly. For requirements engineers, the StRS is the basis of further development of the requirements to populate the SyRS that forms the centerpiece of the FBL of the system. From a project management perspective, the StRS assists in bounding the scope of the project from which refined cost and schedule estimates can be developed.

3.4.3 C3—Define System Requirements

The next process in conceptual design is the definition of system requirements, which starts with the endorsed SNR and elaborates (decomposes/derives) them into system requirements. Figure 3.11 summarizes the activities involved with the system requirements definition process.

The aim of defining system requirements is to determine what the system must do in order to meet the business and stakeholder needs and requirements—that is, to describe the requirements at the system level and be able to relate the functional design back to the BNR/SNR, particularly the StRS. System requirements definition therefore uses elaboration to bridge from stakeholder requirements to system requirements.

During system requirements definition, the following tasks are performed:

- Establish an appropriate requirements framework;
- Perform requirements analysis and allocation—define the major requirements (with associated performance and verification) required of the new system (including maintenance, support, operational, and any other necessary functions) and assign attributes;
- Produce the draft SyRS;
- Determine the technical performance measures (TPMs) to be applied during the systems engineering effort;
- Conduct SRRs.

These activities are outlined briefly in the following sections and described in more detail in Chapter 8.

3.4.3.1 C3.1: Establish Requirements Framework

The first step in defining system requirements is to establish a requirements framework (Figure 3.11, activity C3.1) around which the logical design of the system is to be based. As noted earlier, the RBS is a useful framework for refining the StRS into the detail of the SyRS while also assisting in traceability. Since we have started with a hierarchical decomposition of mission, goals, and objectives, we can use that information to begin our development of a logical hierarchy that can be captured in an RBS.

3.4.3.2 C3.2: Perform Requirements Analysis and Allocation

This activity involves the elaboration of major requirements for the system (with associated performance, verification, and rationale) including operations, maintenance, support, and other necessary functions. When we use the word system from now on, we are using the systems paradigm, because, as noted earlier, there will be more than one SyRS (design to specification) produced for each StRS (integrated set of needs) for the capability system: one for the aircraft, one for the personnel, one for training, and so on. As noted earlier,

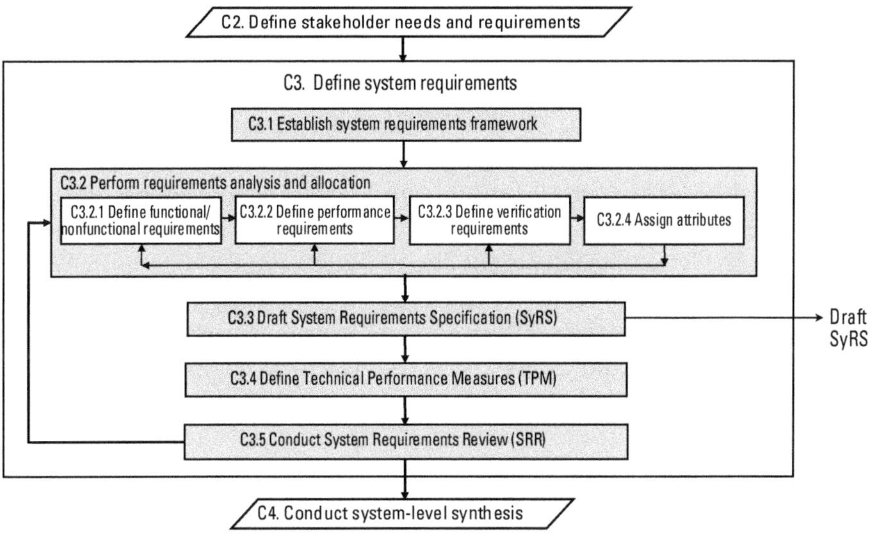

Figure 3.11 Summary of C3—Define system requirements.

the process of requirements analysis is the bridge between stakeholder requirements in the StRS to the system requirements in each of the SyRS.

3.4.3.2.1 C3.2.1—Define Functional/Nonfunctional Requirements

The central step in requirements definition (and in conceptual design, for that matter) is the definition of the requirements (functional, nonfunctional, and constraints) (Figure 3.11, step C3.2.1) for the system. While users are understandably focused on operational requirements, maintenance and support requirements also need to be defined during conceptual design. Human factors considerations such as user interface issues also produce a number of additional requirements.

Defining requirements, an important part of the overall system requirements analysis, identifies the many different functions that the system needs to perform to meet the requirements contained in the endorsed StRS. The analysis identifies all functions, including maintenance functions, that are necessary to maintain or return the system to operational use. At this stage in the development, the concentration is still on what needs to be achieved, not how it should be done. FFBDs are useful tools to assist.

Operational needs and requirements are normally given the highest priority during the early stages of a project because operational aspects tend to be the most tangible and most easily understood. The life cycle approach emphasized in systems engineering also acknowledges that design decisions made early in the acquisition phase can have a large impact on the maintenance and support of the system. To that end, maintenance and support requirements need to be determined and specified during conceptual design, so that the design of the system ensures efficient and effective support during the utilization phase.

Human factors considerations also produce a number of requirements. Requirements for the system are obtained from an analysis of factors such as physical space limitations, climatic limits, eye movement, operator reach, occupational health and safety issues, ergonomics, cognitive limits, and usability. The assessed skill levels of known users also help to constrain the design.

3.4.3.2.2 C3.2.2—Define Performance Requirements

For each requirement, requirements engineers and stakeholders must agree on the performance-related parameters (Figure 3.11, step C3.2.2) that the new system must achieve. Having decided what the system must do, for example, the designer must now determine how well the system is to perform each of those requirements. A good discipline is to ensure that every time a requirement is articulated, at least one corresponding performance statement is made.

Most of the operational functions will have obvious performance parameters associated with them such as speed, accuracy, endurance, and acceleration. Support and other functions also require parametric definition to define completely the requirement. Examples of these parametric requirements include environmental considerations, quality factors, support requirements, and utilization requirements.

3.4.3.2.3 C3.2.3—Define Verification Requirements

The definition of requirements is not complete until verification requirements (Figure 3.11, step C3.2.3) are included. It is good practice to ensure that every time a performance requirement is articulated, a corresponding verification statement is made. It is often difficult to write verification requirements for system-level functions, but the discipline of doing so is important since there is little point in stating a requirement for a function or constraint without consideration of how the function is to be verified. An additional benefit of adhering to the discipline at this stage is that test plans become much easier to write because the tests required at any level can be considered within a framework provided by an aggregation of the verification requirements at that level.

3.4.3.2.4 C3.2.4—Assign Attributes

An attribute is additional information associated with a need or requirement that is used to aid in its definition and management. Need and requirement statements are supported by associated attributes (Figure 3.11, step C3.2.4) that aid in the definition and management of needs and requirements and of sets of need and requirements [24]. Attributes also help to achieve many of the characteristics of needs and requirements defined in the INCOSE *Guide to Writing Requirements* [25].

It is particularly useful to record the rationale behind each requirement, which further assists with the removal of ambiguity and, more importantly, assists in situations where more than one person is involved in the requirements analysis over the course of the project. The rationale explains why each requirement is necessary and the logic behind performance levels assigned to the requirement. This rationale remains attached to a requirement throughout the acquisition phase, providing guidance to lower level designers as to why and how the requirement has been derived and why the performance and verification requirements were selected.

3.4.3.2.5 Requirements Analysis and Allocation

The requirements articulated by the stakeholder are generally at a high level and often tend to be more qualitative than quantitative. For example, our stakeholders require that the new aircraft can operate from any class X airfield. While stakeholders will articulate some additional requirements during requirements elicitation/generation, these broader functions need to be analyzed to identify the lower level requirements necessary to achieve the parent requirement. As discussed in Chapter 2, these lower level requirements are often called decomposed or derived requirements, because they are not stated directly by a stakeholder but are an artifact of the requirements-analysis process. For example, if the parent requirement is to be able to operate from any class X airport, designers would decompose/derive requirements and constraints on the aircraft associated with operation from that class of airfield, which may include requirements such as minimum runway lengths, runway surfaces, and maximum allowable aircraft weight.

Requirements analysis also involves investigation of the relationships between requirements. In addition to the derivation of requirements during requirements analysis and allocation, system designers group and allocate requirements. Allocation has three aspects. First, requirements are allocated to the appropriate group of the RBS. Second, higher level requirements flow down to lower level requirements. Finally, there is the logical-to-physical allocation in preliminary design as the system-level functions are allocated to the physical configuration items.

The RBS as it is developed is useful in supporting the design and communicating with stakeholders. The RBS simply states the function required of the system, however, and it is sometimes difficult to determine how those functions relate to each other. The flow of functions is ideally illustrated by FFBDs, which are also useful tools at this stage, both as analysis tools and as additional mechanisms for communication with stakeholders.

In performing the system requirements analysis, the following tasks are performed:

- Determining the major functions required of the new system (including maintenance, support, and operational functions);
- Assigning the necessary performance-related parameters or requirements to each of the functions, as applicable;
- Articulating any other system-level requirements;
- Determining the TPMs to be applied during the systems engineering effort.

3.4.3.3 C3.3—Draft SyRS

The complete and populated RBS forms the framework for the SyRS, which, when approved, becomes the centerpiece of the system's FBL. At this stage, however, the populated RBS only forms the basis for the draft SyRS (Figure 3.11, activity C3.3), because the requirements collected and analyzed so far have yet to be synthesized into an architectural solution. There will therefore be some degree of latitude allowed in certain requirements and priority labels (such as mandatory, important, and desirable), leaving some flexibility for subsequent design.

The SyRS may take any one of many forms [26] and the most appropriate form depends on the application. The SyRS is perhaps the most important of all systems engineering artifacts, because it becomes the source of reference for all the subordinate specifications that are produced at later stages through the design process. A solid SyRS provides an effective foundation for the remainder of the design and development effort. Errors or omissions in the SyRS flow into the remaining design effort. The later these errors are discovered, the more expensive and time-consuming the rectification becomes.

The endorsed stakeholder requirements in the StRS are expanded in the SyRS to be a set of system-level requirements that, when developed, will result in a system that meets the StRS requirements. However, the SyRS is much more detailed and has transformed the StRS statements in the stakeholders' language into more formal requirements that relate solely to the system. That is, the SyRS therefore only includes statements that relate to the system of interest that is the subject of that SyRS—the aircraft, the facility, the simulator, and so on.

3.4.3.4 C3.4—Define TPMs

Some of the requirements from the set determined in the preceding activities may be considered to be useful as key indicators of system performance (or may represent particular risk)—a sort of "health check." These selected requirements are the TPMs (Figure 3.11, activity C3.4).

IEEE-STD-1220 states that appropriately selected TPMs can be used to assess conformance to requirements, assess conformance to levels of technical risk, trigger development of recovery plans for identified deficiencies, and examine marginal cost benefits of performance in excess of requirements [27]. TPMs, identified early in the system development effort, are continually monitored and tracked throughout the system development as a means of managing the progress and risks associated with system development.

3.4.3.5 C3.5—Conduct System Requirements Reviews (SRRs)

SRRs may be conducted periodically throughout conceptual design to verify and approve versions of system-level requirements (Figure 3.11, activity C3.5). The aim of SRRs is to monitor and approve progressively the system-level requirements that are developed on the way to the FBL. Progressive reviews allow the requirements analysis effort to continue with confidence in the preceding decisions. SRRs may or may not be considered formal reviews.

The SRRs may also review other information such as manufacturing plans, design schedules, and personnel requirements plans that will assist in fine-tuning the PLCs. The number of SRRs depends on the size and complexity of the system. Large and complex systems will require multiple SRRs during conceptual design, whereas simple systems may require very few. It is recommended, however, that at least one SRR is held, even in simple projects, for each of the major activities described above—the development of BNR, SNR, and system requirements—so that business management and stakeholders have some formal involvement in the transition from the BNR (in particular, the BRS) to the SNR (in particular, the StRS) to the SyRS.

3.4.3.6 Project Management, Systems Engineering, and Logistics Considerations

Requirements are captured in the RBS, which when complete, forms the structure of the SyRS. There are, however, a number of additional considerations that need to be captured. These additional issues relate to how the customer expects the system development to proceed and to management issues such as project management, systems engineering, and logistics.

The following list provides a nonexhaustive set of example issues that need to be considered and specified prior to the commencement of system development. As described, most of the following examples will not reside in the RBS (SyRS) but may be more at home in, and will therefore also affect, the WBS and other associated contractual documentation such as a SOW:

- Organizational responsibilities for the project management, systems engineering, logistics management, operation and support should be determined and clearly articulated so that all stakeholder organizations are aware of their respective responsibilities during the different stages of the system's life cycle. Some of these requirements are contained in the contractual documentation, and some are contained in internal customer processes and procedures.
- Project management consolidations such as principal responsibilities, stakeholder engagement, schedule issues, budget issues, risk management, project management milestones, and payment schedules.
- Systems engineering aspects including T&E expectations, configuration management requirements, documentation deliverables, technical reviews and audits, and adherence to certain engineering standards. These requirements need to be considered and included because they will dictate how the design and development effort will be organized and managed. Systems engineering aspects add expense and time to the project but should reduce the technical risks of the system development effort.

3.4.4 C4—Conduct System-Level Synthesis

System design has now progressed to the stage where some of the system-level design decisions can be made. Synthesis (at the conceptual design level) establishes a system configuration that is representative of the final system form. The configuration established at this stage is not assumed to be final as the design is immature and may go through significant changes later in the design process.

Based on the results of the requirements engineering and analysis, a range of architectural options is developed (Figure 3.12, activity C4.1) that represents potential system solutions. The selection of one of these options as the preferred solution requires information to be collected (Figure 3.12, activity C4.2) from the systems engineering process, life cycle costs, quality assurance, T&E, maintenance, integrated logistics support, and so on. The potential solutions are then evaluated (Figure 3.12, activity C4.3), which requires the development of suitable evaluation criteria as well as an evaluation framework. Any discrepancies in the systems engineering products are noted) and fed back to previous activities. The preferred solution is then chosen (Figure 3.12, activity C4.4).

The system synthesis activities described in activities C4.1 to C4.4 in Figure 3.12 can also be illustrated by Figure 3.13, in which architectural

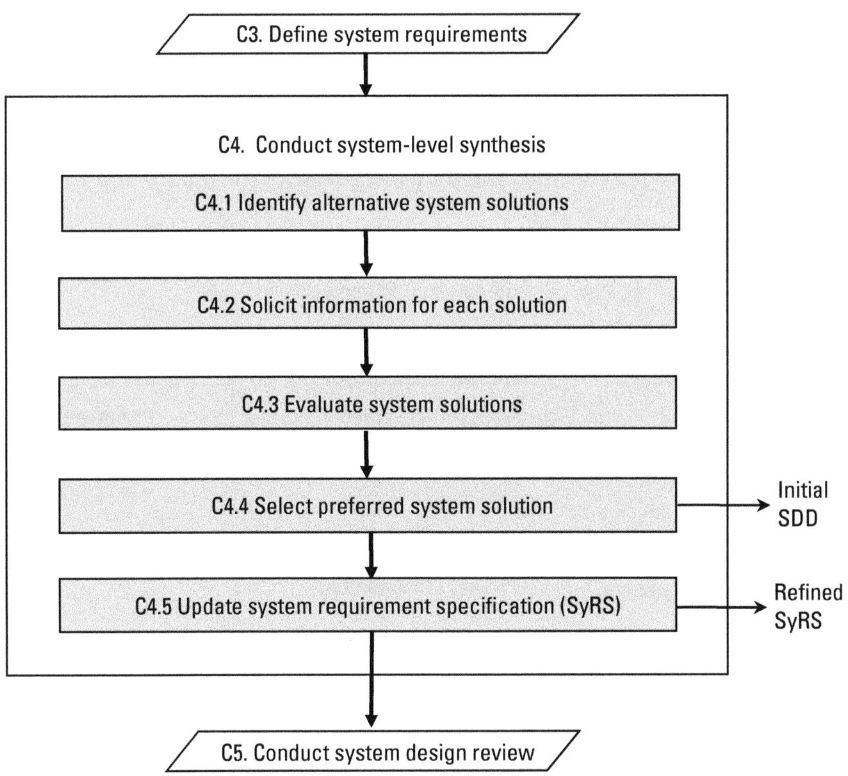

Figure 3.12 Summary of C4—Conduct system-level synthesis.

options are developed; they are evaluated in the context of an evaluation framework using selected criteria with input from the stakeholders involved; and a preferred system architecture is developed.

In some applications of the systems engineering process, the evaluation of alternatives may take the form of evaluation of responses to a request for tender (RFT). In that case, potential contractors are provided with a draft SyRS and draft contractual documentation (including a draft SOW) upon which they base their proffered system-level solutions. The customer then evaluates the tender responses (which represent feasible system-level designs) before selecting the preferred system-level solution. In the process, the draft SyRS is further refined to make it more suitable as the system's initial FBL and to take into account the range of feasible alternatives. As feasible alternatives are presented, the customer may need to revisit the StRS and seek endorsement from stakeholders for proposed changes. Some of the endorsed requirements may not be feasible within the approved schedule or cost constraints.

The final activity in process C4 (Figure 3.12, activity C4.5), produces the refined SyRS. The draft SyRS necessarily contains margins of acceptable performance and requirement priorities. Potential system solutions have now been investigated (at least in broad terms), and the preferred approach to the problem has been determined. With this approach in mind, the draft SyRS must be refined to remove the ranges of acceptable performance and to reflect

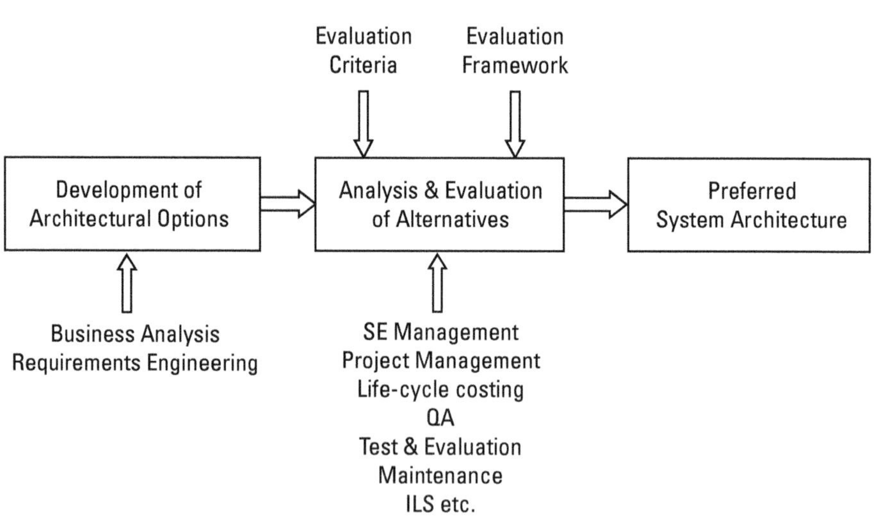

Figure 3.13 System synthesis activities—selection of the preferred system-level solution.

exactly what is being offered by the preferred system solution and the minimum acceptable level of performance. The main products of synthesis are the refined SyRS and a broad system solution for achieving the specification requirements.

The SyRS is perhaps the most important of all systems engineering documents because it becomes the source of reference for all the subordinate specifications that are produced during later stages of the design. If the SyRS is solid, it forms an effective foundation for the remainder of the design and development effort. Conversely, errors or omissions in the SyRS flow into the remaining design effort. The later these errors are discovered, the more expensive and time-consuming the rectification will be.

3.4.5 C5—Conduct SDR

At the end of conceptual design the SDR provides the following:

- A formalized check of the logical design with respect to the specified requirements;
- A formalized communication of the intended design approach to the major players in the design effort;
- A means by which interface issues can be raised, discussed, and resolved;
- A formal record of design decisions and acceptance.

Figure 3.14 summarizes the activities involved in the SDR process.

SDR occurs once the entire set of requirements has been placed in the SyRS. It is essential in all reviews to establish the entry and exit criteria for the review—that is, it must be clear what needs to be in place before the review begins as well as what must have been achieved before the review can be considered as complete. Consequently, the review cannot begin until the entry criteria have been met (Figure 3.14, activity C5.3). Part of that confirmation is to ensure that all of the necessary documentation has been prepared and reviewed (Figure 3.14, activity C5.1) in advance of the review and that all parties have agreed to the agenda for the review (Figure 3.14, activity C5.2).

The principal purpose of the review is to examine each requirement to ensure that it is met by the preliminary system architecture determined during system-level synthesis—that is, the review confirms the system-level solution (Figure 3.14, activity C5.4). If all is satisfactory in this regard, the SyRS is approved and baselined as part of the initial FBL. It is rare that SDR ends with all outstanding issues resolved during the review, since many will take some time to resolve. Rather than hold up subsequent design, action items are

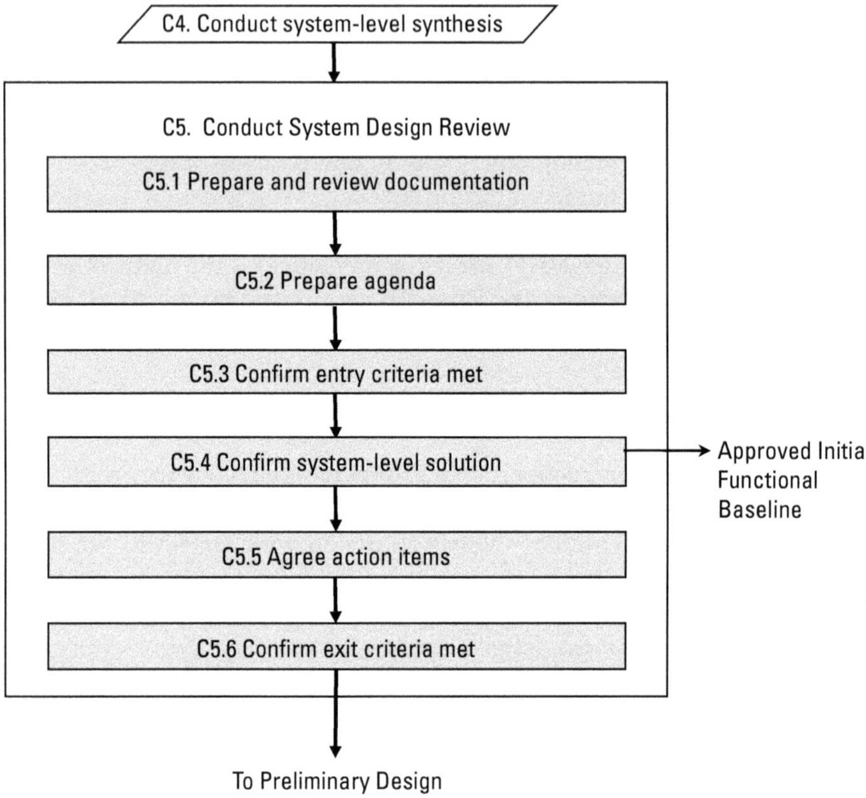

Figure 3.14 Summary of C5—Conduct system design review.

agreed on (Figure 3.14, activity C5.5) with agreed-upon timelines to account for any outstanding action from the review (providing, of course, that resolution of those issues will not affect the remainder of the design). These actions are completed in parallel with the early preliminary design activities and are reviewed for completeness in conjunction with a later review or audit. Before the review can be considered to be complete, the exit criteria must be met (Figure 3.14, activity C5.6).

The delineation between conceptual design and preliminary design is often not clear. SDR aims to provide some demarcation by performing the following functions:

- *Evaluation of the system-level design proposal:* This evaluation investigates the activities of conceptual design to estimate the ability of the proposed system-level designs to meet the system-level requirements (remembering that these requirements include operational and support requirements and that the system design includes hardware, software, personnel, and facilities). This may involve a review of the major subsystems and their role in achieving system performance.

- *Approval of the SyRS:* The SyRS is the document that needs to reflect the system-level requirements approved during SRR. New system-level requirements may need to be added to the SyRS, and errors or conflicts in existing requirements may need to be clarified prior to the final approval of the SyRS. It is recommended that the acceptable margins determined during SRR will be replaced by a more acceptable performance level for each requirement. It is also recommended that the priorities (essential, important, and desirable) are removed so that each statement in the specification becomes an unambiguous requirement.
- *Establishment of the initial FBL:* Through the approval of the SyRS, the initial FBL for the system is established—this is the baseline from which the remainder of the design will follow.

3.5 Example Systems Used as Case Studies

Throughout the following chapters we use a number of examples, wherever possible, to illustrate and reinforce the key concepts as we introduce them. There is no intent to capture the entire design process with these examples. Rather, the systems have been chosen as convenient examples that can be readily recognized by readers from a wide variety of disciplines and specialties. That is, readers are not forced to become domain experts in a particular field just to understand the illustration. Through these examples, the majority of readers can immediately understand the system context and business, stakeholder, and system needs; subsequent requirements; interface issues; technical performance measures; the logical-to-physical translation; broad trade-off analyses; and the physical configuration items involved in the final design.

It should also be noted that there is no suggestion that the aggregation of examples throughout the text represents an adequate or complete design for any of the example systems; the available space prohibits the inclusion of sufficient detail regarding the context, which would also obscure the general lessons provided. Additionally, an illustration often leaves an aspect of an artifact unfinished to illustrate the application of CDM, rather than to complete the design.

3.5.1 Medical Center Example

ACME Medical Services has just acquired a block of land with the intention of developing a medical center called the ACME Medical Center.

The block of land is located at the outer edge of one of a major suburban shopping mall (Figure 3.15). The block had been occupied by a hotel complex that was torn down by the previous owner with an intention to redevelop. The block is on a corner of the shopping mall bounded on two sides by the mall itself. It is relatively flat and has been cleared of any rubble. Services such as gas, electricity, water, and telecommunications had been provided to the old hotel, and access to each is still available at the perimeter of the site. Ample car parking for patients is available around the shopping mall, and there is no intention to provide special parking spaces for the medical center. The shopping mall has a body corporate that has developed guidelines for buildings in the area, covering factors such as the overall look of the building and location of signage. The local council zoning regulations limit the building to two floors.

ACME intends to provide medical [general practitioner (GP)], dental, physical therapy, and pharmacy services. Its feasibility study has identified that a profitable mix would comprise a doctor's surgery supporting three GPs, a dentist's surgery for two dentists and two dental assistants, a physical therapy

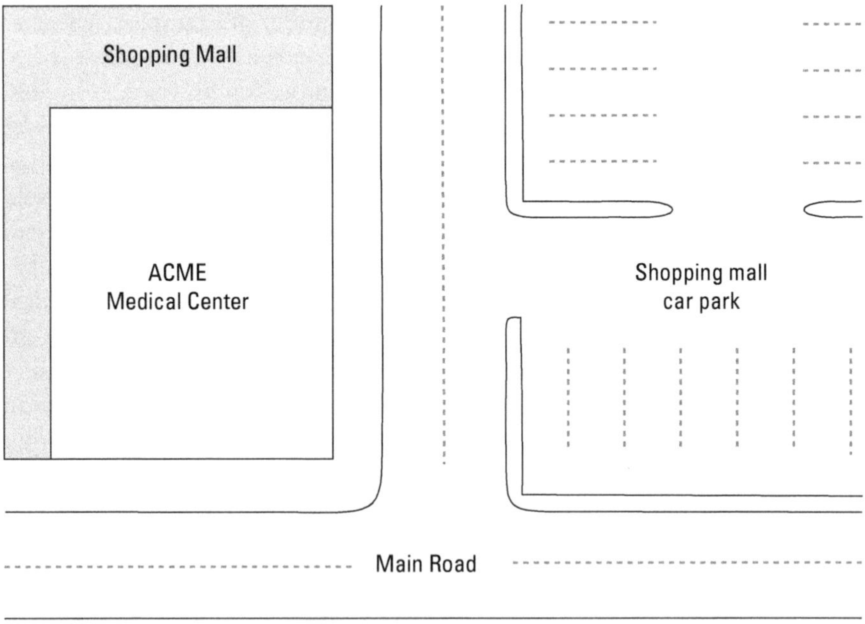

Figure 3.15 Location of the ACME Medical Center.

center supporting three physical therapists, and a pharmacy with a basic dispensing service provided by one pharmacist and an assistant. Services offered by the GPs, dentists, and physical therapists are only available to Medical Center patients; the pharmacy is to be available to Medical Center patients, as well as to general customers who are not patients but wish to use the pharmacy.

All professional and general staff are employees of ACME, which finances and manages the venture. The center will be managed by a center manager, who will have an assistant manager.

ACME would like a center that provides a comfortable (but not extravagant) environment for patients. Waiting rooms should be roomy and comfortable. Security is important, as are the maintenance and logistics aspects of the center. However, ACME intends to make the operational aspects of the center invisible to the patients, thereby not intruding on patients' experience of the center. Patients will park in the shopping center car park.

ACME is an employer with a strong record of caring for its employees. The accommodation for staff should be comfortable, with provision of a kitchenette and restroom facilities that are separate from the patients. Private parking is not required for center staff; like the patients, they will park in the shopping mall car park.

This is the major example used throughout Chapters 4 to 8.

3.5.2 Replacement Aircraft Example

An aircraft operator (ACME Air) has identified a business need for a class-leading, medium-sized aircraft to replace the aging platform that it currently operates over domestic routes and some short international routes.

3.5.3 Small Business Local Area Network Example

A small organization is looking to install a local area network within its leased premises.

3.5.4 Domestic Security Alarm Example

Another division of ACME Industries, ACME Alarms, has a business need to develop a domestic security alarm. The company proposes to sell the alarm into the domestic market to compete in price and functionality for all forms of domestic dwellings such as houses, condos, and apartments. The alarm is to be capable of being installed by the customer and must be able to operate in a back-to-base monitored mode as well as a stand-alone mode. ACME expects the alarm to be designed, developed, and manufactured in house as it has an existing product in the marketplace, a car alarm. The company therefore has a design team, a technical regulatory framework for engineering, and a production line. It would like to reuse as many of the car alarm parts (as much of its production line) as possible in the domestic security alarm.

3.5.5 Defense Long-Range Communications System Example

A country's defense force requires a long-range communications system to support connectivity between the defense force's headquarters and deployed forces, which may be as far away as 1,200 miles.

3.5.6 Cockpit Voice Recorder Example

After a recent safety review, the federal government has directed the installation of cockpit voice recorders (CVRs) into all civilian rotary wing aircraft. ACME Air therefore needs to install CVR into its current fleet of Whirlybird medium-lift helicopters.

3.6 Summary

Chapters 1 to 3 have provided the context within which CDM activities are undertaken. The following chapters provide greater detail on those activities: Chapter 4 (Identify Major Stakeholders and Constraints), Chapter 5 (Define Business Needs), Chapter 6 (Scope System and Define Business Requirements), Chapter 7 (Define Stakeholder Needs and Requirements, and Chapter 8 (Define System Requirements). Chapter 9 focuses on writing needs and requirements statements as part of the recording of text-based BNR, SNR, and system requirements in the SyRS.

Review Questions

1. Briefly describe the key aspects of CDM: top down, use of intuition and bottom-up analysis, use of iteration, and traceability.
2. Draw a diagram to illustrate the five CDM processes (C1—Define business needs and requirements; C2—Define stakeholder needs and requirements; C3—Define system requirements; C4—Conduct system-level synthesis; and C5—Conduct SDR.

3. Draw a diagram to illustrate the CDM process of C1—Define business needs and requirements. Briefly describe each of the major activities in the process.
4. Draw a diagram to illustrate the CDM process of C2—Define stakeholder needs and requirements. Briefly describe each of the major activities in the process.
5. Draw a diagram to illustrate the CDM process of C3—Define system requirements. Briefly describe each of the major activities in the process.
6. Draw a diagram to illustrate the CDM process of C4—Conduct system-level synthesis. Briefly describe each of the major activities in the process.
7. Draw a diagram to illustrate the CDM process of C5—Conduct SDR. Briefly describe each of the major activities in the process.

Endnotes

[1] EIA/IS-632, *EIA Interim Standard—Systems Engineering*, Washington DC: Electronic Industries Association, 1994.

[2] MIL-STD-499B, *Military Standard—Systems Engineering—Draft*, Washington DC: US Department of Defense, 1994.

[3] IEEE-STD-1220, *IEEE Standard for Application and Management of the Systems Engineering Process*, New York: IEEE Computer Society, 2005.

[4] EIA-632, *Processes for Engineering a System*, Washington, DC: Electronic Industries Association (EIA), 2003.

[5] ISO/IEC/IEEE 15288, *Systems Engineering—System Lifecycle Processes*, International Standards Organization, 2023.

[6] INCOSE, *Systems Engineering Handbook: A Guide for System Lifecycle Processes and Activities* [Walden, D.D., et al., (eds.)], INCOSE-TP-2003-002-05, Hoboken, NJ: John Wiley & Sons, 2023.

[7] Wheatcraft, L. S., M. J. Ryan, and T. E. Katz, *Needs and Requirements Manual: Needs, Requirements, Verification, Validation Across the Lifecycle*, Hoboken, NJ: John Wiley & Sons, 2024.

[8] MIL-STD-499B, *Military Standard—Systems Engineering—Draft*, Washington DC: US Department of Defense, 1994.

[9] IEEE-STD-1220, *IEEE Standard for Application and Management of the Systems Engineering Process*, New York: IEEE Computer Society, 2005.

[10] EIA-632, *Processes for Engineering a System*, Washington, DC: Electronic Industries Association (EIA), 2003.

[11] EIA-632, *Processes for Engineering a System,* Washington, DC: Electronic Industries Association (EIA), 2003.

[12] Jones, B. L., and M. J. Ryan, "ISO/IEC 15288(E)—Visual Enhancement for Effective Communication," *Systems Engineering/Test and Evaluation Conference SETE2011,* Canberra, Australia, May 2–4, 2011.

[13] International Council on Systems Engineering (INCOSE), INCOSE-TP-2003-002-05, *Systems Engineering Handbook* (fifth edition), Hoboken, NJ: John Wiley & Sons, 2023.

[14] Wheatcraft, L. S., M. J. Ryan, and T. E. Katz, *Needs and Requirements Manual: Needs, Requirements, Verification, Validation Across the Lifecycle,* Hoboken, NJ: John Wiley & Sons, 2024.

[15] International Council on Systems Engineering (INCOSE), INCOSE-TP-2021-003-01, *Guide to Needs and Requirements,* 2022.

[16] International Council on Systems Engineering (INCOSE), INCOSE-TP-2010-004-04, *Guide to Writing Requirements,* 2024.

[17] International Council on Systems Engineering (INCOSE), INCOSE-TP-2021-006-01, *Guide to Verification and Validation,* 2022.

[18] International Council on Systems Engineering (INCOSE), INCOSE-TP-2024-146, *Guide to Security Needs and Requirements,* 2024.

[19] Wheatcraft, L. S., M. J. Ryan, and T. E. Katz, *Needs and Requirements Manual: Needs, Requirements, Verification, Validation Across the Lifecycle,* Hoboken, NJ: John Wiley & Sons, 2024.

[20] Miller, G. A., "The Magical Number Seven, Plus or Minus Two: Some Limits on Our Capacity for Processing Information," *The Psychological Review,* 63, 1956, pp. 81–97.

[21] Simon, H. A., *The New Science of Management Decisions,* Englewood Cliffs, NJ: Prentice Hall, 1977.

[22] Van Gundy, A. B., *Idea Power: Techniques and Resources to Unleash Creativity in Your Organization,* New York: American Management Association, 1992.

[23] IEEE-STD-1233, *IEEE Guide for Developing System Requirements Specifications,* New York, NY: The Institute of Electrical and Electronic Engineers (IEEE), 1998.

[24] Wheatcraft, L. S., M. J. Ryan, and T. E. Katz, *Needs and Requirements Manual: Needs, Requirements, Verification, Validation Across the Lifecycle,* Hoboken, NJ: John Wiley & Sons, 2024.

[25] International Council on Systems Engineering (INCOSE), INCOSE-TP-2010-004-04, *Guide to Writing Requirements,* 2024.

[26] Sources of information on the content of the SyRS can be found in the following:

Data Item Description (DID), *System/Subsystem Specification (SSS),* DI-IPSC-81431, DD Form 1664, US Department of Defense, April 1989.

ECSS-E-10A, *Space Engineering—System Engineering*, Noordwijk: The Netherlands: European Cooperation for Space Standardization, 1996.

MIL-STD-961E, *Department of Defense Standard Practice for Defense and Program-unique Specifications—Format and Content*, Washington DC: US Department of Defense, August1, 2003.

MIL-STD-490A, *Military StandardSpecification Practices*, Washington, DC: US Department of Defense, 1985.

[27] IEEE-STD-1220, *IEEE Standard for Application and Management of the Systems Engineering Process*, New York: IEEE Computer Society, 2005.

4

Identify Major Stakeholders and Constraints

4.1 Introduction

As discussed in Chapter 3, conceptual design begins with identification of the major stakeholders and of the constraints that apply to the system and its development. As illustrated in Figure 4.1, the steps involved in C1.1—Identify major stakeholders and constraints include the identification of major stakeholders, followed by the identification of business, project, external, and design constraints.

4.2 C1.1.1—Identify Major Stakeholders

Stakeholders are the owners of the problem at the business management and operations levels and are the people from whom we elicit stakeholder needs and requirements and then from whom we obtain the input necessary during requirements analysis to assist in tasks such as trade studies. Their input to the system life cycle is crucial. Careful selection of appropriate stakeholders is therefore fundamental to the success of the project. It follows that if we choose a different set of stakeholders we will most likely end up with a different set of system requirements and, as a result, a different system. It is essential that

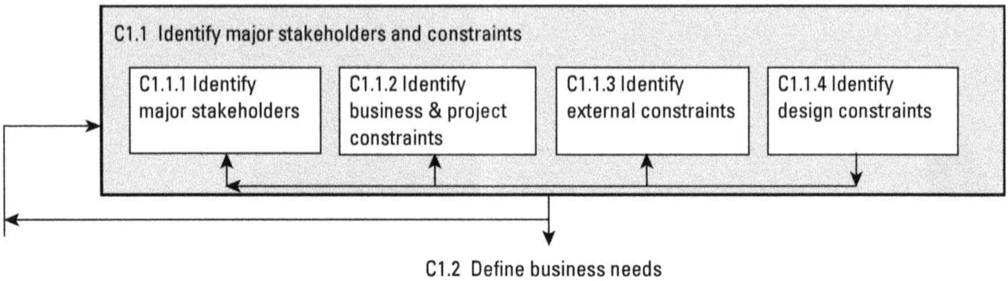

Figure 4.1 Steps involved in C1.1—Identify major stakeholders and constraints.

major stakeholders are nominated by business management—it is not reasonable for business management to pass their needs and requirements to the operations level and to let those at that level decide among themselves who are to be the stakeholders of the system.

4.2.1 Who Is a Stakeholder?

Stakeholders invariably include managers at the operations level and may include maintainers, logisticians, and perhaps clients—but the selection of stakeholders warrants considerably more thought. In project management, a stakeholder is defined as someone who has a stake in the project—that is, someone who is affected by the system in some way or who can affect the system in some way [1]. This definition is commonly used in requirements engineering; for example, Young defines a stakeholder as anyone who has an interest in the system or in its possessing qualities that meet particular needs [2], or as any group, individual, or organization that is affected by or can affect the project [3]. As an example, Bittner and Spence identify the users, the development team, investors, sponsors, authorities, customers, and even the developers of third-party software, as stakeholders [4]. Hull, Jackson, and Dick define a stakeholder as any person or organization that has an opinion or a responsibility for, or who may be influenced or affected by the proposed system [5]. Kotonya and Sommerville define a stakeholder as anyone who is affected by the system in some way [6].

While these definitions may be a place to start, the identification of stakeholders cannot involve simply listing those who can affect either the system or the project, or those who perceive themselves to be affected by either. In most systems, this is not a useful definition since it is often difficult to find someone who is not affected by the system in some way, or able to affect it. By this definition, in fact, even in a simple system such as an automatic teller machine (ATM) network for a bank, there may be millions of stakeholders. This is also true in committee-based organizations such as public service organizations, where the number of potential stakeholders is almost limitless, particularly if taxpayers are included.

We, therefore, cannot progress with such a definition of the people from whom we elicit requirements; the reasons are outlined as follows:

- First, if there are so many, we cannot possibly have the time to engage with everyone who will have an involvement with the new system (even if we only restrict ourselves to representatives of selected groups).
- Second, it assumes that anyone affected by the system is a source of requirements. Our competitors will be affected, for example, but we are not going to ask them what they would like to see in the new system, particularly since we are planning their demise. Similarly, potential intruders are definitely going to be affected by our domestic alarm, and we will certainly take them and their skill sets into account during the design, but they are not stakeholders.
- Additionally, such a broad definition would mean that anyone who wants to influence requirements should be able to—we would therefore be at the mercy of any stakeholders who want to contribute requirements, whether they are useful or not.
- Finally, the new system will have disparate effects on stakeholders—they will not be affected equally, and they should not therefore have equal rights in expressing requirements. No complex system can be optimum to all parties concerned; nor can all functions be optimized [7]. Stakeholders must therefore be prioritized, or we will be left with an intractable problem with no design space available.

For these reasons, the identification of stakeholders involves much more than simply listing all those who could be considered to have a stake in the new system. While we must consider anyone who is affected by the new system when we are considering requirements, they are not automatically stakeholders. For example, when considering the design of an ATM system, the armed guards who distribute the cash around the system are clearly important actors within the system; although their requirements place considerable constraints on the design of the system, they are not stakeholders. Similarly, the tellers whose employment may be threatened by the proliferation of ATMs are clearly affected by the system, but they are not necessarily stakeholders.

Section 4.2.4 examines these issues through further detailed examples of stakeholder identification.

4.2.2 A Better Definition of Stakeholders

Rather than defining a stakeholder as someone who can affect or be affected by the system, a stakeholder is more usefully defined as someone (either individually or as a representative of some organization) who has a *right* to influence the outcome of the system, rather than someone who is simply affected by the system. In other words, stakeholders are not just those involved in the system or affected by a problem or opportunity that results from the system; stakeholders are those with a right to affect the outcome of the problem-solving activity.

Alexander and Stevens provide a similar definition of a stakeholder as someone who has a justifiable claim to be allowed to influence the requirements—although they go on to include people whose lives are affected by the system, managers who are concerned to see the system succeed, and even regulators such as local and state governments [8].

Hence, even our better definition does not allow us to identify our stakeholders automatically. If a stakeholder has a right to influence the requirements, we need to identify what or who gives them that right. Even then, as discussed in Section 4.2.1, we need to examine candidate stakeholders more carefully. To do so, we may consider the following questions.

- Do all stakeholders have equal rights?
- If not, who decides which stakeholders have higher priority?
- What do we do if stakeholders do not agree?
- If a group of people is considered to be a stakeholder, do they all have a voice, or is a spokesperson to be elected/nominated?
- How do we discount requirements collected from a stakeholder who is clearly confused and whose contributions are unenlightening?

These issues are addressed by the formal process outlined in Sections 4.2.3 and 4.2.4.

4.2.3 Identifying Major Stakeholders

The number of tasks involved in the C1.1.1—Identify major stakeholders step are listed as follows and detailed in Sections 4.2.3.1 to 4.2.3.11.

- Identify a business owner that is able and willing to conduct/endorse the activity of identifying stakeholders and to assist in stakeholder management;
- Identify all candidate stakeholders (anyone affected by the system or able to affect it);

- Evaluate candidates and select stakeholders (individuals or groups of stakeholders by type);
- Understand roles, responsibilities, and interrelationships of stakeholders and stakeholder groups;
- Nominate the principal stakeholder;
- Identify a stakeholder representative (an individual) for each stakeholder group;
- Prioritize stakeholders;
- Form the stakeholder board;
- Develop stakeholder management strategies;
- Prepare the stakeholder management plan (SMP).

4.2.3.1 Identify a Business Owner

In the context of a particular system development, there must be a business owner (most likely at the business management level) who is responsible for (and accountable/answerable for) the system development and its associated project. This individual (or perhaps committee) must have sufficient authority to make decisions with regard to the project and the system development. Additionally, they must be able to endorse the SMP, to appoint stakeholder representatives, and to agree to stakeholder priorities. They must also have sufficient authority to settle disputes between parties. Without such an owner of the process, very few of the necessary trade-offs will be able to be made during the conceptual design process, and we will be left with no design space. With this broad range of responsibilities, the responsible business owner is also commonly referred to as the project champion [9]; here we call them the business owner. The business owner may not have sufficient time to devote to the system design, so they could nominate one of the major stakeholders to be the principal stakeholder, to whom the business owner's accountabilities are delegated.

In this key role, the business owner (or the delegated principal stakeholder):

- Nominates stakeholders;
- Allocates stakeholder rights and priorities;
- Manages the elicitation/elaboration process;
- Chairs the stakeholder board (or delegates the chair);
- Arbitrates stakeholder conflict and makes hard decisions where necessary;

- Provides the conduit between the business (the business case) and the project (particularly the conceptual design effort);
- Negotiates resources on behalf of the project;
- Protects the project from any vagaries of the business environment;
- Manages stakeholder and business expectations.

4.2.3.2 Identify Candidate Stakeholders

The first task in identifying stakeholders is to identify the candidate stakeholders—that is, to list all those actors who may have, or may be perceived to have, or who may perceive themselves to have, a stake in the project. In accordance with the commonly used definition of a stakeholder, we should identify and assess anyone who is either affected by the outcome of the system or who is able to affect it. Where possible we should identify them by type—that is, we should group them in accordance with their common interest and common perceived stake.

We should recognize that we may not be able to identify all candidate stakeholders at the outset; like most aspects of conceptual design, our ability to identify stakeholders will improve as we understand the system better as we progress through the design. The steps that follow should therefore be applied iteratively.

4.2.3.3 Evaluate Candidates and Select Stakeholders

Once we have a list of candidate stakeholders, we evaluate each to determine whether they are a genuine stakeholder, or whether they are simply an actor within the system, or simply a constraint placed either on the way in which the system is to operate or how it is to be developed. It must be emphasized that anyone on our list of candidate stakeholders is important, and we must engage with them. The crucial difference lies in what we make of their contribution. Genuine stakeholders have the right to influence the system requirements, so it is from them that we allow opinions about what the system should do and how it should look and perform. Others such as operators and maintainers may be stakeholders in some circumstances, but, since it is not necessarily their business, we are normally interested in their requirements as constraints, rather than considering them as stakeholders.

If we want to include all actors as stakeholders rather than constraints (perhaps at least for political reasons), we might consider prioritizing stakeholders so that we can manage them better. It may also be useful for these actors to consider themselves as stakeholders rather than constraints, even though we consider them more as the latter when we take into account their requirements. See the discussion in Section 4.2.3.6.

4.2.3.4 Understand Roles and Responsibilities of Stakeholders

Once we have defined the stakeholders and stakeholder groups, we must understand the roles and responsibilities of each individual and group and how they interact. This understanding is essential to guide the conceptual design process in aspects such as requirements elicitation/elaboration, and in trade studies—it is also an essential precursor to the prioritization (categorization) of stakeholders.

4.2.3.5 Nominate the Principal Stakeholder

The business owner may be able to be available for the entire conceptual design process. In many cases, however, it would suit their purposes to nominate a principal stakeholder to whom one or more roles can be delegated. For example, if the business owner is not going to chair the stakeholder board, they would nominate the principal stakeholder to be the chair. The principal stakeholder should be someone in the organization as senior as possible and supportive of the venture. Additionally, the principal stakeholder should have the same desirable qualities as the stakeholder representatives, as outlined in Section 4.2.3.6.

4.2.3.6 Identify Stakeholder Representatives

We cannot engage with all stakeholders of a certain type, because we don't have enough time, and, even if we did, we would be at the mercy of such a large number of opinions. (Even if they are similar in background and emphasis, each person will provide us with a slightly different view.) We must therefore identify appropriate representatives who can speak on behalf of that stakeholder community—it is from these representatives that we elicit requirements and from whom we seek input necessary during requirements analysis and negotiation. Since their input to the system life cycle is crucial, appropriate representatives must be selected with care.

In addition to being able to contribute to the requirements elicitation process, it is very important that each representative is supportive and powerful (at least within their community). This individual must be accepted by the group as their spokesperson so that they are able to speak with authority on behalf of the group.

Ideally, a suitable representative of each stakeholder group needs to do the following:

- Understand the business domain;
- Understand this particular application area within the business domain;
- Understand the specific problem;

- Understand systems engineering;
- Understand the conceptual design process;
- Understand engineering;
- Understand the associated technology;
- Faithfully represent the requirements of their stakeholder community;
- Be supportive of the system and the requirements definition/analysis process;
- Be powerful, at least within their stakeholder community;
- Have sufficient time to be able to contribute when required.

Clearly, the success of the conceptual design process (particularly elicitation and elaboration) will depend heavily on the careful selection of stakeholder representatives. Even then, since there are not likely to be too many stakeholders that possess all of the above attributes, a period of training and education will be required at the beginning of every project to ensure that stakeholders are aware of the conceptual design processes to be implemented.

4.2.3.7 Prioritize Stakeholders

Stakeholder management works best when everyone involved in the process understands their role in the design process—that is they understand the importance of their opinion in the scheme of things. It is almost impossible to progress with a large number of stakeholders, each of whom has equal rights to the others; consensus becomes increasingly difficult as the number of stakeholders increases. Further, it is not possible to develop an optimum system by combining optimum inputs. Stakeholders should therefore be grouped into categories, and the priority and the rights of each category should be made explicit in the SMP.

In particular, requirements analysis is greatly aided if stakeholders' inputs are prioritized, because conflicting requirements can then be traded off more easily—design spaces are much more readily utilized if some dimensions have more weight than others. Additionally, the priority of stakeholders will assist the requirements engineers and business analysts in prioritizing their elicitation and elaboration activities, particularly early in the process.

4.2.3.8 Form the Stakeholder Board

The stakeholder board provides the mechanism by which stakeholder needs and requirements are elicited, crystallized, traded off if necessary, endorsed, and then managed throughout the life of the system. The board is formed from the selected stakeholder representatives, each of whom has a voice commensurate with the priority allocated to them. The board is chaired by the business owner, who is responsible for leading the conceptual design effort, or they may delegate that responsibility to the principal stakeholder. There should always be an even number of stakeholder representatives on the board so that any deadlock can be broken by the casting vote of the chair (should the business environment be one that requires consensus).

4.2.3.9 Develop Stakeholder Management Strategies

Once all stakeholders have been identified and grouped, and representatives have been identified and prioritized, strategies can be developed to maximize stakeholder involvement in the conceptual design process. Largely, this involves creating a favorable environment by recognizing that, while there are some technical aspects of conceptual design, it is fundamentally a process dominated and driven by humans (with all their associated preconceptions, prejudices, and agenda)—it is therefore often less concerned with what we can do and more concerned with what we want to do. These "soft" aspects must be accommodated and managed if the system development is to succeed.

Consequently, stakeholder management is arguably even more important than requirements management; yet it is much more difficult. Requirements can be stored and manipulated in a database—managing stakeholders is more akin to herding cats! For example, consider the following:

- Stakeholders may not always be supportive; for example, stakeholders who currently use a manual process are not likely to support the automation or outsourcing of functions that will see their positions made redundant.
- Some stakeholders may also resist change because of the additional effort that will ensue; given the rate of change in many organizations in the modern world, some stakeholders are simply suffering change fatigue.
- Some stakeholders may have a hidden agenda; that is, they may be interested in the failure of this project because they feel that it is diverting resources and organizational focus away from another project with which they have much more sympathy.
- Certain stakeholders have a stranglehold over other individuals and ideas within an organization—these gatekeepers can control the way in which a project progresses and therefore must be actively managed as part of the conceptual design process [10].
- Engineers and software developers may resist certain requirements because they are too hard or too time-consuming to implement.

- Project managers with a predefined scope and budget may resist any requirement that appears to threaten on-time delivery or impinge upon budget or cash flow.

Stakeholder management is one of the 10 knowledge areas in the PMI's *Guide to the Project Management Body of Knowledge* [11]. Interestingly, it is only a relatively recent addition having not been included in the *Body of Knowledge* until version 5 in 2013; prior to that the only reference to stakeholders was to their identification as the first act of Project Communications Management. In the current versions of the *Guide*, however, Project Stakeholder Management is the tenth knowledge area comprising the following tasks:

- Identify stakeholders;
- Plan stakeholder engagement;
- Manage stakeholder engagement;
- Monitor stakeholder engagement.

The *Guide to the Project Management Body of Knowledge* lists four models for identifying stakeholders: grid methods (power/interest, power/influence, or impact influence); stakeholder cube; salience model; and directions of influence [12]. Here, by way of illustrating a helpful technique, we make use of the stakeholder management matrix in Figure 4.2, proposed by McLucas [13].

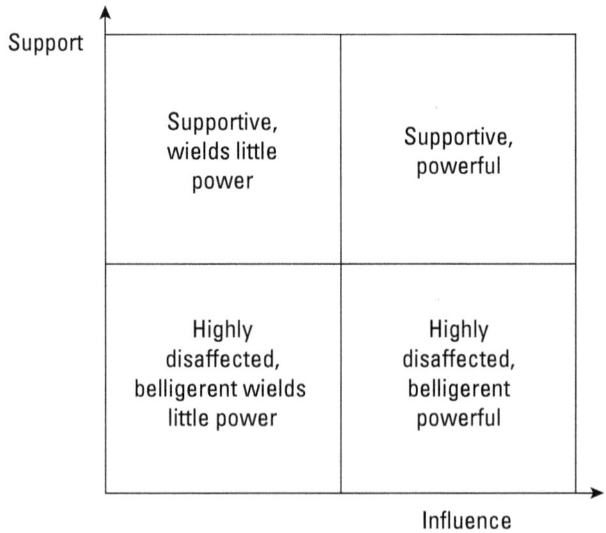

Figure 4.2 Stakeholder management matrix (after McLucas).

Stakeholders are placed within the matrix in accordance to their support and influence and are grouped into four broad quadrants: supportive, wields little power; supportive, powerful; highly disaffected, belligerent, wields little power; and highly disaffected, belligerent, powerful.

The aim of stakeholder management strategies, therefore, is to do as much as possible to ensure that stakeholders are all supportive and powerful (as illustrated in Figure 4.3). If stakeholders are supportive but not powerful, they must be given power where possible—perhaps by assigning them control of the project budget or by moving the entire project under their control. Similarly, if a stakeholder is powerful but not supportive, they must be either persuaded of the project's merit, or they (or perhaps the project or budget) must be moved so that they have little or no power with respect to the system's development. Again, the business owner or principal stakeholder has an important role to play here—they must plan and implement these strategies.

4.2.3.10 Prepare SMP

As a minimum, the SMP should do the following:

- Identify the business owner;
- List stakeholders by category (priority) and type;
- Identify the stakeholder representative for each category;

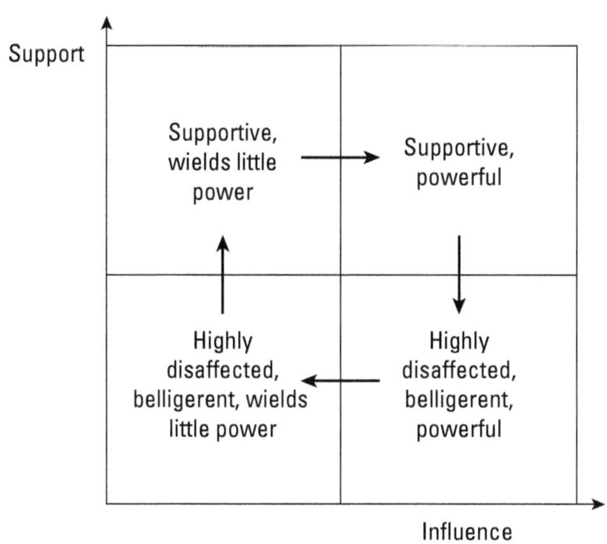

Figure 4.3 Stakeholder management matrix, illustrating generic stakeholder management strategies (after McLucas).

- Identify stakeholder roles and responsibilities;
- Nominate the principal stakeholder (if not the business owner);
- Nominate the stakeholder board and identify the chair;
- Nominate stakeholder management strategies. (This may be in a separate document or in an annex available only to business management.)

4.2.3.11 Layers of Stakeholders

It must be recognized that the identification of stakeholders is often not as simple as outlined, because there may be a number of layers of stakeholders. For example, one or more of the stakeholders may also have their own stakeholders who must be satisfied (who may also have stakeholders that must be satisfied); while the members of this latter group are at arm's length to the project, their influence is often felt, and they may have to be considered in the SMP.

4.2.4 Identify Major Stakeholders: Examples

Several examples in this section illustrate the steps associated with the activity C1.1.1—Identify major stakeholders. The purpose of doing so is to reinforce that stakeholders are not always self-evident by simply naming the SOI; nor do potential stakeholders have innate rights. We must take care early in conceptual design to identify candidate stakeholders and to analyze them critically for their right to be involved in the conceptual design process (and, if so, to what extent).

4.2.4.1 Example: Your Own Home

You and your partner are building a home for yourselves and your two children.

> *Identify business owner:* In this case there would, no doubt, be a team approach, and you and your partner would be the business owners (assuming that the mortgage is in joint names). Of course, there may be circumstances in which one of you may take a principal role, even if both financial contributions are equal. There may even be circumstances in which only one of you holds the full mortgage but still prefers a team approach. Alternatively, the holder of the full mortgage could demand full rights, because they are providing the funding. There may even be circumstances in which one partner holds the full mortgage but is happy to delegate full rights to the other. This is why it is so important to spend time analyzing stakeholders—it is not possible at the start of a project to presume in advance the arrangements of rights, to know who the business owner is, and even to know who is the principal stakeholder.

Every situation will be largely unique, even in such a simple example as building your own home.

Identify candidate stakeholders; evaluate candidates and select stakeholders. The following may be considered candidate stakeholders:

- *You and your partner.* As noted, there are any number of ways that you and your partner may arrange rights. Let's assume that you and your partner are the joint mortgage holders, in which case it is difficult to imagine that any other candidate stakeholder has the right to dictate to you any aspect of the system design. In this case, the business owners (you and your partner) are clearly stakeholders and clearly the principal stakeholders. As discussed above, however, it may be possible for one partner to be the business owner and for the other partner to be the principal stakeholder, or any other combination.
- *Your children.* You will certainly take into account the wishes of your children because it is their home too, but you cannot afford to meet all their perceived requirements. They will therefore have rights that are significantly moderated by your desires (and budget). Of course, in some cases, you may give them limited stakeholder rights, such as the right to choose the color of their room, but you are not likely to give them stakeholder rights when deciding on the number of rooms you can afford, or on the inclusion of a swimming pool. Although you want them to be happy, you will discuss with them what they would like and provide what you can, presumably with some negotiation. This is why we say that requirements practice is largely a political activity not an engineering one; solving these types of conflict is a matter of leadership, not requirements engineering—conceptual design without leadership is fundamentally flawed from the outset.
- *In-laws who visit and may stay occasionally.* You are also unlikely to give your in-laws full stakeholder rights to dictate any requirements of the house. However, should an aged parent be occupying the house with you, you may give them similar rights to the children with regard to the layout of their room or similar. Additionally, should you have an in-law who is limited in their physical ability in some way, you might place increased emphasis on access requirements—although this does not make them a stakeholder. In some cases (abnormal, of course), you may not like your in-laws and may deliberately design the house without a spare room so that they cannot come and stay with you—that is, you may deny them any rights and deliberately exclude them from the system. Of course, there may also be circumstances

where this harsh exclusive view may change. If an in-law has made a significant financial contribution to the purchase of the house, they may have the right to dictate requirements and become a stakeholder (although, if their donation was purely philanthropic, they may well waive that right).

- *Friends who visit and may stay occasionally.* Depending on your relationship with your friends and in-laws, friends may have a similar priority to in-laws. They are, however, most likely to play only a minor role and would not be characterized as stakeholders. If you have many friends who visit and stay regularly, you may have the need for a guest bedroom, but its design and inclusion will be your choice on their behalf—they will not become stakeholders.
- *The bank that holds the mortgage.* Despite the fact that the bank will invariably hold one or more mortgages over the property and therefore will have significant power over you and the house, it is not a stakeholder—it does not have the right to dictate any aspect of the house design. Since the bank does dictate the budgetary limit of your aspirations, it quite clearly provides a number of external constraints on the project and the resultant house; however, the bank is not a stakeholder, and you will not engage with it again during the project provided that you do not default on your mortgage or need to apply to exceed its current limit. The bank's principal input will be as a source of constraints.
- *Local government.* As with the bank, local government regulations will place significant constraints on the house design and its location on the block, the utilities it will access, and so on. The ability to impose regulatory constraints does not make local government a stakeholder, however. Its principal impact will be as a source of constraints.
- *Neighbors.* Although neighbors are generally constraints, not stakeholders, in this case they will be stakeholders with limited rights. Of course, neighbors must be taken into account because they may cause you difficulty in the building approval process if some aspect of the design impinges on their community rights. Note, however, that it is the community expectations of you (the local governmental regulations that constrain you) that account for any neighbor's right to have an opinion about the system requirements. In that sense, then, they can only have issues with your building project if it breaches some local regulation or guideline—that is, they can show that you have not met a regulatory constraint. The fact that a neighbor is a stakeholder in the community does not make them innately a stakeholder

in your project. However, when we analyze the neighbors as candidate stakeholders, it becomes obvious that they would normally have the right to specify the fence in common with your boundary (and would normally be required to bear half of the cost of its construction). So, this does mean that the three neighbors on your three boundaries are stakeholders with limited rights. However, provided that your building conforms to local regulations (as constraints), neighbors have no more rights than to specify and pay for half of the common fence. Note also that, should the neighbors wish only to agree to pay half of the minimum fence cost, a more expensive fence can be erected if the additional cost is borne by you (provided that the neighbor agrees). All that being said, it is now up to the business owner as to how they wish to handle this category of stakeholder/constraint. If they have a close relationship with a neighbor, they will no doubt want to continuously update them on the project's progress. If, however, they have had an adversarial relationship over many years, the business owner may choose to conduct all communications through more formal channels. Again, conceptual design is a political process that requires business leadership to be successful.

- *The architect and draftspersons.* The architect has a considerable interest in the success of the project in that their reputation will suffer if the building looks awful when completed. They will also be legally liable for any flaws in their work. However, this interest in the success of the project does not make them a stakeholder; rather their level of skill and ability, their availability, their interest, and their enthusiasm are constraints.
- *The builder and subcontractors.* Similarly, builders and their subcontractors are sources of constraints, not stakeholders. Their motivation and their skills must be taken into account, and they will also be legally liable for any flaws in their work, but they have no particular right to dictate requirements. However, for example, you may be purchasing your home from a developer/builder and have arranged some form of discount on the purchase price in return for allowing the builder to use your home as a display home for a short period—in that case, the builder may become a stakeholder with limited rights.

Identify roles, responsibilities, and interrelationships. Although you and your partner are the only genuine stakeholders in this venture, you may wish (even if only to increase their enthusiasm for the impending move to a new house) to provide the children with some stakeholder rights.

Similarly, you may wish to provide in-laws with limited rights. In that case, the roles responsibilities, and interrelationships within the family will be obvious and straightforward to describe.

Nominate the principal stakeholder. In this case, you or your partner could be nominated as the principal stakeholder. However, that may not be necessary, and you could simply both act as the business owners.

Stakeholder representatives. You and your partner would naturally jointly share the role as principal stakeholder, although one of you may waive this right in most circumstances and defer to the opinions and preferences of the other. Assuming that you have decided that the children will share a single room, you may choose one as the representative, or you may decide to make them equally responsible for their decisions—that is, they must agree on the requirements you have allocated to them (such as the color and layout of the room) and therefore each be a stakeholder. Should they have separate rooms, then you would most likely allow each child to be a stakeholder (albeit limited) in their own right. Similarly, you may wish to engage with one or both in-laws as limited stakeholders. You will also have to negotiate with three neighbors to reach agreements regarding the fencing separating adjoining blocks of land.

Prioritize stakeholders. In this case, the following stakeholder categories are useful:

- *Category 1.* You and your partner will be Category 1 stakeholders, either jointly, or one partner may shoulder more of the load.
- *Category 2.* The children may each be Category 2 stakeholders with limited rights as defined at the beginning of the project. Similarly, Category 2 rights may be extended to in-laws.
- *Category 3.* The three neighbors may be Category 3 stakeholders with limited rights over their adjoining fence.
- *Category 4.* You may wish to refer to the architect and builder as Category 4 stakeholders to encourage them to have some ownership of the project, although they will largely be constraints. Similarly, the bank and local government will be sources of constraints.

4.2.4.2 Example: A Small Business Local Area Network

A small organization is looking to install a local area network within their leased premises.

Identify business owner: In this case the owner or manager of the small business is clearly the business owner—the person who is ultimately responsible for the system and the business it serves.

Identify candidate stakeholders, evaluate candidates, and select stakeholders: The following may be considered candidate stakeholders for the sake of this example:

- *The business owner/manager.* The business owner is clearly a stakeholder who has the right to influence the system. In many cases, however, they may not have sufficient time to devote to the project so they may nominate a business manager as the principal stakeholder.
- *The finance manager.* The finance manager has a crucial role in managing the organization's finances, particularly cash flow, and will no doubt have requirements with regard to expenditure spread, financial guidelines, and milestones. While these constraints must be taken into account, the financial manager is not automatically a stakeholder and does not necessarily have the right to articulate requirements. Once the project budget is set, however, they will become an essential source of constraints over budget, financial milestones, and cash flow.
- *The operations manager.* Similarly, the operations manager will impose constraints regarding timing of work, disruption to existing business and cutover dates, but they are not automatically a stakeholder. They could, however, be nominated by the business owner to be the principal stakeholder if there is not a business manager or if the business manager does not have sufficient time to devote to the project.
- *Employees.* While they will undoubtedly be affected by the system, employees are not naturally stakeholders. Employees have no native right other than to a safe workplace and fair working conditions; they do not have the right to dictate how the business allocates its funds. However, it is important that employees are comfortable with and have some ownership of the new system so they will have views on its operation. Again, leadership is important, employees will be kept up-to-date with the progress of the project, their opinion will be sought regarding some aspects of the new system, and they may be given the right to dictate some requirements (such as, perhaps, the location of printers or their choice of computer brand). In that case, they would become stakeholders with limited rights.
- *The building manager.* The building manager must be consulted, but the requirement to consult and any subsequent requirements are

constraints. The building manager will be the guardian of rules and regulations that are constraints that are to be followed—they personally are not a stakeholder and can only repeat the regulations, not augment them.

Identify roles, responsibilities, and interrelationships. In this case, the principal stakeholder may wish to include the employees as limited stakeholders in which case they may choose to hold occasional meetings to engage with them. Alternatively, they may ask the employees to nominate a representative to participate in stakeholder meetings.

Nominate the principal stakeholder. The business owner would be the principal stakeholder unless they have delegated the role to the business manager or the operational manager.

Stakeholder representatives. In this case, the only group is that of the employees and a suitable representative would have to be chosen. Alternatively, to ensure that all employees feel that that their opinions have been heard, an employee working group may be formed to vocalize any issues.

Prioritize stakeholders. The following stakeholder categories may be used:

- *Category 1.* The only certain stakeholder is the business owner who is the Category 1 stakeholder. However, they may delegate to the business manager or operations manager as the principal stakeholder.
- *Category 2.* Unless the business owner has delegated the principal stakeholder role to them, the financial manager and the operational manager would be Category 2 stakeholders.
- *Category 3.* The business owner or principal stakeholder may also wish to make a suitable employee representative a Category 3 stakeholder.
- *Category 4.* The business building manager may be nominated as a Category 4 stakeholder, if the principal stakeholder thought that was a useful way to engage. They remain, however, largely a source of constraints.

4.2.4.3 Example: ACME Medical Center

In the case of the ACME Medical Center, there are a number of actors who can be considered to have a stake in the system or at least have some role to play in its development or operation.

Identify business owner: In this case the business owner or manager will be the manager who has been appointed by the ACME Medical Services board of directors to be the project champion who is accountable for the project.

Identify candidate stakeholders, evaluate candidates, and select stakeholders: The following may be considered candidate stakeholders for the sake of this example:

- *ACME Medical Services.* If we apply the test of who has a "right" to influence the design of the system, we see that it is really only ACME Medical Services, as the business owner, that has an undeniable right to influence the system. The business owner will therefore be the accountable person in ACME Medical Services who has been given the responsibility to serve as the project champion and the principal stakeholder.
- *Professional staff (GPs, dentists, physical therapists, and pharmacists).* The professional staff may be granted rights in some instances, but largely their requirements will be constraints on the way in which the system is built or on the way in which it is operated. That is, the system must facilitate their workflow and ensure that their workplace is safe and secure, but they have no right to tell ACME how to run their business. A further complication is that, even if ACME wishes to allocate some stakeholder rights to professional staff and general staff, such staff may not yet have been employed. In that case, access will be required to professional staff in another of ACME's medical centers.
- *General staff.* As with professional staff, general staff are not stakeholders in the sense that they would have the right to be a source of requirements. They have the right to a safe workplace and to appropriate work conditions, but they have no right to tell the organization what they would like the Medical Center to do, or how they think it should look. As with professional staff, ACME may wish to allocate some rights to general staff, but those personnel will not yet have been hired so ACME would have to make staff available from other medical centers.
- *Patients/customers.* Patients and customers are clearly important to the Medical Center, but there is no need for ACME to ask them what they would like in terms of services. ACME will know that from their other medical centers, and their business case will already have determined the mix of services required to be profitable in this new location. In any case, ACME is well aware that patients would like free care available without any delay—it is up to ACME to decide how much to charge for services and how to meet the demand. Having said that,

ACME would like to keep prospective patients and customers aware of the progress of the Center, even if it is purely a marketing activity.

- *Local community.* The local community is important to ACME, because it is the principal source of their patients and customers. ACME would, therefore, like to keep the community aware of the progress of the Center, and so conduct probably the same activity that it would for patients and customers.
- *Shopping mall body corporate.* An early investigation is required to determine if the new Center is part of the shopping mall's body corporate. It appears not to be, but that should be confirmed because the body corporate has a number of rules that will be constraints if it is necessary to follow them. Note that, even if the Center is not part of the body corporate, ACME may ask designers to consider the relevant rules and follow them if practicable—that is, ACME may voluntarily take on the constraints imposed by the body corporate so that the Center fits in with the surrounding shopping mall. Alternatively, ACME may want the Center to stand out. In either case, information is required to facilitate ACME's decision, which will then provide a number of business constraints.
- *Shopping mall tenants.* The shopping mall tenants are not stakeholders, but ACME may wish to keep them informed of the Center's progress.
- *Local government.* Although the local government has regulations that must be met, they are not stakeholders in the Medical Center's development or in its operation. As a test of their status as stakeholders, the local council representative can have no requirements regarding the development or operation that are not incorporated in council regulations—they cannot add any of their own requirements. That is, the council is a source of constraint; it is not a stakeholder.
- *State and federal health departments.* As with the local municipal government, state and federal government departments are sources of constraints. They cannot insist on any requirements that are not incorporated in regulations and laws.
- *Professional medical associations.* The professional medical associations will be sources of constraint with regard to any standards or regulations that must be followed. They will not be stakeholders and will probably not be engaged in any way (unless there is a regulatory requirement to do so). ACME will no doubt have a relationship with the professional associations, should one be required, since they already manage a number of medical centers.

- *Staff unions.* The staff unions will not be stakeholders and presumably will not need to be engaged with if this Center conforms to the same arrangements as ACME's other centers.
- *Construction industry.* The members of the construction industry are constraints, not stakeholders. Their motivation and their skills must be considered, and they will also be legally liable for any flaws in their work, but they have no right to dictate requirements since they do not have the right to dictate how ACME spends its shareholders' funds.
- *Utilities/services and other health providers.* At this point, utilities/services and other health providers are really interfaces and should be included in that consideration when we come to it. In many respects, however, it does not matter greatly in which category actors are placed, as long as they and their requirements are identified. Having thought of them here, we can readily move them to our consideration of interfaces.

Identify roles, responsibilities, and interrelationships. The roles and responsibilities are again quite clear in this case. There is a business owner and a nominated principal stakeholder. Any interrelationships among professional staff members and between professional and general staff are well known and can be observed, if necessary, in ACME's other centers.

Nominate the principal stakeholder. In this case, the principal stakeholder will be the manager who has been appointed by the ACME Medical Services board of directors as the project champion who is accountable for the project.

Stakeholder representatives. The only groups are those of the employees, and a suitable representative would have to be chosen, probably one from each of the professional staff and the general staff groups. However, this is unlikely to happen because none of the professional staff or general staff will have been hired at the time of conceptual design. In that case, if needed, a member of ACME's head office may act as representative of all staff.

Prioritize stakeholders. The four main categories of stakeholder are described as follows. (Please note that the remaining candidate stakeholders are properly interfaces, and we can move them to that area for consideration.)

- *Category 1.* In this case, there is really only one Category 1 stakeholder (ACME Medical Services, the business owner). Then, since a stakeholder has to be a person, ACME may appoint a principal stakeholder (perhaps a senior person in the responsible business unit).

- *Category 2.* Professional staff and general staff might be considered Category 2 stakeholders, with whom we would engage to elicit their expectations as actors in the system (some of which expectations ACME Medical Services may agree should be turned into requirements). In that case, ACME must nominate a suitable representative. An ACME employee could, of course, act as the representative of both types of staff.
- *Category 3.* We may decide that we should engage with patients and customers and the local community in a manner that encourages them to feel a sense of "ownership" of the Center. (Their satisfaction is, after all, essential to the success of the venture.) In this case, we should include them as Category 3 stakeholders. This category of stakeholders would not be expected to be a source of requirements, but the SMP would identify the communications activities undertaken to keep them informed of the progress of the project, perhaps through community consultation evenings. Additionally, ACME may wish to make shopping mall tenants Category 3 stakeholders and invite them to the community consultations.
- *Category 4.* Since we would like to keep shopping mall tenants informed during the development, we may call them stakeholders and allocate them to Category 4. Again, this category of stakeholders would not be expected to be a source of requirements, but the SMP would identify the communications activities undertaken to keep them informed of progress of the project. Perhaps they could be invited to attend the community consultation evenings along with the Category 3 stakeholders.

4.2.4.4 Example: Domestic Security Alarm

In the case of the in-house development by ACME Alarms of a Domestic Security Alarm, there are ACME managers and employees who have a stake in the system or at least have some role to play in its design, development, manufacture, and operation.

Identify business owner. If we apply the test of who has a right to influence the design of the system, we see that it is really only the system owner, ACME Alarms, that has an undeniable right to influence the system. The business owner will be the accountable person in ACME Alarms who has been given the responsibility as the project champion and the principal stakeholder.

Identify candidate stakeholders, evaluate candidates, and select stakeholders. The following may be considered candidate stakeholders in this example:

- *ACME Alarms.* The accountable person in ACME Alarms who has been given the responsibility will be the project champion and the principal stakeholder. In this case ACME may wish to establish a stakeholder board comprising other senior personnel who will have a significant interest and the right to affect the system based on the accountabilities of their positions in the company: chief operating officer, chief engineer, production manager, distribution manager, and sales manager. Of course, by virtue of their positions in the company, they will each be principally accountable for a subset of the alarm's requirements.
- *Residents.* The residents who will buy the alarm come from a big tent, and an individual's opinion is of interest but not sufficient to define the features required by the market. If ACME's business development and marketing people are not intimately aware of the market need, they could establish a market research focus group. So, although residents are the target market who need to be delighted by the product, they are not stakeholders.
- *Competitors.* Competing alarms will certainly be considered in designing this system, as meeting the standard set by alarms already in the marketplace will be a source of constraints: If the alarm is to be successful, it will need to match or improve on its competitors' features. Competitors are not stakeholders, however, and they would not really be aware of the ACME's product development until ACME's alarm begins taking their market share.
- *Intruders.* As with competing companies, intruders will certainly be taken into account by the alarm, but they are not stakeholders. The skill sets of the intruders and their physical shape will be constraints on the system.
- *Security monitoring firms.* Security monitoring firms will need to accredit the alarm for inclusion in the suite of alarms that they can monitor. There will be certain conditions to be met if the alarm is to be suitable to be monitored. Although ACME Alarms will need a business relationship with one or more monitoring firms, the conditions the alarm needs to meet would be better considered as constraints. Of course, it may be possible for ACME to partner with a monitoring firm, in which case the monitoring firm has a pecuniary interest and therefore a right to be a stakeholder.
- *Insurance companies.* Home insurance companies will often give a discount to residents if they have an alarm installed. The minimum

requirements for such an alarm must be features of the design for this alarm, so should be considered as constraints. It is not likely that ACME will need a business relationship with the insurance companies.

- *Retailers.* ACME Alarms may choose to sell the alarm through a limited number of retailers (with whom they may already have a relationship for the sale of their car alarm). Such retailers are not stakeholders but will be the source of constraints related to the size of the packaged alarm to sit on shelves, the need for marketing with QR codes, and so on.
- *Police involvement.* The police will act as they would when patrolling the area. There is no expectation that they would respond to the alarm, although there is nothing to stop them if they wished to investigate. There is no need to engage with the police in either the design of the alarm or with regard to the regulations and laws that govern its operation. Police are therefore neither stakeholders nor source of constraints.
- *Neighbors/community groups.* Provided the alarm meets all laws and regulations, neighbors and community groups are not sources of requirement, nor do they represent any constraints.
- *Regulations.* Any applicable regulations will be constraints, and any regulators may need to be engaged with to ensure that those constraints are understood. However, regulators will not be stakeholders.
- *Service providers.* Service providers (such as utility providers) will be sources of constraint, not stakeholders.
- *Installers, repairers, suppliers of parts.* The support concept has yet to be written, but at this stage it would seem that the skill sets and availability of installers are constraints. If the concept is for faulty items to be returned to ACME for repair, then other repairers will not really be relevant. In any repair context, suppliers will be sources of constraint.

Identify roles, responsibilities, and interrelationships. In this case a number of senior personnel will have a significant interest and the right to affect the system based on the accountabilities of their positions in the company: chief operating officer, chief engineer, production manager, distribution manager, and sales manager. ACME may wish to establish a stakeholder board that is comprised of those individuals and chaired by a nominated senior manager as the principal stakeholder.

Nominate the principal stakeholder. The principal stakeholder will be the accountable person in ACME Alarms who has been given the responsibility as the project champion.

Stakeholder representatives. ACME may wish to request points of contact with the insurance companies, retailers, and security monitoring firms to act as stakeholder representatives. These representatives will be engaged by those undertaking conceptual design and may also be invited to attend one or more of the meetings of the stakeholder board (perhaps those that sign off on the BNR and SNR).

Prioritize stakeholders. The four main categories of stakeholder are described as follows. (The remaining candidate stakeholders are more precisely described as interfaces, and we can move them to that area for consideration.)

- *Category 1.* The business owner, ACME Alarms, is the only Category 1 stakeholder. Since a stakeholder has to be a person, ACME may appoint a principal stakeholder and delegate accountability to them. In this case, ACME may wish to establish a stakeholder board comprising the other accountable senior personnel: chief operating officer, chief engineer, production manager, distribution manager, and sales manager. Of course, they will each be principally accountable for a subset of the alarm's requirements, but ACME may wish them to meet as an equal board to ensure that there is consensus in the company. The principal stakeholder can be empowered to arbitrate and break any deadlocks.
- *Category 2.* The security firms and insurance companies are largely sources of constraint, but ACME management may wish to have them closer to the design process to ensure that business arrangements remain solid.
- *Category 3.* The residents are important in that the alarm must be an attractive acquisition, but they are not stakeholders. Their desires and predilections are constraints on the features of the alarm, however, so they must be taken into account. ACME may decide to do so by utilizing their business development and market staff to act as the voice of the customer, or they may combine those perspectives with a series of activities engaging potential customers through focus groups and workshops. It is up to ACME to decide to what extent such activities are conducted.
- *Category 4.* Retailers, installers, repairers, and suppliers of parts will be engaged early in the conceptual design in order to ascertain the extent to which they constrain the design.

4.2.4.5 Example: CVR

In the case of the CVR, there are a number of candidate stakeholders who have a stake in the system or at least have some role to play in its development or operation. Please note that this example is somewhat artificial, because, in reality, ACME Air would most likely simply procure an OTS CVR and install it. However, we have treated it here as a development project, because it is illustrative of a number of projects in which the arrangements of stakeholders are not as clear-cut as in our previous examples.

> *Identify business owner.* In this case the business owner will be ACME Air, which will be represented by the accountable manager who has been given the responsibility to be the project champion and the chief stakeholder. ACME will presumably be required to fund the development and installation.
>
> *Identify candidate stakeholders, evaluate candidates, and select stakeholders.* The following may be considered candidate stakeholders, for the sake of this example:
>
> - *Aircraft owner/operator.* The principal stakeholder will be the accountable member of ACME Air who has been given the responsibility to be the project champion. ACME Air may also wish to nominate other stakeholders from its operational and support staff.
> - *Aircrew.* Aircrew, in this case, are similar to employees in a number of our previous examples. That is, while the pilots are loyal employees, they do not really want the oversight of the CVR, which may not be their friend should an incident occur. They will therefore be engaged with to ensure that they understand why the CVR must be installed and to obtain their input regarding the placement of the CVR in the cockpit so that it does not interfere with their operations. Aircrew are therefore largely constraints but can be considered to be stakeholders with limited rights. A suitable representative must be nominated.
> - *Regulators/regulations.* Any applicable regulations will be sources of constraints that must be met, and any regulators may need to be engaged with to ensure that those constraints are understood. However, that does not mean that regulators will be stakeholders.
> - *Air incident investigators.* The purpose of the CVR is to record information that will ultimately be examined by investigators of air incidents, whether near misses or crashes, to assist them in determining the causes and the related activities of the incident. In that sense then, almost all

of the functions of the CVR (including recording duration, quality of recording, and recording which sources) will be dictated by them, not the business owner (who is installing the CVR to respond to a mandate) nor the aircrew (who would probably not appreciate the CVR's monitoring of their performance). Because of the mandate, however, ACME must allocate stakeholder rights over a considerable portion of the system and the air incident investigators will be the stakeholders with the highest priority since the CVR really serves them, despite the fact that ACME will be paying for its development and installation.

- *Legal.* A legal opinion certainly must be sought with regard to aircrew privacy and perhaps other issues. Such an opinion may be a source of constraints, but that does not make legal representatives stakeholders.
- *Project management organization.* The manner in which the installation project will be conducted will be outlined in the acquisition concept when it is developed. If the project management is to be conducted in-house, the project manager will be a stakeholder but only with limited rights. For example, they will not have the right to dictate requirements. Should outside project management be engaged, then the project manager will be a stakeholder, again with limited rights.
- *Aircraft OEM.* The aircraft OEM is an important source of information. ACME will engage with them in some detail to ensure that the installation of the CVR is in accordance with the OEM's installation guidelines and, more importantly, is in concert with current and future configuration management plans for the platform. It would not be useful for ACME's current installation of the CVR to interfere with some future modification or upgrade.
- *OT&E organization.* Before a modification plan is approved, the OT&E organization will need to ensure that any updated aircraft are fit to return to service. The representative of the OT&E will be a stakeholder with limited rights, although the advice will be taken in the context of the overall efficiency of the life cycle activities.
- *Public/passengers.* The public and any passengers are served by the presence of the CVR, but they are not stakeholders, nor constraints.
- *Installers, repairers, maintainers, suppliers of parts.* The support concept has yet to be written, but at this stage it would seem that the skill sets and availability of installers are constraints, whether they are in-house or outsourced through contract. Since this is a development project, the support concept will also address repair and the source of spare parts. Installers, repairers, and suppliers will be sources of constraint.

Identify roles, responsibilities, and interrelationships. In this case, the roles and responsibilities line up with the major operational and support roles within the organization, so they are quite clear.

Nominate the principal stakeholder. The principal stakeholder will be the accountable person in ACME Air who has been given the responsibility of being the project champion.

Stakeholder representatives. ACME will nominate a member of the aircrew to assist, or they may rely on the chief of operations to act on the pilot's behalf. A representative must be identified within each regulator, and one must be sought from the air incident investigators. The project manager will be in continuous contact with a representative from the OT&E organization to ensure that the essential validation activities are scheduled appropriately.

Prioritize stakeholders. The three main categories of stakeholder are described as follows. (The remaining candidate stakeholders are more properly defined as sources of constraint, and we can move them to that area for consideration.)

- *Category 1.* The business owner, ACME Alarms, is the only Category 1 stakeholder with full stakeholder rights. Since a stakeholder has to be a person, ACME may appoint a principal stakeholder and delegate accountability to them. In this case, however, almost all the requirements for the CVR will come from the air incident investigators, so ACME may provide them with full rights over the performance characteristics of the CVR and include them as a Category 1 stakeholder.
- *Category 2.* If ACME Air has nominated a member of the aircrew, an operational staff member, and a support staff member as stakeholders, they will be Category 2 stakeholders with rights likely limited to their operational and support roles. The OT&E representative may also be included as a Category 2 stakeholder with rights limited to their area of expertise.
- *Category 3.* Installers, maintainers, repairers, and suppliers of parts will mostly be sources of constraints but may be classed as Category 3 stakeholders.

ACME may wish to establish a stakeholder board comprising a representative of aircrew, an operational staff member, a support staff member, and a representative of the OT&E organization. They may wish to include the regulators and air incident investigators on the stakeholder board, or they may wish to engage with them in a separate forum.

4.2.4.6 Example: ACME Timber

Before leaving our consideration of stakeholders, it is worth revisiting the wood-felling example discussed in Section 3.3.1.1. We showed in that example that it was important for business management to take a top-down approach to the development of needs and requirements. We saw that when the business did not communicate clearly the mission for the chainsaw that a good systems approach to the elements of capability was necessary but not sufficient to meet the business needs. Indeed, had the bottom-up approach been undertaken on the assumption that the purpose of the chainsaw was to cut down more wood, the business would have suffered.

Note also that from a stakeholder point of view that the identification of stakeholders can't exist until the business has undertaken its consideration at the business management level. For example, in the first instance where we saw that the chainsaw's purpose was to cut down more wood, then we had stakeholders at the operations level at the wood-felling team, at the operations management level, at the wood-clearing level, at the sawmill level, and at the marketing outlet level. However, in the second case where the purpose of the chainsaw was to cut down wood more efficiently, we would have the same stakeholders at the wood-felling team level, the operations management level, and the wood-clearing level, but the sawmill operators and managers would not need to know about the chainsaw, because their operations are not affected—since the same number of trees are being felled. Similarly, marketing would not need to know about the chainsaw, because they are not affected. So, in the first instance the sawmill operators and the marketing people are stakeholders; in the second they are not even involved in the project. Consequently, it is clear that the top-down approach is important not just for the articulation of the correct mission but also because we can't properly identify stakeholders involved that top-down perspective. If we approach the design bottom-up, our consideration of stakeholders will be inappropriate.

4.3 C1.1.2—Identify Business and Project Constraints

Before focusing on the detail of the desired system, it is essential to identify the business and project constraints that are relevant to the system and its acquisition. This analysis provides essential information about the development environment for the system and begins the top-down approach to system development.

It is perhaps a fine point, but the difference between a stakeholder and a constraint is that we must take into account requirements given to us by stakeholders, but we should always examine constraints carefully to determine whether or not the business will allow them to remain constraints or whether they can be removed. (See the note of caution at the end of this chapter.) As a general rule, stakeholders are people, and constraints are inanimate (e.g., regulations and laws).

Business constraints include any policies, procedures, standards, or guidelines that are set by the organization to guide system development and procurement. These constraints can include partnering relationships with other companies, contracting policies, and so on. For example, an aircraft company may have a policy of buying engines only from a particular manufacturer so that through-life support issues are simplified across the entire fleet—a constraint such as this must be articulated early as it has a significant effect on the procurement.

Project constraints include the resource allocations to the project as well as any externally imposed deliverables and acquisition timeframes. Many companies have business-wide standards for processes such as quality assurance and systems engineering, and these methodologies guide the way projects can operate. Additionally, the business may require the project to report progress in a particular way or to implement particular metrics, tools, and documentation procedures.

4.3.1 Example: ACME Medical Center

Business constraints. In the ACME Medical Center example, business constraints might include the following:

- *Budget.* As in any other business ventures, ACME will be able to obtain funding at a particular rate, including financial and cash flow constraints as required to return on investment. These constraints will then drive project milestones.
- *Schedule.* ACME will also place constraints on the time by which the Center is to be in service, including completion milestones for the project that line up with funds availability.
- *Conformance to ACME templates.* ACME will have a number of templates that the Center design will need to conform to, such as for branding, color schemes, and building styles.

- *ACME SOPs.* Since ACME runs a number of existing medical centers across the country, it will presumably have SOPs that govern the way those centers operate, which will be the way it wants this new center to operate. If there are SOPs, ACME will invariably require the new center to conform to them—the SOPs will therefore be constraints.
- *Existing support procedures and arrangements.* For efficiencies across the organization, ACME will also have arrangements and procedures for such support as for pharmaceutical supplies, for general supplies, and for personnel recruitment. These existing arrangements must be acknowledged and accommodated as constraints.
- *Existing partnering arrangements.* The project must also accommodate any existing partnering arrangements for construction and so on. Conformance to these existing partnering arrangements will be a source of constraints.
- *Enterprise standards.* ACME as an organization will be accredited to and must be bound by the requirements of a number of enterprise-level standards such as those for quality insurance, project management, systems engineering, and risk management.
- *Contracting procedures.* ACME will also have some form of procurement manual that will identify contracting procedures to be used within the organization in terms of sole-source arrangements, tendering requirements, contract negotiations, financial delegations, and so on.
- *ACME policies and procedures.* ACME will also want to ensure that its accounting procedures, HR policies, and any other major policies and procedures are complied with.

Project constraints. Project constraints would include the following:

- *Budget.* Budget (as well as cash flow) for the project will be inherited from the overall business constraint.
- *Schedule.* The broad structure of the schedule will be inherited from the major payment and completion milestones required by ACME.
- *HR constraints.* ACME will most likely have capped the number of staff in the project team or have placed limitations on their employment.
- *Project management standards.* The project manager will be required to comply with project management standards, and other project management procedures, methodologies, and tools.
- *Location.* There is little space around the center location for the storage of building supplies and for the delivery and pouring of concrete. Considerable negotiations may be required with the shopping mall

owners to be able to use a portion of the mall parking lot during development. This use may also come at a cost, which must be incorporated into the project budget.

- *Access.* In addition to the physical access required to the mall parking lot during construction, there will be considerable disruption to the traffic on surrounding roads. Discussions with local officials will identify a number of constraints related to traffic control.
- *Availability of contractors.* The project schedule will depend on the availability of contractors to complete the work. Reliability of effort from contractors may come at a cost that will need to be incorporated into the budget. Any period of unavailability must be accommodated in the schedule.
- *Weather.* The region's weather will have two main effects. First, the general pattern of the weather in the building months will lead to a number of project days lost due to bad weather—the project schedule will need some contingency to accommodate these delays. The second effect will result from the climate of the region—building in the depths of winter in cold, wet climates is obviously going to be slower than in more sunny climes.

4.3.2 Example: Domestic Security Alarm

Business constraints. Business constraints for the Domestic Security Alarm might include the following:

- *Target range of residences/environments.* ACME's business case will target the alarm at a specified range of residences. The range will not be unnecessarily limited so they may could these needs in terms of a minimum capability that can scale up to a maximum size (maximum number of sensors, for example).
- *Competitors.* ACME's business case will have determined which alarms would be competing for market share. The competitors and alarms themselves will not necessarily be constraints, but they will be sources of constraints, such as the minimum features required for an alarm to be competitive, the target market price, and the time to market. The constraints are discussed in their own right later in this section.
- *Design and production constraints.* ACME may mandate that the design and/or production are to be undertaken in house. Additionally, they may even mandate that faulty items are to be returned to the company for repair.

- *Schedule issues.* There must be some timeliness associated with getting to market, depending on ACME's business case, competitors, and the state of the market. ACME will therefore provide a date by which the first alarms are to be on the market and the rate(s) at which alarms are delivered after that.
- *Utility (range of features required).* A market survey and advice from the business development staff will help ACME to identify the minimum features (and perhaps appearance) required for an alarm to be competitive as well as additional features that would result in a market-leading alarm that will immediately attract market share. Such features include the following:
  - Ease of use;
  - Acceptable false alarm rate;
  - Back-to-base or stand-alone (or both).
  - Back-to-base connectivity: phone line, fiber, and mobile telephone network;
  - Transportability.
- *Design-to-budget.* A market survey and a business development staff also will help ACME to determine the price sensitivity of the market in conjunction with the trade-off of price for various alarm features. ACME will therefore set a design-to market price for the alarm when it arrives at retail shelves.
- *Target cost to build alarm.* Similarly, ACME will set a target cost for the nonrecurring engineering required to design and build the alarm for two reasons. First, the difference between the total development cost and the income from alarm sales over a given period will represent the return on investment over that period. Second, the difference between the design, production, and distribution costs of an alarm and the income from selling an alarm at retail will represent the profit per alarm. Both measures of profit will be useful to ACME.
- *Target through-life support costs.* In addition to the target cost to build, ACME will take a life cycle view and may mandate an approach to support as well as a target cost for that support. ACME will consider the skill sets of the installers, maintainers, and repairers as well as packaging, delivery, and storage requirements, which will then be placed as constraints on the project and the system.
- *Enterprise standards.* ACME as an organization will be accredited to and must be bound by the requirements of a number of enterprise-level standards such as those for quality assurance, project management, systems engineering, and risk management.

Project constraints. Project constraints include the following:

- *Schedule.* The project manager will take the overall business constraints regarding timing for the first items delivered to the market and the subsequent delivery date. They will then develop a project schedule within those overall constraints, identifying major milestones.
- *Budget and cash flow.* The project manager will then use the schedule to develop the cash flow required to fund the particular project activities—an early burst of funding may be required to place deposits on long–lead time items and upgrades to the production line. Subsequent funding might become more uniformly distributed to finance design work and the ensuing steady state of production.
- *HR constraints.* ACME will most likely have capped the number of staff in the project team or placed limitations on their employment.
- *Project management standards.* The project manager will be required to conform to project management standards, and other project management procedures, methodologies, and tools.

4.3.3 Example: CVR

Business constraints. Business constraints for the CVR might include such issues as the following:

- *Project cost.* Although the need to install a CVR has been dictated by government, ACME Air will have an opinion on total project cost.
- *Support.* Again, although the installation has been directed, ACME Air will have a view on support costs and place constraints on factors such as spares holdings.
- *Schedule.* The business will mandate (or have mandated on them) a date by which installations on all aircraft are to be complete.
- *Availability of platforms.* Given that ACME Air would strongly prefer to remain operational throughout the modification process, they may well mandate the rate at which aircraft can be made available. If there is sufficient time before the completion date, they may mandate that the installation occurs in conjunction with a planned maintenance period. These issues will be elaborated in the acquisition, deployment, and support concepts.
- *Training.* ACME may well state that they want the CVR installation to have minimal impact on the training of aircrew and ground crew and so will place a restriction on any additional training required.
- *Procedures.* ACME would also like minimal variation to existing

procedures and would like the operation of the CVR to be transparent, at least during flight operations.

- *Diversity.* ACME Air will no doubt dictate that the same CVR is to be installed on all aircraft in the fleet. They may also dictate that this project is to standardize use of the CVR on another aircraft type, increasing the project scope and budget.
- *Privacy.* Aircrew unions or operators may raise concerns over voice recording as a privacy issue.
- *Liability.* Recording cockpit audio may increase legal exposure in the event of an incident, depending on how recordings are used in litigation. Given that the need to install a CVR is government-directed, this may not be a concern, but the issue must be examined.
- *Contracting.* ACME Air may also mandate the way in which the CVR is procured and installed. That is, they would mandate whether the procurement contract is to be for the equipment, installation, and support, or whether the CVR is to be procured under contract but then installed by ACME personnel.

ACME Air may have an opinion on these constraints at this point in the conceptual design; however, they may not develop a firm view until the development of the operational, acquisition, deployment, operational, and support concepts.

Project constraints. In addition to HR constraints and the need to follow enterprise standards, project constraints would include the following:

- *Availability of platforms.* The project schedule will depend heavily on the ACME Air need for operational availability as well as the final date by which all installations are to be completed.
- *Availability of CVRs.* In addition to the availability of platforms, the availability of CVRs (how quickly they can be developed along with any ancillaries) will affect the overall project schedule.
- *Availability of installation personnel.* The delivery schedule will also be constrained by the availability of installation personnel. For example, if there are only two teams available, only two aircraft will be able to be modified simultaneously.

4.4 C1.1.3—Identify External Constraints

In addition to business-imposed constraints, there are wider external constraints on the development of the system. It is useful to consider these constraints from several perspectives, discussed as follows:

- *Business environment.* The system will no doubt be affected by changes in the broader business and economic environment, particularly those related to cost, pricing, availability, and licensing. While the business may buffer the project against the harshest of these changes (including changes in exchange rates and interest rates) the volatility of the business environment must be considered by the project management plan. Most of these changes will have little impact on requirements, although the effect of component availability and pricing may affect proposed designs.
- *Conformance to laws and regulations.* A number of requirements for the system will come from the need to conform to national and international laws and regulations. Conformance to laws is binding within a national or international legal construct; regulations are normally provided by governing bodies within the application domain of the development. While the relevance of any specific law or regulation will depend on the nature of the system being developed, such constraints might include the frequency of operation of a radio system, the use of hazardous materials, disposal (recycling) regulations, and air-worthiness regulations.
- *Compliance with standards.* Industry standards provide similar constraints to laws and regulations, except that compliance with any individual standard may be at the discretion of the developer, unless the standard is mandated by the business or by the contract for the development, if one is in place. Constraints can be placed by standards across areas as diverse as systems engineering processes, design procedures, radio frequency emission levels, building codes, and building materials.
- *Ethical considerations and social responsibility.* Despite any formal constraints placed on the system development by laws, regulations, and standards, system developers have a moral and ethical responsibility to the owners and users of the system, as well as to the community.
- *Interoperability and interfacing requirements.* Since it is rare that a system would stand alone, interoperability and interface considerations must be considered during development. External interfaces are normally dictated by existing systems, although interoperability may be required with other systems under development. For software, for example, a common interoperability and interface constraint for a software application is to be able to run within a particular operating system. Chapter 6 discusses further the process for identifying the system boundary and specifying the external interfaces.

- *Operating environment.* The system will have to exist within an operational environment that will provide constraints in terms of temperature, humidity, radiation, and robustness to shock.

Again, an important aspect of top-down design is to understand these constraints before considering system-level and lower level requirements.

4.4.1 Example: ACME Medical Center

External constraints for the ACME Medical Center would include the following:

- *Compliance.* There will be a considerable number of constraints on the system because of the need to conform to building codes and regulations, local government regulations, body corporate requirements, health regulations, and health authority requirements.
- *Procedural interfaces.* Constraints will be drawn from interface requirements with service providers such as health insurance providers, providers of specialist medical services, radiology and imaging, hospitals, and blood services companies.
- *Physical interfaces.* The Center has a physical interface as a building to the shopping mall walls and has interfaces to sidewalks and to the shopping mall parking lot.

4.4.2 Example: Domestic Security Alarm

External constraints for the Domestic Security Alarm would include the following:

- Environmental constraints (operation/disposal);
- Neighbors and local community in terms of noise pollution;
- Legal liability;
- Domestic power supply;
- Interoperability with existing commercial communications infrastructure (commercial phone line);
- Avoidance of interference with other uses of the commercial communications infrastructure (particularly telephone and modems);
- Avoidance of interference with other household electrical systems;
- The need to act ethically (that is, the system is not to harm intruders or homeowners);
- Interface requirements for back-to-base monitoring systems;

- Meeting regulations (e.g., building regulations, safety, and environmental factors (including noise and RF emission and disposal).

4.4.3 Example: CVR

External constraints for the CVR would include the following:

- *Government directive.* The government directive to install CVR must also come with some constraints. It may well dictate such aspects as duration of recording, specific standards and procedures to be applied, and the date by which all installations are to be complete.
- *Certification.* In addition to any government mandates, installations must meet aviation authority certification regulations in order to attain a supplemental type certificate (STC).
- *Other regulatory requirements.* An approved engineering order or installation drawings must be developed and approved by a certified design organization or licensed engineer to meet environmental qualification standards, functional and data interface standards, and electrical and EMI/EMC compliance. A certificate of approval would be required from the relevant regulatory authority—in the United States, from an FAA inspector of a Designated Airworthiness Representative (DAR). An air flight manual supplement may be required if the CVR affects aircraft operations. Further, the CVR installation must be included in the Aircraft Maintenance Program (AMP); maintenance instructions are derived from the CVR manufacturer's Instructions for Continued Airworthiness (ICA).

4.5 C1.1.4—Identify Design Constraints

Design constraints include those factors that directly affect how the system is designed and the way the design is conducted. Of course, a number of business, project, and external constraints (such as budgets, regulations, and standards) will flow down and be inherited as design constraints.

Typical design constraints include the state of the art of relevant technologies, the skill sets of available engineers and tradespeople, as well as extant methodologies and tools to assist in the design, development, construction, and production of the system. Additionally, bounds such as all-up weight may be a design constraint for an aircraft system if it is to land on certain classes of airfield.

4.5.1 Example: ACME Medical Center

In the case of the ACME Medical Center, design constraints might include the following:

- *Location.* The location for the Medical Center is on a corner block, which will limit the number of facing walls, perhaps having an impact on the ability to separate entrances.
- *Location of current utilities.* ACME may conduct a study and determine that the cost of moving the current utility services (i.e., water, power, and gas) is prohibitive, and it may dictate that the current locations are to remain and that the Center is to be designed around them.
- *Location of the parking lot.* There is to be no bespoke parking lot, so the shopping mall parking lot is to be used. This may place design restrictions on the accessibility of the Center from the parking lot and may drive design constraints such as the location of the main entrance (which will then drive the internal layout).
- *Operations and support must be unobtrusive.* The need to ensure that operations and support functions are not visible to patients means that they can't use the same door as the patients, and the internal functions must be separate from the flow of patients into and throughout the Center.
- *Patient privacy.* The internal layout and functions of the Center must be such that patients are triaged and interviewed without being overheard and possibly even seen by other patients.
- *Staff privacy.* Staff facilities must not be visible to patients.
- *Comfort requirements.* The degree of comfort required by ACME will place design constraints in terms of the nature of the waiting areas, entertainment system, and play areas for children.
- *Maximum numbers of staff and patients inside the Center.* The maximum numbers of people inside the Center will drive a significant number of design constraints such as the fire egress arrangements, the size of the air conditioning units, and the size of ablutions.
- *Security.* The need for security must in some way interfere with the flow of patients and their access to services and the delivery of those services by staff. Careful design must address how to maintain the desired level of security while not having a significant impact on Center operations.
- *Height restriction.* The local government regulations limit the height of the building to two stories. This regulation needs to be checked—presumably, it relates to building height, not number of floors, but

nonetheless it does restrict (along with the area of the block) the internal capacity of the building.

- *Pharmacy access.* Pharmacy access is required for patients from the inside of the Center, as well as by customers from the outside of the Center.
- *Wheelchair access.* The building codes will ensure that the building is accessible to wheelchairs and other aids, so such access may not be considered at all since conformance to regulations will pick up this constraint. However, the Medical Center would need to recognize that the users of mobility aids may well be using them for the first time and, therefore, lack skill in their use; subsequently, they may need addition room externally and internally.
- *Shopping mall body corporate.* As noted when discussing stakeholders, ACME may wish to serve as a good neighbor and conform where possible to the shopping mall body corporate rules.

4.5.2 Example: Domestic Security Alarm

Design constraints for the Domestic Security Alarm would include the following:

- Company's ability to design and manufacturing capability/capacity (facilities, tools, and skill sets).
- Requirement to use certain in-house (partner) components or technologies.
- Company's technical regulatory framework.
- Cost to design.
- Target market cost.
- Need to utilize OTS products (current technologies) wherever possible.
- Useful authentication technologies (e.g., keypad and biometrics).
- Safety/suitability of various technologies, tools and processes.
- Robustness, size, and weight.

4.5.3 Example: CVR

Design constraints for the CVR would include the following:

- *Size, weight, and power.* It is not likely that the CVR will add significantly to the weight of the aircraft and there should be sufficient power available, but those factors will need to be confirmed. The size, however, will need more consideration, as a suitable space must be found in the cockpit.

- *Environment.* The CVR has to operate in the environment of the aircraft cockpit and will need to conform to the full range of environmental conditions, including temperature, humidity, and pressure.
- *Interfaces.* In addition to environmental interfaces, the CVR will need to interface to the aircrafts' DC bus, and most likely to a specified fuse.
- *Operational.* Almost all the operational aspects of the CVR will come from the air incident investigators in terms of the frequency range, quality, and length of recording; the environments within which the CVR must survive in terms of aircraft crash; mechanisms for retrieving data; and so on. These constraints will become requirements.

4.6 Some Caution When Considering Constraints

Having identified business, project, external, and design constraints, work should not progress until each constraint is tested and taken on knowingly into the next activity. That is, we must convince ourselves that each constraint is inviolate and must remain as a constraint. It doesn't necessarily follow that a current constraint will remain so without question. Before moving on, we should therefore pause to consider what might allow us to remove the constraint if doing so facilitates the progress of the project.

In many cases the constraint will be intransigent; for example, if a new vehicle is to operate on public roads, it must be roadworthy and therefore conform to the appropriate regulations. In many other circumstances, however, the constraint may be able to be lifted to the advantage of the project; for example, budgetary and schedule constraints may be eased by negotiation with the business or enterprise (assuming a good case can be made), or the restrictions imposed by a business-mandated standard may be able to be removed if it can be shown to be unnecessarily prescriptive or potentially to result in significant cost increases without sufficient attendant advantages.

The cautionary note is therefore that a constraint isn't so just because some stakeholder or regulator said it is, or because it always has been so in the past. It behooves those undertaking conceptual design to investigate the source of the constraint to ensure that the precise nature of the constraint is known. Further, if a constraint places some undesirable bound on the project, it should not just be accepted without question, but we should investigate the possibility of removing the restriction if that is possible and to our advantage.

Review Questions

1. Define a stakeholder and explain why a stakeholder is not simply someone who has a stake in a system.
2. Briefly describe the nine tasks involved in the step C1.1.1—Identify major stakeholders.
3. Briefly describe the nine tasks of the chair of the stakeholder board.
4. Conduct the C1.1.1; Identify major stakeholder tasks for a space telescope project (similar to the Hubble telescope, for example) with the following stakeholders:
   - Astronomers.
   - Manufacturing company;
   - National space agency;
   - Launching company;
   - Operating company;
   - Politicians.
5. Briefly describe business and project constraints.
6. Briefly describe external constraints.
7. Briefly describe design constraints.
8. Briefly explain why constraints should be tested before carrying their effects into the next activity.

Endnotes

[1] Project Management Institute Standards Committee, *A Guide to the Project Management Body of Knowledge*, Upper Darby, PA: Project Management Institute, Version 6, 2017.

[2] Young, R., *The Requirements Engineering Handbook*, Norwood, MA: Artech House, 2004.

[3] Young, R., *Effective Requirements Practices*, Boston, MA: Addison-Wesley, 2001.

[4] Bittner, K., and I. Spence, *Use Case Modelling*, Boston, MA: Addison-Wesley, 2003.

[5] Hull, M. E. C., K. Jackson, and A. J. J. Dick, *Requirements Engineering*, London: Springer, 2011.

[6] Kotonya, G., and I. Sommerville, *Requirements Engineering*, Chichester, UK: John Wiley & Sons, 2000.

[7] Maier, M., and E. Rechtin, *The Art of Systems Architecting*, Boca Raton, FL: CRC Press, 2009.

[8] Alexander, I. F., and R. Stevens, *Writing Better Requirements*, London: Addison Wesley, 2002.

[9] Leffingwell, D., and D. Widrig, *Managing Software Requirements: A Unified Approach*, Reading, MA: Addison-Wesley, 2000.

[10] Weisbord, M. R., *Productive Workplaces*, San Francisco: Jossey-Bass, 1987.

[11] Project Management Institute Standards Committee, *A Guide to the Project Management Body of Knowledge*, Upper Darby, PA: Project Management Institute, Version 6, 2017.

[12] Project Management Institute Standards Committee, *A Guide to the Project Management Body of Knowledge*, Upper Darby, PA: Project Management Institute, Version 6, 2017.

[13] McLucas, A., *Decision Making: Risk Management, Systems Thinking, and Situation Awareness*, Canberra, Australia: Argos Press, 2003.

5

Define Business Needs

5.1 Introduction

Having identified major stakeholders and constraints, business management must understand their needs (and then requirements). As illustrated in 5.1, we begin with the C1.2—Define business needs activity which involves defining: the mission, goals, and objectives for the system; the preliminary operational scenarios for its use; the upper level measures by which we will judge success of the system development; and the PLC.

5.2 The Need for a Mission Statement

It is essential that the development of the system begins with a concise statement of the mission, which is then further elaborated by statements of the system-level goals and objectives. Good problem definition is not a new notion—as demonstrated by the following words of wisdom from the past:

- "A problem is half-solved if properly stated." (John Dewey);
- "A problem well-stated is a problem half-solved." (Charles Kettering);
- "A problem clearly stated is a problem half-solved." (Dorthea Brande);
- "Every problem has in it the seeds of its own solution." (Norman Vincent Peale).

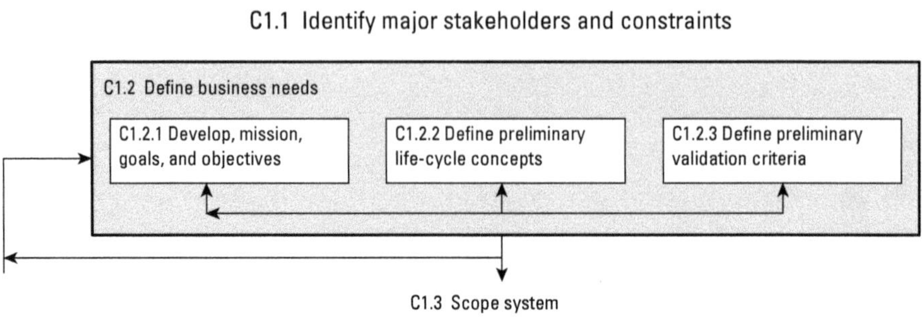

Figure 5.1 Steps involved in C1.2—Define business needs.

Or, as put in the negative by Robert Mager:

> . . . if you're not sure where you're going, you're liable to end up someplace else" [1].

Once the mission statement is formulated, it is then expanded and qualified by short declarative statements of the system goals. Goals are normally relatively broad statements, each of which spawns a number of more-specific objectives.

As shown in Chapter 2, requirements are represented in a hierarchy, first in a logical hierarchy in the BRS, StRS, and SyRS, and then in a physical hierarchy in the configuration items in the development specifications. We also saw that traceability is assisted by a hierarchical numbering of requirements.

Backward traceability from children to parent requirements leads to a useful observation related to heredity, in that the highest level requirements also ought to be traceable to some higher statement. As illustrated in Figure 5.2, it follows therefore that there is a single parent statement that serves as the highest level requirement statement, back to which all subsequent requirement statements should be able to be traced. This then is the principal role of the mission statement in systems/requirements engineering—at the system level, the system mission statement is the parent statement that provides the seminal point for conceptual design and that is the ultimate system requirement [2] from which all other requirements and designs flow [3].

Once again it is clear that the system mission statement performs an essential role since its contents must be sufficient to start the system design. While it is elaborated further by the high-level system requirements that are gathered next, the system mission statement cannot be at odds with them, nor can it contain information that is not then elaborated on by at least one subsequent system requirement.

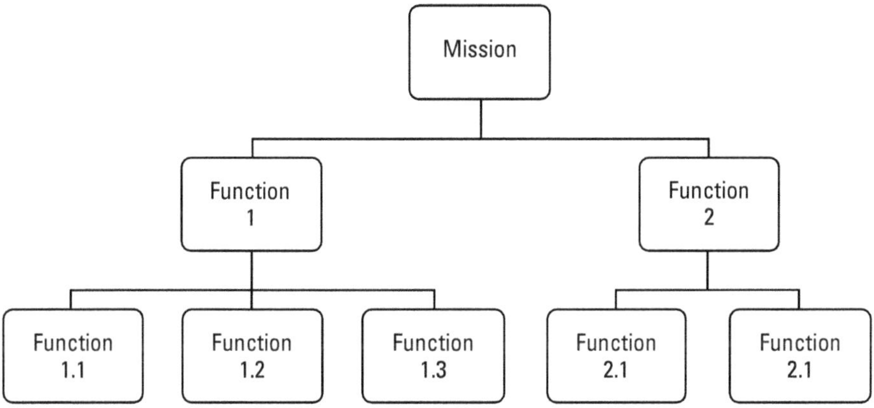

Figure 5.2 The system mission statement as the common antecedent of all subsequent goals, objectives, needs, and requirement statements.

That is, in the hierarchy, the system's mission statement provides the base statement from which the high-level requirements for the system are drawn. The system mission statement therefore must be a complete description of the system and contain concepts pointing to every function (and subsequently, every element) of the system.

We should also note here that the system mission statement does not exist in isolation and is not the starting point for the analysis of requirements. As illustrated in Figure 5.3, the system mission statement should be able to be traced to a higher project mission statement and then perhaps to a business mission statement. For example, in the case of the ACME Medical Center, ACME Medical may have an enterprise mission to increase profit to shareholders (at an appropriate level and within an appropriate timeframe) that is decomposed into three business missions: to increase existing income streams, to add complimentary income streams, and to improve efficiency across the business. The first of these business missions (increase existing income) might be decomposed into three projects to develop new Medical Centers, one of which is to be developed on the site of the old Canberra Hotel. The hierarchy of mission statements might then be the following (with our project statement and mission statement in bold):

- *Enterprise Mission:* Increase profit to shareholders . . .
  - *Business Mission 1:* Increase existing income streams . . .
    - ***Project 1.1 Mission:* The Medical Center Project shall develop a Medical Center on the Canberra Hotel location . . .**

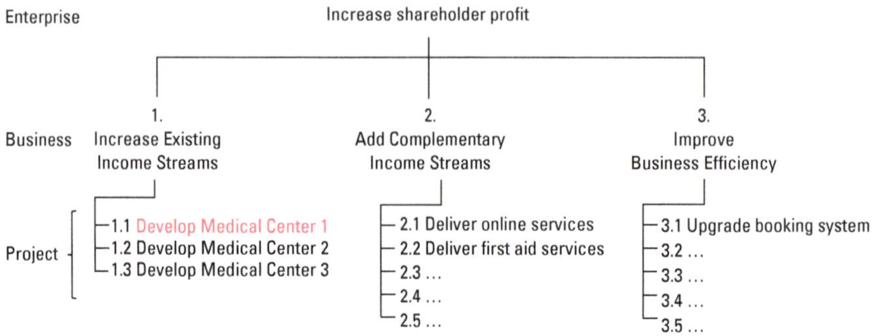

Figure 5.3 Hierarchy of missions from the enterprise to the business to the project level.

- ***System 1.1 Mission:* The Medical Center System shall deliver Medical Services . . .**
  - *Project 1.2 Mission:* Develop a Medical Center at . . .
    - *System 1.2 Mission:* Deliver Medical Services at . . .
  - *Project 1.3 Mission:* Develop a Medical Center at . . .
    - *System 1.3 Mission:* Deliver Medical Services at . . .
  - *Business Mission 2:* Add complementary income streams . . .
    - *Project 2.1 Mission:* Develop online services
      - *System 2.1.1 Mission:* Deliver online services
      - *System 2.1.2 Mission:* Deliver . . .
      - . . .
    - *Project 2.2 Mission:* Develop first aid training
      - *System 2.2.1 Mission:* Deliver first aid training
      - *System 2.2.2 Mission:* Deliver . . .
      - . . .
    - *Project 2.3 Mission:* Develop . . .
      - *System 2.3.1 Mission:* Deliver . . .
      - *System 2.3.2 Mission:* Deliver . . .
  - *Business Mission 3:* To improve business efficiency . . .
    - *Project 3.1 Mission:* Upgrade booking system
      - *System 3.1 Mission:* Deliver upgraded booking services
  - . . .
  - . . .
  - . . .
    - . . .

Note that the relationships between conceptual design and systems engineering and project management starts at this point—the enterprise mission and business missions form the start point for the development of the project mission statement and the scope statement required by the project manager for the project and for the system that the project will bring into being. The project mission statement is a parent of the system mission statement, which is the start point for the elaboration conducted by the system/requirements engineers. It should also be noted that, for ease of explanation, the hierarchy above is reasonably sparse in that each project tends to relate to a single system (that is, each mission statement tends to have a single subordinate system mission statement). A project may, of course, bring a number of systems into being. In that case, as shown for the projects under Business Mission 2, there may be a number of system mission statements for a given project mission statement.

5.2.1 Desirable Characteristics of a Mission Statement

To extend the hereditary relationship discussed in Section 5.2, it follows that the parent mission statement (and any intervening goal, objective, and need statement) should have the same desirable characteristics as those requirement statements that are its ultimate descendants. As discussed in Chapter 2, requirements should be necessary, singular, correct, unambiguous, feasible, appropriate-to-level, complete, conforming, and verifiable [4]. It follows, therefore, that the seminal system statement—the system mission statement—should have the same characteristics as the individual requirement statements that are its descendants. Specifically, the mission statement should be described by the following adjectives:

- *Necessary:* It should be obvious from the system mission statement that the system is necessary—that is, for example, that it does not already exist in some form (either as a complete system, or in the combination of the functionality of a number of other existing systems).
- *Singular:* The system mission statement cannot be a combination of mission statements for two or more systems. That is, each system must have its own unique mission statement.
- *Correct:* The system mission statement must contain only those elements that pertain to the system—that is, since it defines the system scope, each element must be within that scope. Therefore, each of the elements in the mission statement must be examined carefully for correctness to confirm that it should be included.

- *Unambiguous:* The system mission statement must remain within the intuition of the participants, and all must share a common understanding of it. It must therefore be unambiguous and convey the same meaning to each reader, regardless of their background. Ambiguity in system definition and subsequent requirements has been [5] and continues to be [6] a significant obstacle to project success.
- *Feasible:* The system must be considered feasible in the judgement of all participants. If the system mission statement is not feasible, it is not likely that the requirements derived from it will be feasible.
- *Appropriate-to-level:* As is expected of all subsequent requirements that flow from it, the system mission statement must be written at the system level and be an appropriate antecedent of any statement at any lower level in the functional hierarchy. To be appropriate at the system level, the system mission statement must be solution-independent—it must describe the problem, not the solution.
- *Complete:* The system mission statement must be complete at the system level. It must have a word or phrase for every important aspect of the system and be able to be traced back to by any statement at any lower level in the functional hierarchy. In Chapter 2 we discussed the writing of requirements and showed that each must have an associated rationale recorded to describe the reason for its existence. To be complete, the system mission statement should also end with ". . . in order to . . ." so that a system-level rationale is evident in the mission statement.
- *Conforming:* As with each subsequent requirement statement, the parent system mission statement must conform to a standard formal structure as defined by the organization, or perhaps in accordance with an external standard.
- *Verifiable:* Chapter 2 also describes why each requirement must be verifiable, in that it can be proved that the system meets or possesses the requirement. Even at the high level of abstraction of the system mission statement, it must also be possible to define the tests that will be used to demonstrate that the system is successful. While such a demonstration is an act of verification for individual requirements, for the mission statement, of course, such proof is validation.

5.2.2 Guidelines for the Format for a Mission Statement

Of these desirable properties for a system mission statement, the properties of necessary, correct, feasible, and verifiable are related to the content of the statement and must be considered carefully during its drafting (addressed in the next section). The properties of singular, unambiguous, appropriate to level, complete, and conforming are similarly a matter of the content but are also assisted by a focus on an appropriate format for the statement. The attainment of these latter characteristics is greatly assisted if the following guidelines are used for the format of the system mission statement:

- *It should comprise a single sentence:* To ensure that the mission statement is singular, it must be able to be expressed in a single sentence. In good written English, a sentence may be defined as encapsulating a single thought—if we want to express two thoughts, a second sentence is normally required. The discipline of writing a single sentence to describe the system mission therefore keeps us at the level of abstraction that defines the system as a "single thought," which is the level of abstraction needed at this stage in the project to define a unique endeavor. If we feel the need to start a second sentence, we run the risk of adding a second "thought," which may well be describing a second project.
- *It should contain no conjunctions:* In addition to being expressed in a single sentence, the mission statement should not contain any conjunctions if it is to be singular. The statement cannot state, for example, that the purpose of the system is to "do this . . . and . . . do that . . . ." The presence of the "and" in the statement implies that there are two purposes and therefore ultimately two systems. Having said that, we can of course have conjunctions in phrases and adjectival clauses. For example, we can have a clause in a mission statement proposing a system that is ". . . comfortable and secure . . ." but we cannot have a system that is "a medical center and riding school . . .," because two (very different) systems are implied. Note that we could have a system mission statement that is for a doctor surgery and a dental surgery and a physical therapy practice . . .," but only because we can (and therefore should) logically group those functions into a mission statement for a "medical center." As a general rule, therefore, we should not have any conjunctions because, if elements of the same type are able to be joined by "and," there is a group noun or adjective that can replace them. If not, then the mission statement does not describe a system with a singular purpose.
- *It should contain no more than five to seven concepts:* The mission statement cannot contain any more than five-to-seven key elements if it is to remain within the intuition of the reader [7] and to remain unambiguous.

- *It should include an "in order to" clause:* The justification, or rationale, for the subject of the mission statement can be made explicit by the inclusion of an ". . . in order to . . ." clause at the end of the system mission statement. This clause then ties the mission statement at each level back to the parent mission statement and ultimately to the business case for the system.
- *It should avoid physical terms:* To ensure that the mission statement is independent of the physical implementation means, the elements of the mission should always be couched in logical, not physical, terms. The mission statement must not imply any particular physical solution.
- *It should rely on iteration:* It does not follow that a perfect mission statement can be written in a single pass of the following process. Participants' and designers' understanding of the system must improve as the design process progresses—regardless of when it occurs, this increased understanding must be able to flow back into previous artifacts. A temporary lack of understanding must not be allowed to halt the design process—if designers are unsure of any particular element, they must continue and come back to the stumbling block when a better understanding has been developed.

5.3 C1.2.1—Develop Mission, Goals, and Objectives

Figure 5.4 describes the iterative approach to developing the mission, goals, and objectives. First the mission is drafted through a constrained bottom-up (middle-out) process in which all important need elements are identified and then crafted into a draft statement. The draft mission statement should be short and expressed in a single sentence, although it should have a word or phrase for every important aspect of the system. While business managers often find it difficult to state the mission in a single, short sentence, the project is doomed to failure if the owners of the concept cannot describe it succinctly at the outset.

This process is not conducted in a single pass but requires iteration. This iteration is essential since each pass will invariably increase the understanding of all participants and allow them to frame the problem more appropriately at each stage.

Once participants are satisfied with the draft mission statement (or are at least sufficiently happy to continue), the mission is elaborated into its constituent elements, which are then grouped into draft goal statements, again in

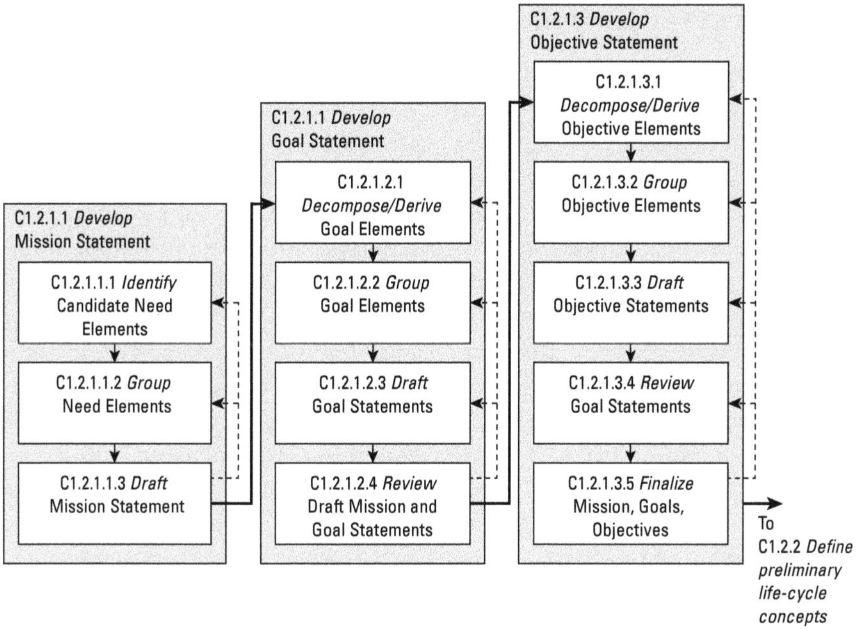

Figure 5.4 The development of mission, goals, and objectives through the subordinate steps of C1.2.1—Develop mission, goals, and objectives.

an iterative fashion. The goals are then reviewed and, if necessary, the mission statement is revised.

The process then continues in a similar vein to develop objectives for each draft goal. Again, the objectives are defined iteratively, with the goals and mission statement being revisited as necessary throughout the process.

At the completion of the C1.2.1—Develop mission, goals, and objectives activity, the participants should be convinced that they have completely described the system (at least at this level of abstraction, and within their current level of understanding).

5.3.1 C1.2.1.1—Develop Mission Statement

As illustrated in Figure 5.4, the first draft of the mission statement involves three main activities: identify candidate need elements, group need elements, and draft mission statement, which are iterated to produce the draft mission statement. Sections 5.3.2 to 5.3.4 discuss each of these activities in more detail, illustrating each activity using the ACME Medical Center example.

5.3.2 C1.2.1.1.1—Identify Candidate Need Elements

Crafting a complete balanced mission statement in one (top-down) pass is very difficult for most participants, particularly if the system is large, complicated, has a number of interfaces, and/or is a new system (that is, it contains elements with which participants have no direct experience). It is therefore productive to ask participants to simply list (in a bottom-up manner) the key words or phrases that are candidates to be considered for inclusion in the mission statement. Once each candidate need element has been assessed for suitability, they are grouped and regrouped into five to seven elements at a consistent level of abstraction. Consequently, the participants' natural tendency to think in a bottom-up manner can be supported, but the output is useful because it contributes to the top-down design—supporting bottom-up aggregation within a top-down construct (thereby remaining within the participants' intuition).

The first step in drafting the mission is therefore to identify candidate elements of the statement by simply listing them in a bottom-up manner. These elements can be obtained through a workshop with key business managers and participants, through individual interviews with those same parties, or through any other appropriate requirements-engineering technique (as discussed in Chapter 2). Note that when participants are considering these high-level elements, they are in fact articulating the major functions of the system (at least as they understand them at that point in time). The inclusion or exclusion of an element here will significantly change the design, but the impact of the change can be considered at a useful level of abstraction that is within the intuition of the participants and their current level of understanding.

The following list illustrates how we might initially list the need elements for our example project for the ACME Medical Center. (See the background for the example project at the end of Chapter 3.) Note that, for convenience, this list starts at the project mission statement level.

- Redevelop the Canberra Hotel location;
- Build ACME Medical Center;
- Deliver GP services;
- Deliver dental services;
- Deliver physical therapy services;
- Deliver pharmacy services;
- Provide a secure environment;
- Provide a comfortable environment;
- Provide privacy;
- To the local community.

Since the set of project mission, goals, and objectives is to be developed iteratively, any reasonably comprehensive set of candidate need elements represents an appropriate place to start the iteration. Having made a start, we iterate our understanding and converge to an agreed-on mission through a process of review and test. Briefly, we have two main tests—whether every mission element we have identified should be included and whether other important elements have not been included. When conducting those tests, we consider whether the system would be any different if this mission element were to be included or omitted.

For example, a mission statement might contain a need element that refers to "users." We might consider whether the definition of user to any lower level of detail would change the nature of the fielded system. If the system is to be used by users of all ages, ethnic backgrounds, languages, and abilities, then we should not define it any further in order to imply to lower level designers that all users need to be accommodated. If, however, we only require support for adult, English-speaking users, then we should reword that need element to say so explicitly, because the resultant system will be identifiably different as a result.

In the case of the ACME Medical Center, we review the need elements above and note that we have included most of the major concepts. We test each of them and note that "redevelop" is the right verb since the site already exists. Similarly, it is important that we include reference to the current location because the project's mission is to redevelop that site, not to find the best site on which to develop a medical center. That the system is a medical center is also important, so we should use that term explicitly (rather than "hospital," "outpatient center," or any other form of medical facility). The requirements for comfort and security are important as they were stated directly in our brief for this exercise (see Section 3.5.1) and will subsequently drive a number of major requirements. We could describe each of the medical services but, since we can use the term "medical services" as a collective phrase for those services at this level, we should use the highest level of abstraction (although we should keep track of exactly what services ACME is seeking, since they will no doubt flow from the business case for the Center).

We might also consider whether the services should focus on the local community as suggested by the brief in Section 3.5.1—this could potentially be a significant design driver in terms of the nature of the services provided by the Center. We don't have enough understanding of the ACME business case to conclude any deliberations here, but ". . . to the local community" is a good example of a phrase that is often stated in a project brief but must be carefully considered before being included in the mission and subsequent design;

we must understand what ACME means by that phrase before we include it. Do they mean that services will only be provided to those from the local community or mostly provided to those from the local community—that is, will our design have to incorporate some identification process that prohibits services from being provided to people who are not from the local area? If so, what is the definition of local? Do they mean local in a geographical or in a demographical sense? In either case, how do they propose to ensure that the constraint is met by the fielded system? It is, of course, likely that the phrase is more of a marketing slogan than it is a design driver, so it must be understood before it is included. In this case, the phrase should probably be dropped from the design, as it is doubtful that ACME intends to limit the range of patients (noting, however, that the company will probably still use the slogan in marketing the Center as part of the local community).

Our second test—the negative test (that is, to identify what is missing)—is much harder to apply and requires considerable experience on the part of the requirements engineers. Here, for the purposes of the example, two major elements are deliberately missing from the project mission statement for the ACME Medical Center in order to show how they might be picked up in the next stage of iteration.

5.3.2.1 C1.2.1.1.2—Group Need Elements

Having identified and reviewed the candidate need elements, we now group them so that we have elements at a similar level of abstraction. We are looking for five to seven grouped need elements from which we will form the draft mission statement. This aggregation will also assist in reducing the length of the sentence. The content of the groups should not be discarded, however, as those terms will no doubt be explicit in the subsequent goals. The following list illustrates how our candidate need elements might be grouped into three need elements (labeled NE and numbered to assist in traceability) that are of a similar level of abstraction:

- NE1. Develop Canberra Hotel location into ACME Medical Center.
  - Redevelop the Canberra Hotel location.
  - Build ACME Medical Center.
- NE2. Deliver a selected range of medical services.
  - Deliver GP services.
  - Deliver dental services.
  - Deliver physical therapy services.
  - Deliver pharmacy services.

- NE3. Provide a secure and comfortable environment.
  - Provide a secure environment.
  - Provide a comfortable environment.
  - Provide privacy.

5.3.2.2 C1.2.1.1.3—Draft Mission Statement

Our understanding of the need elements is now sufficiently mature to proceed, so we pause with that level of understanding and form the elements into a single concise sentence—the draft mission statement:

> *. . . to develop the Canberra Hotel location into a medical center (ACME Medical Center) that provides a selected range of medical services in a secure and comfortable environment.*
>
> > Are we sure of the range of medical services?
> > What about profitability?
> > We should add rationale: in order to . . .?

Note that we do not have to be completely confident that we have captured absolutely every element of the mission at this stage, because we have a chance to revisit the sentence in the next iteration after we have given more detailed consideration to the goals that flow from the mission. In fact, we should not at all be confident that we have captured every element of the mission at this stage, since we cannot assume that the original statement of a problem is necessarily the best, or even the right one [8].

With regard to purpose, we should always include a short ". . . in order to . . ." clause at the end of the mission statement describing why the system is to exist—this is an important step because this portion of the mission statement is in fact a system-level rationale, which begins the good requirements-engineering practice of recording a rationale for each requirement. The following iteration of the mission statement illustrates that the rationale in the case of ACME Medical Center will no doubt be able to be tied back to the business case and will most likely be couched in terms of the required return on investment:

> The Medical Center Project shall develop the Canberra Hotel location into the ACME Medical Center that delivers a selected[1] range of medical services in a secure and comfortable environment, in order to attain a nominated profit.[2]

Notes:

1. Medical services to include doctor's surgery supporting three GPs, a dentist's surgery for two dentists and two dental assistants, a physical therapy center supporting three physical therapists, and a pharmacy that provided a basic dispensing service provided by one pharmacist and an assistant.
2. Profit level of *x*% is to be attained within *y* years (confirm with business case).

There is often tension at this point between the need to be concise and the desire to include in the mission statement every aspect of the project/design that is important (at least at the appropriate level of abstraction)—we need to be as specific as possible so that the detail is not missed. For example, for the ACME Medical Center, the nature of the medical services provided by the center is critical to the success of the business case, but we do not necessarily have to list all the medical services in the mission, because it would invariably become too long and not a useful driver of the project. As illustrated above, one solution to this dichotomy is to provide a footnote in which the detail is described. Alternatively, and probably more usefully in the long run, a project glossary can be developed, in which the term "Medical_Services" can be explicitly defined in the context of this particular project—a glossary term is capitalized when used in requirements to show that, although the term may have a general meaning or a number of meanings, it has a specific meaning in the context of this system. The underscore symbol is used to ensure that there is no confusion that the two words could be two separate glossary entries. In fact, in addition to Medical_Services, we could usefully create glossary terms for Medical_Center and Nominated_Profit and rewrite the mission using those terms:

> *The* ***Medical_Center_Project*** *shall develop the Canberra Hotel location into the ACME* ***Medical_Center****, which delivers* ***Medical_Services*** *in a secure and* ***comfortable*** *environment, in order to attain a* ***Nominated_Profit***

We should also recognize that the Medical Center is more than just the building and is in fact a system (the Medical_Center_System) comprising the Medical_Center and all the other necessary inputs to that capability: people, infrastructure, supplies, and so on. We could amend the term from now on or simply acknowledge in the glossary that the term Medical_Center implies the system. We will discuss the great utility of the project glossary in more detail in Chapter 9 when we consider the discipline of writing requirement statements.

The draft project mission can be confirmed within the group of participants that developed it, or, more usefully, it can be communicated to a wider group and used to elicit comment that can inform subsequent revision. Because the mission is effectively a single-sentence description of project scope, early communication facilitates a shared understanding among all participants.

Additionally, it may not be possible to define one or more of the key elements of the mission precisely. For example, those developing the mission statement for a new piece of equipment may not have had enough information to decide immediately whether it is better to lease or buy. So, they might choose to use a more general verb, such as "acquire" in the first instance. This lack of certainty of the direction of the project identifies a critical design decision that must be taken (sooner or later) and would likely be the trigger for the commissioning of a trade study, the results of which can be returned to refine the mission statement by the inclusion of a more precise verb. The trade study is recorded as part of the design since it will provide the rationale for the inclusion of particular requirements as well as assist in traceability between levels of design.

5.3.2.3 Iteration

Note that the development of the draft mission statement is an iterative process that allows participants to diverge and converge in their thinking [9]. First, an iterative process supports participants employing their natural bottom-up processes to select and group key need elements. These are then formed into a sentence that is tested and adapted accordingly. Once a balanced statement has been drafted, need elements are then decomposed from it and the mission statement is redrafted if required. Importantly, during this process, participants have been considering the system in its entirety but at a level of abstraction that is within their intuition and suitable to their current understanding of the system.

5.3.3 C1.2.1.2—Develop Goal Statements

As illustrated in Figure 5.4, drafting of the goal statements has four main activities: decompose/derive goal elements, group goal elements, draft goal statements, and review mission and goal statements.

5.3.3.1 C1.2.1.2.1—Decompose/Derive Goal Elements

Having satisfied ourselves that we have a draft mission statement that is a suitable start point for the design (recognizing that we will invariably return later and modify the mission statement as we understand the system of interest better), we can now begin the process of identifying the requirements that are explicit in the draft mission statement (that is, the mission statement can be decomposed into those requirements), and those requirements that can be inferred (that is, derived) from it. So, we can decompose the draft mission into six elements (highlighted in bold font in the draft mission statement, and numbered GE1 to GE6 to make it easier to identify them). From the draft mission, however, we can also derive two other requirements (GE7 and GE8) that have not been explicitly stated—we need to be able to manage and support the Medical Center and its services once they have been provided. We have now begun the logical decomposition/derivation process that continues throughout the remainder of the design:

> *The* ***Medical_Center_Project*** *shall develop the* ***Medical_Center*** *on the Canberra Hotel location to deliver* ***Medical_Services*** *in a* ***secure*** *and* ***comfortable*** *environment, in order to attain a* ***Nominated_Profit***

The following functions are decomposed from the need:

- GE1. Redevelop the current site into the medical center
- GE2. Provide the medical center infrastructure
- GE3. Deliver the selected range of medical services
- GE4. Provide a secure environment
- GE5. Provide a comfortable environment
- GE6. Attain the nominated profit level

The following functions are derived from the need:

- GE7. Manage the medical center and its services
- GE8. Support the medical services

5.3.3.2 C1.2.1.2.2—Group Goal Elements

Once we have decomposed/derived the required goal elements from the draft mission statement, we can begin to group these elements into goals. Note that this is the start of the allocation and grouping process. We cannot have a large number of goals or else the design will be unnecessarily complicated at this very high level—it is desirable to have approximately seven goals so that the design at this level stays within our intuition.

In this case, our groups are fairly obvious so that we can develop five goals that logically group the elements that flow from the mission. The goal phrases are brief so that they are easy to discuss—it is good practice to use verb- or noun-based expressions that are two or three words long. (Verb-based, three-word phrases are probably the most useful.) We then number the goals to begin the formal traceability process:

1. Develop Medical_Center.
   GE1. Redevelop the current site into the medical center
   GE2. Provide medical center infrastructure
2. Deliver Medical_Services.
   GE3. Deliver a selected range of medical services
3. Manage Medical_Center.
   GE7. Manage medical center and its services
4. Support Medical_Center.
   GE8. Support medical center and its services
   GE4. Provide a secure environment
   GE5. Provide a comfortable environment
5. Attain Nominated_Profit.
   GE6. Attain nominated profit level

5.3.3.3 C1.2.1.2.3—Draft Goal Statements

Once we have established the goal statements and formally numbered them, we revisit them to confirm (at our present level of understanding) that the goals are balanced and complete. If necessary, we adjust them and also revisit the draft mission statement to confirm that it is still appropriate (and adjust it if necessary). We may again choose to confirm the draft mission and goals by communicating them to the broader community of participants, not just those immediately involved in their development.

Once iteration is complete, each phrase should be rewritten as a sentence. Note that one of the advantages of the three-word, verb-based structure is that the verb and object of the statement are already identified—we simply need to add the subject to create the full sentence. The characteristics of a good goal statement and the principles for its construction are similar to those discussed previously for the mission statement:

1. Develop Medical_Center.
   The Medical_Center_Project shall develop the Medical_Center_System.
2. Deliver Medical_Services.
   The Medical_Center_System shall deliver the Medical_Services.
3. Manage Medical_Center.
   The Medical_Center_System shall manage the Medical_Center.
4. Support Medical_Center.
   The Medical_Center_System shall support the Medical_Center.
5. Attain Nominated_Profit.
   The Medical_Center_System shall attain the Nominated_Profit.

5.3.3.4 C1.2.1.2.4—Review Draft Mission and Goal Statements

Now that the draft goals have been developed, it is useful to review the draft mission statement to ensure that all goals are explicit in the statement, which will ensure that each goal is directly traceable to the mission. Once again, the draft mission and goal statements can be confirmed within the group of participants that developed it or, more usefully, it can be communicated to a wider group and used to elicit comment that can inform subsequent revision.

> The **Medical_Center_Project** shall develop the **Medical_Center** on the Canberra Hotel location to deliver **Medical_Services**, with appropriate **management** and **support**, in order to attain a **Nominated_Profit**.
>
> Glossary entries:

1. Medical_Center: The ACME Medical Center, which is to be developed to operate on the Canberra Hotel location, including the building and its elements of capability (e.g., building, facilities, equipment, and personnel).
2. Medical_Services to include a doctor's surgery supporting three GPs, a dentist's surgery for two dentists and two dental assistants, a physical therapy center supporting three physical therapists, and a pharmacy that provided a basic dispensing service provided by one pharmacist and an assistant.
3. Nominated_Profit level of x% is to be attained within y years (confirm with business case).

5.3.4 C1.2.1.3—Develop Objective Statements

As illustrated in Figure 5.4, drafting of the objective statements has three main activities: decompose/derive objective statements, group objective elements, review goal statements, and finalize mission, goals, and objectives.

5.3.4.1 C1.2.1.3.1—Decompose/Derive Objective Elements

Our next task is to elaborate the objectives for each goal by continuing the decomposition and derivation process. Again, a bottom-up process can be used within a top-down framework.

Starting with Goal 1, we decompose/derive from it the objective elements required to be performed to achieve that goal. For example, in order to provide ACME Medical Center, we will need (at least) to clean the site and acquire appropriate approvals; design, build, fit out, and commission the building; and then provide appropriate facilities for staff, equipment, and utilities. Of course, each of these functions will require other lower level functions, each of which will require further lower level functions, and so on. This, therefore, is the beginning of our top-down hierarchical elaboration of the elements of the mission statement into a set of requirements that will become the SyRS. Again, the objective elements are numbered—1OE1 is the first objective element of Goal 1, 2OE1 is the first objective element of Goal 2, and so on.

1. Develop Medical_Center.
   - 1OE1 Engage building designer.
   - 1OE2 Conduct building design.
   - 1OE3 Approve building design.
   - 1OE4 Clean building site.
   - 1OE5 Acquire appropriate approvals.
   - 1OE6 Provide site security.
   - 1OE7 Engage building contractor.
   - 1OE8 Build Center building.
   - 1OE9 Provide utilities interfaces.
   - 1OE10 Provide medical facilities.
   - 1OE11 Provide staff facilities.
   - 1OE12 Provide customer facilities.
   - 1OE13 Install medical equipment.
   - . . . .

We then continue the hierarchical elaboration for Goal 2 and the subsequent goals.

5.3.4.2 C1.2.1.3.2—Group Objective Statements

Once we have developed a first draft of the objectives for each goal, we need to group and allocate them to provide a balanced structure of objectives. Any redundancy or overlap here will result in redundancy and overlap in lower level requirements if we continue to elaborate those requirements without first grouping and allocating objectives here. The following list illustrates how we might begin to group the objective elements of Goal 1—again, we number the objectives hierarchically (for example, Objective 1.2 is the second objective of Goal 1).

1.1 Design Medical_Center.
- 1OE1 Engage building designer.
- 1OE2 Conduct building design.
- 1OE3 Approve building design.

1.2 Prepare for build.
- 1OE4 Clean building site.
- 1OE5 Acquire appropriate approvals.
- 1OE6 Provide site security.

1.3 Build Medical_Center.
- 1OE7 Engage building contractor.
- 1OE8 Build Center building.
- 1OE9 Provide utilities interfaces.

1.4 Fit out Medical_Center.
- 1OE10 Provide medical facilities.
- 1OE11 Provide staff facilities.
- 1OE12 Provide customer facilities.
- 1OE13 Install medical equipment.
- 1OE14 Install general equipment.

1.5 Commission Medical_Center.
- . . .

As illustrated in the next list, the objective elements are not discarded once we have the goals as they form the next level in the elaboration process.

1.1 Design Medical_Center.
- 1.1.1 Engage building designer.
- 1.1.2 Conduct building design.
- 1.1.3 Approve building design.

1.2 Prepare for build.

1.3 Build Medical_Center.
1.4 Fit-out Medical_Center.
1.5 Commission Medical_Center.

The hierarchical numbering of the goals and objectives continues (and begins the hierarchical numbering system that will end in the SyRS requirements at the end of the process)—the first level of decomposition, derivation, and grouping is shown here for Goals 2–5.

2. Deliver Medical_Services.
   - 2.1 Deliver GP services.
   - 2.2 Deliver dental services.
   - 2.3 Deliver physical therapy services.
   - 2.4 Deliver pharmacy services.
   - 2.5 Deliver medical support.
3. Manage Medical_Center.
   - 3.1 Manage center administration.
   - 3.2 Manage center finances.
   - 3.3 Manage center staff.
   - 3.4 Manage center contracts.
   - 3.5 Manage support services.
4. Support Medical_Services.
   - 4.1 Provide financial services.
   - 4.2 Provide logistic support.
   - 4.3 Provide administrative support.
   - 4.4 Provide comfortable environment.
   - 4.5 Provide secure environment.
5. Attain Nominated_Profit
   - 5.1 Implement business strategy.
   - 5.2 Understand customer needs.
   - 5.3 Maintain customer satisfaction.
   - 5.4 Undertake center marketing.
   - 5.5 Engage local community.

The decomposition, derivation, grouping, and allocation of objectives in the preceding lists are illustrative of only the beginning of the process, which would continue until a complete set of balanced objectives has been derived. Should any element of the design require more detailed consideration, trade studies would be commissioned to inform any design decisions and to provide subsequent rationale for the requirements.

5.3.4.3 C1.2.1.3.3—Draft Objective Statements

Again, the objective statements are initially brief (three-word, verb-based phrases) to allow them to be readily understood—once iteration is complete, each should be written as a full sentence. The characteristics of a good objective statement and the principles for its construction are similar to those discussed previously for mission and goal statements.

5.3.4.4 C1.2.1.3.4—Review Goal Statements

Once we have established the objectives, we revisit the goal statements to confirm (again, at our present level of understanding) that the goals are balanced and complete. If necessary, we adjust them and revisit the draft mission statement to confirm that it is still appropriate (and adjust it if necessary).

5.3.4.5 C1.2.1.3.5—Finalize Mission, Goals, and Objectives

We may again choose to confirm the draft mission, goals, and objectives by communicating them to the broader community at the business management and business operations levels, not just those immediately involved in their development. Once we have confirmed the artifacts, we finalize the mission, goals, and objectives into a formal set from which traceability of requirements can be conducted.

5.3.5 RBS Versus WBS

Chapter 2 introduces the RBS (the functional hierarchy) as a framework within which to capture requirements. The term is deliberately chosen to differentiate this structure from the well-known project management document known as the WBS [10]; the RBS is grouped logically (functionally), and the WBS is structured by physical work packages (including physical system elements) and contains other project-related work. The logical groupings of the RBS are then allocated to the physical groupings of the physical system elements in the WBS.

As illustrated in Figure 5.5, the hierarchy of project mission, goals and objectives can be summarized using an RBS. Figure 5.6 shows how the system mission, goals and objectives (the system RBS) can be extracted from the project RBS. Note that the elements of the system mission persist for as long as the system is in service—those goals associated with the project requirements are only relevant for the duration of the project that provides the Medical Center. Figure 5.7 shows just the system RBS once the project is completed.

Figure 5.8 illustrates the relationship between the RBS and the WBS for the ACME Medical Center at the project and system level. The RBS contains

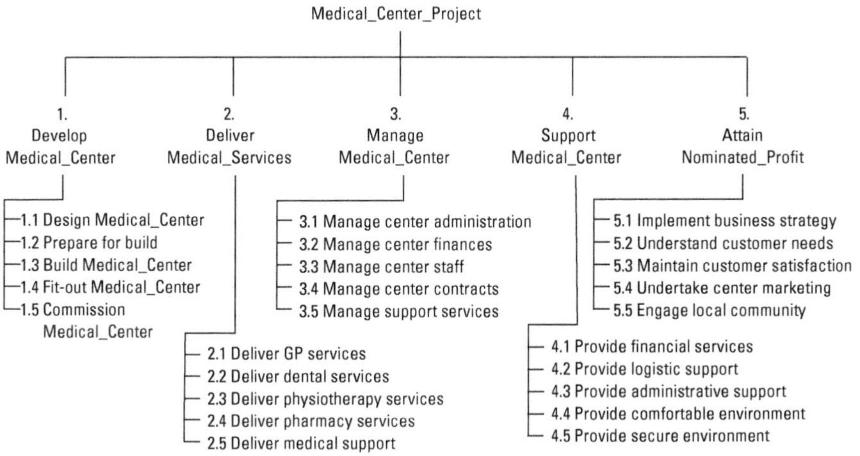

Figure 5.5 Project RBS for the Medical_Center_Project to deliver the Medical_Center.

the functions required to deliver the Medical Center and its services. The WBS contains all the work that needs to be undertaken to produce those functions. The RBS provides a functional description and the WBS provides a physical description—the process of mapping the functions in the RBS to the physical work in the WBS is called allocation.

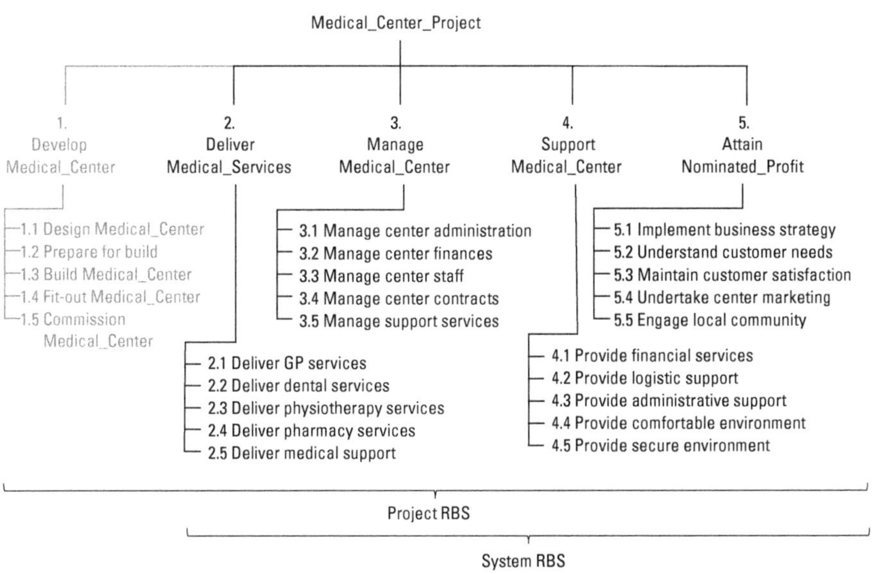

Figure 5.6 System RBS as a subset of the Project RBS for the Medical_Center.

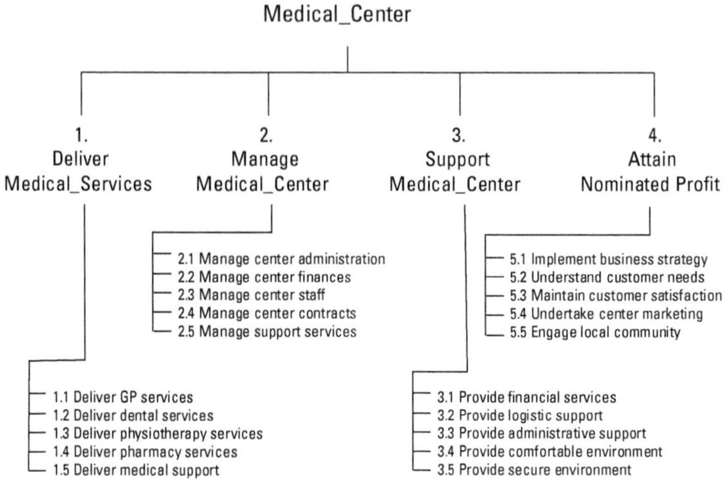

Figure 5.7 System RBS for the Medical_Center once the Medical_Center_Project is complete and the Medical_Center is in operation.

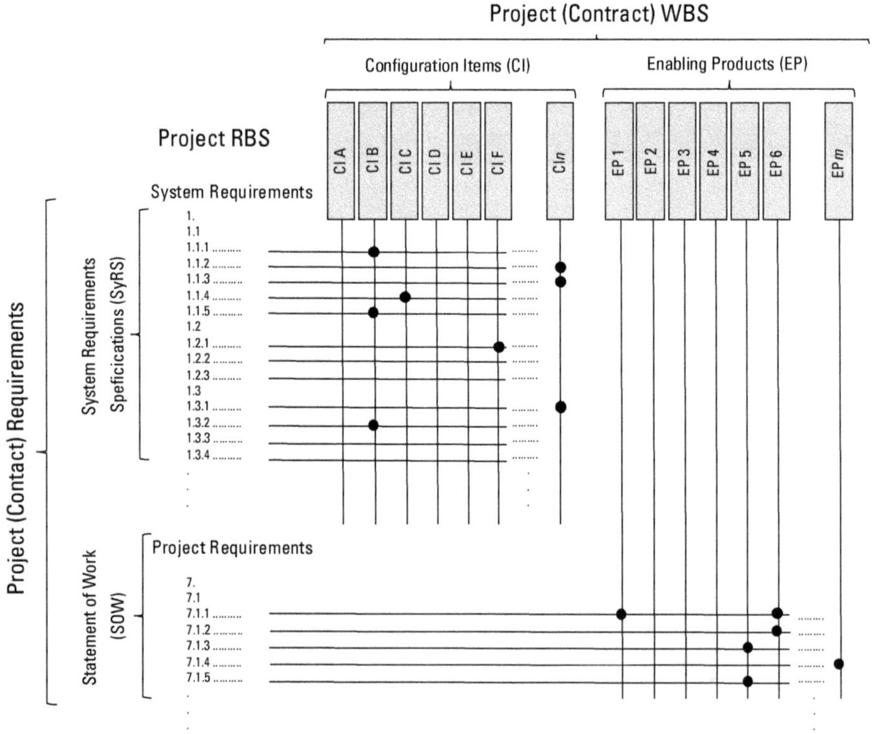

Figure 5.8 The relationship between the RBS and WBS at the project (contract) and system levels.

Note that both the project requirements and the system requirements map onto the work packages in the WBS: The project requirements in the RBS map onto the work packages that relate to the enabling products produced by the project; the system requirements map onto the CIs that are delivered as part of the system. Project requirements are contained in the SOW, and system requirements are contained in the SyRS. Once the delivery contract has been signed, the project requirements will therefore begin with, "The Contractor shall . . ." whereas the SyRS requirements begin, "The Medical_Center shall . . ."

5.4 C1.2.2—Define Preliminary Life-Cycle Concepts

As discussed in Chapter 1, it has been common in the past to consider systems only in the light of the issues associated with the acquisition phase and to pay little attention to through-life support. Project teams have typically focused on the acquisition phase of the project and on the development of a system that meets the operational participant requirements while minimizing cost and schedule. This has often led to larger-than-expected costs in the utilization phase to be met from budgets that are insufficient to keep the system in service. A life cycle focus requires a focus on the entire capability, not just on the material system. A capability-system focus takes into account all constituent elements of the system including operation, maintenance and support, and retirement or disposal.

Early in the acquisition phase, therefore, business management must give some guidance on the life cycle concepts related to the acquisition, deployment, operation, support, and retirement of the system. While the systems engineering processes that follow will ensure a life cycle focus, it is important that the business management focus on the major cost drivers that will affect the utilization of the system. There are a number of life cycle–related tradeoffs at a business-case level.

5.4.1 Preliminary OpsCon

The preliminary OpsCon (as with all PLCs) is developed by business management and contains as much detail as they wish to communicate regarding (and sometimes constraining) the solution, which is to be refined in subsequent activities by stakeholders at the business operations level. The preliminary OpsCon is commonly the first concept developed and takes primacy among the five life-cycle concepts, because it drives the principal needs and requirements for the system. However, development of the life-cycle concepts is iterative, and the operational concept is tempered by the subsequent support concept.

5.4.1.1 OpsCon Content

There are a number of excellent references that provide guidance on the development process, content and layout of an OpsCon. Perhaps the principal guidance is in ANSI/AIAA G-043A-2012 [11] and ISO/IEC 29148 [12], which suggest that OpsCon contain the following sections:

- *Scope:* This section provides an overview section describing in broad terms the organizational context, the proposed capability system, and the system's intended purpose within that context.
- *Glossary and referenced documents:* A glossary is essential to ensure that terms are used consistently throughout the OpsCon (and other LC). It is also critical to keep a record of the source of information in the OpsCon in case of conflicting and ambiguous requirements, or some disagreement between participants.
- *Proposed system operational overview:* This section identifies the needs (business and then participant) for the system, including the major mission(s), how these will be performed, techniques used, and any constraints. Each mission is investigated to determine common needs in order to develop an RBS based on the system mission, goals, and objectives. The section addresses proposed operational environments as well as any operational policies and constraints on the system. The section also includes personnel requirements derived from the types of personnel involved in the system, as well as consideration of operational support issues (in concert with the other life-cycle concepts).
- *Operational scenarios:* A description of possible operational scenarios should be provided to help describe the likely events facing the system. (These are often summarized in the body of the OpsCon and outlined in detail in an annex.) See Section 5.4.1.2.
- *Existing system(s):* This section outlines the existing system(s) to the extent that they are relevant to the acquisition of the new system, including the current major mission(s), how these are performed, major participants, and any current constraints.
- *Nature of (and justification for) proposed changes:* The existing system(s) (to the extent relevant to the acquisition of the new system) are examined to identify the shortcomings that lead to the needs for the proposed system. There are many types of justifications for a proposed

system: There may not be a current system in service; the current system may be beyond economical repair; there may be new operational needs for which the current system was not designed; and so on.

- *System scope:* This section includes a description of the system context diagram, external interfaces, significant constraints, system states and modes, and any architectural considerations that need to be addressed.

5.4.1.2 Operational Scenarios

The examination of operational scenarios begins with a description of the general operational environment for the system to identify all the environmental factors that may have an effect on the operation of the system. Specific operational scenarios are then described to depict the full range of circumstances under which the system is required to operate. It is not necessary to describe every possible scenario, but all types of operations must be represented. Scenarios also need to represent all perspectives.

These scenarios, often called use cases, vignettes, or user stories [13], provide valuable guidance to the system designers and form the basis of major events in the acquisition phase such as acceptance testing of the system as it is introduced into service—that is, use cases become test cases. Despite any more detailed technical verification and validation procedures, the system's fitness for purpose (and thence the basis for its common-law warranty) is fundamentally related to its ability to perform in accordance with the operational scenarios defined at this stage.

In many cases it is also useful to define the various modes of operation for the system. Designers need to understand whether the system exists in different modes even if it is as simple as the difference between the fully operational mode or the training mode. Complex systems may have their requirements stated in a number of modes. For example, a modern fighter aircraft may have modes defined for air-to-air combat, ground attack, reconnaissance, maritime operations, and nontactical flights. Each mode must be associated with the particular conditions (e.g., mission, operational, environmental, and configurational) that define it.

Users tend to think of system operation in terms of its normal use under ideal conditions—care has to be taken to define exception conditions. For example, a pilot of a combat aircraft will naturally describe a number of modes and states during which adversary aircraft are engaged and destroyed but will need some prompting to describe what happens when the pilot's aircraft is damaged, and the pilot must eject. At every stage, therefore, in each scenario, we must ask the question, "What could go wrong here?" and use that information to generate requirements.

The value of a good set of operational scenarios cannot be understated, for two reasons. First, the use cases are a form of virtual prototyping that allows designers to "walk" through the system functions and performance with the users with the aim of garnering the attendant system-level requirements. Second, as noted earlier, the use cases that are so useful at the beginning of system design become test cases at the end of the acquisition phase. That is, if the system can perform as required within the operational scenarios, the users will be satisfied.

In this activity, the operational scenarios are preliminary in that they are relatively high-level and provide broad guidance from the business management perspective in terms of how the system is to fit into the business and the organizational context. The detail of each scenario is to be fleshed out by the participants at the business operational level when the SNRs are developed.

Preliminary operational scenarios are developed using the following general guidelines (see Chapter 7 for more detail):

- First, the general operational environment for the system is described to identify all the environmental factors that may influence the operation of the system. (These may or may not have been identified in the development of the mission, goals, and objectives.) This scenario is sometimes called "A day in the life of . . ." because it describes a typical day or cycle of the system as it is used by all involved. This general scenario is then used as the context for more detailed scenarios.
- It is not possible to develop all possible scenarios from all possible perspectives for all possible viewpoints, so consideration must be given to the way a suitable (minimal) subset of scenarios is to be developed. While the smallest number of scenarios is sought, all the system requirements must be covered, and all relevant actors must be described.
- Specific operational scenarios are then described to depict the full range of circumstances under which the system is required to operate. It is not necessary to describe every possible scenario, but all types of operation must be represented. There must be at least one scenario that represents the perspective of each major user of the system.
- Where necessary, the exception conditions of each scenario are outlined. For example, in a scenario that describes the use of a movie theater by patrons, care must be taken to identify how the scenario would vary in exceptional conditions such as an emergency in the theater (perhaps due to a fire or power blackout) or a disruption such as a faulty projector.

- There must be as little duplication between scenarios as possible. To minimize repetition, it is often helpful to group scenarios in a hierarchy where the lower level scenarios inherit the properties of those at the level above.
- Scenarios also tend to be more useful and more easily interpreted if a standard structure is used as a broad template for each scenario.
- Scenarios must be consistent in the use of terms for functions and actors—this consistency is greatly assisted by the development of a project/system glossary. See Section 9.2.2 for more detail on the use of a glossary.

5.4.1.3 Preliminary Acquisition Concept

Based on the business operational needs for the proposed system, business management defines the proposed acquisition concept. Any business needs for acquisition are articulated and may include the following:

- Budget and schedule;
- Preferred (and perhaps prohibited) supply sources;
- Type of desired contract arrangements;
- Relevant existing contractual arrangements;
- Any relevant existing acquisition support arrangements;
- Program/project management considerations;
- Reporting issues;
- Relationship to other system acquisitions within the organization.

This preliminary acquisition concept is continually refined—first, after the feasibility analysis in Activities C1.4.1 to 1.4.3; then, after participants have begun to consider their needs and requirements; and then, at every major review as the design continues.

5.4.1.4 Preliminary Deployment Concept

In this concept, business management outlines the issues surrounding the deployment of the system (to the extent that they are concerned with the introduction into service of this system). For example, business management may dictate the nature of transition arrangements between an existing and replacement system, since business continuity is at risk during this period. Additionally, the preliminary deployment concept may address issues such as transition between facilities, training of operational personnel (and possibly engaging new staff and retiring others), and transition support arrangements.

5.4.1.5 Preliminary Support Concept

The preliminary support concept will capture the needs of business management for the support of the system throughout its life cycle. Issues addressed would include relevant support policies and procedures; any relevant existing support arrangements; the support environment; desired maintenance levels and cycles; and the anticipated impact of the proposed system on support facilities, equipment, personnel, and training.

The support services that are provided (by what is sometimes called the support system) during the utilization phase include the following. (The constituent capabilities that provide those support services are acquired during the acquisition phase.)

- *Operating support capability:* This capability encompasses the support elements of operating facilities system operators, support equipment, operator manuals and technical data, operating support procedures, and operating support information systems.
- *Engineering support capability:* Engineering support encompasses the support elements of engineering facilities, engineering personnel, engineering support and test equipment, engineering technical data, engineering processes, engineering information management system, and software support.
- *Maintenance support capability:* Maintenance support is the capability that includes the support elements necessary to develop, establish, and integrate a maintenance support system capable of sustaining a system throughout its life. The capability comprises the elements of maintenance facilities, maintenance personnel, maintenance support and test equipment, maintenance technical data, maintenance processes, and maintenance information management system.
- *Supply support capability:* Supply support encompasses the following support elements: supply facilities, supply personnel, supply support equipment, supply technical data, supply processes, supply information management system, spares, and packaging.
- *Training support capability:* The training support capability includes the support elements training facilities, training personnel, training equipment, training materials and other technical data, training processes, and training information management systems.

It should be noted that the consideration of support concepts must be made in the context of achievement of the capability required to meet operational scenarios (which is why the preliminary operational scenarios are developed first). In particular, the support aspects of a system must not be optimized by themselves without consideration of the operational use of the system. For example, the lowest cost through-life support for an aircraft is achieved by minimizing fuel consumption and reducing wear and tear that requires maintenance—in short, by not operating it all. Clearly, therefore, through-life support must be optimized based on the desirable operational scenarios, not independent of them. Customers must also recognize that through-life support optimized for the contracted operational scenarios may not necessarily be optimal for any other scenarios, should they change their operational context in the future.

5.4.1.6 Preliminary Retirement Concept

The preliminary retirement concept addresses the retirement phase, which is the final stage in the system life cycle. Functions associated with phase-out and disposal include transportation and handling, decomposition, and processing of the retiring system. If considered early, disposal and phaseout issues will form some of the criteria against which the system is designed. It is important, however, that system designers focus on retirement, rather than the more limiting issues of disposal—planning for disposal is important, but a system may retire from a number of life cycles before its ultimate disposal.

As illustrated in the example in Figure 5.9, a system may be developed from conceptual design in an initial life cycle and then delivered, deployed, operated, and supported throughout that life cycle. If the organization that owns the system changes focus and moves away from the business area for which the system was acquired, the initial owners may sell the system to a second organization that, having completed conceptual design and preliminary

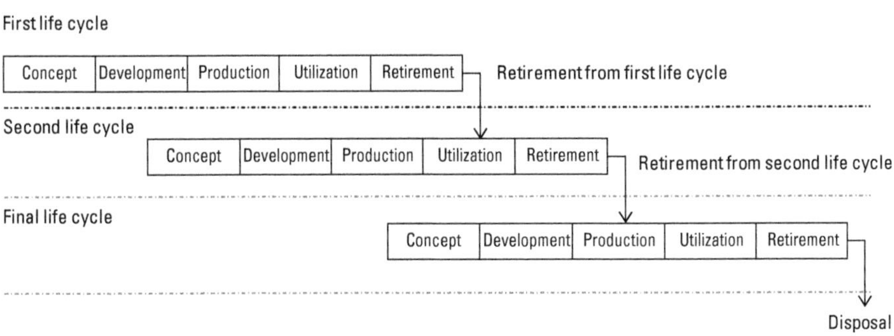

Figure 5.9 A system may be retired from a number of life cycles before final disposal at end of life.

design in that life cycle, has decided to meet business requirements with an OTS solution, albeit secondhand. As illustrated in Figure 5.9, the system then enters a second life cycle, perhaps with a number of its elements (the operating personnel, for example) replaced. When the second organization wishes to replace the system (perhaps because it has become costly to maintain), the system may be sold to a third organization that wishes to meet its systems design with a modified-OTS solution based on a refurbished secondhand system—again, a number of system elements may be added or removed and other elements added. As illustrated in Figure 5.9, the third organization acquires the system, refurbishes it, and then fields it in a third life cycle. In this example, Figure 5.9 illustrates that the system is uneconomical to repair at the end of the third life cycle and is disposed of as scrap.

A focus on designing for more than one life cycle is important because, if designers assume that there is only one life cycle and design just for disposal of the system products at end of that life cycle, then that disposal method may well be the only option available at the end of that life cycle (or, at least, other options may have been constrained). Yet, for minimal environmental impact, system design should address the system's existence in as many life cycles as possible (on the basis that the longer a system exists, there is lower cost and lower environmental impact resulting from the need to develop replacement systems). A good system design should therefore not only achieve longevity of use in the first life cycle of the system but should aim for longevity of use in all life cycles until disposal is the only viable option remaining [14].

In reality, retirement issues are often not considered until they are imminent. Very little is written about formal methods for the inclusion of retirement and disposal in the early systems engineering activities. Perhaps as environmental and ecological constraints become more stringent, retirement and disposal will begin to play a more prominent part in the system life cycle. Indeed, environmental considerations such as emission control and material recycling are already playing an important role in constraining the design, construction, operation, and disposal in the aircraft and motor vehicle industries.

When considering the retirement aspects of the system during its design, it is useful to undertake three broad tasks [15]: to identify the reasons for potential retirement, to identify potential retirement methods for the system, which then allows conceptual designers to identify design issues that may arise from the consideration of each retirement method. These three tasks are described as follows:

- *Identify reasons for potential retirement:* There can be any number of reasons why the system may require retirement:

- The system may have reached the end of its life and no longer be usable and/or supportable.
- The system may still largely be useful, but one or more critical components may be unusable or unsupportable.
- The system may still be perfectly useful, but it (or a significant subsystem of it) is being retired for one of the following reasons.
  - The business need for the system may have disappeared (for example, the owner's business direction may change, or the owner may be retiring and winding up the business);
  - The business need is still valid, but the business owner is forced to retire the system (due to issues such as cash flow, the operating license is revoked, scandal, public pressure, or disputes between business partners).
  - The system (or one of its critical components) may be damaged beyond economical repair (by accident, natural disaster, act of war, or vandalism).
- *Identify potential retirement methods for the system:* Once the reasons for potential retirement have been identified, each can be examined for potential retirement methods for the system through the following actions (although not all may apply to all reasons for retirement):
  - It could be sold as a secondhand item.
  - It could be traded in on a replacement system.
  - It could be redeployed as a training aid.
  - It could be destroyed and disposed as waste.
  - It could be disassembled—that is, broken up and sold as working subsystems, assemblies, or components.
  - It could be scrapped—that is, broken up and sold as subsystems, assemblies, or components for their value as scrap.
  - Placed in storage—that is, mothballed because none of the other methods are currently acceptable or tenable.
- *Identify design issues that arise:* Once the reasons for potential retirement and the potential retirement methods have been identified with the stakeholders, designers can identify any design issues that may arise, including the following:
  - Cost of refurbishment for resale or trade-in;
  - Desirability of salvage or material recovery for resale;
  - Cost of transportation for retirement/disposal;
  - Observance of recycling regulations;
  - Avoidance of the use of potentially hazardous/toxic materials;
  - Environmental impact;

- Cost of disposal/destruction/uninstallation;
- Need for specialized personnel/equipment required for disposal;
- Observance of any caveats on disposal (such as might be placed on military equipment by international arms trade agreements or that might be placed on high technology by national sanctions);
- Time taken to arrange disposal;
- Availability of design data and drawings to support disposal.

By way of illustration, the following is a brief example of the conduct of the three tasks outlined above when considering the preliminary retirement concept for the ACME Medical Center.

Reasons for potential retirement of the ACME Medical Center and potential retirement methods. Identifying first the reasons why the business may be retiring (including disposing of) the building, and noting for each the potential retirement methods:

- Business is booming and the current facility is too small:
  - Relocate the Center to another location.
  - Sell or lease the building as a general purpose building.
- Business is unprofitable for any one of a number of reasons:
  - Close the Center.
  - Sell or lease the building as a general purpose building.
- ACME wishes to reduce the number of medical centers it operates (owner is retiring or the business strategy has changed):
  - Sell the Center (services and building) as an ongoing medical services business.
  - Retain the building and lease the medical center business to another operator.
  - Sell or lease the building as a general-purpose building.
- External influences force the business to close/move (such as regulation changes, zoning changes, catastrophic damage due to fire or earthquake, and so on):
  - Close the Center.
  - Sell or lease the building as a general-purpose building.
- Some portion of the Center is unprofitable (one of the services, for example):
  - Reduce the Center footprint by collapsing to a smaller area (perhaps to only one of the two floors) and lease the remaining capacity.

- Reputation damage (some incident at the Center has meant that patients no longer attend in sufficient numbers):
  - Close the Center.
  - Sell or lease the building as a general-purpose building.
- Building damaged beyond economical repair (by fire or flood, for example):
  - Close the Center.
  - Clean the site, and either sell the site or rebuild on it.

Identify potential retirement methods for the ACME Medical Center and design issues that arise. Note that there are a number of potential retirement methods that are common for various reasons for retirement. Accordingly, the following lists the potential retirement methods and then identifies design issues that arise from each:

- Close the Center:
  - Equipment must be capable of being removed and installed in a manner that doesn't require building modifications to get it out.
  - Staff contracts must allow for termination or redeployment.
- Relocate the Center to another location:
  - Equipment must be capable of being relocated without disturbance to current building. For example, large equipment should be on the bottom floor, and perhaps a large plate glass window on that floor could be readily removed to allow egress, or, alternatively, the main entrance could be made sufficiently large. This needs to be included as an interface.
  - Center must be capable of providing uninterrupted reduced services as it transitions from old site to new.
- Sell the Center (services and building) as an ongoing medical services business:
  - Building must be designed to suit the market as well as ACME's needs—that is, it can't be so specific to ACME as to be less desirable to potential purchasers.
  - Any equipment leases must be transferrable.
  - Staff contracts should allow for termination or redeployment or transfer to the new lessor.
- Sell or lease the building as a general-purpose building:
  - Building must be designed to be quickly repurposed into a general-purpose building such as an office building.

    - Building must be attractive as a medical center as well as a general-purpose building (in terms of such factors as external look and feel and access points).
- Retain the building and lease the medical center business to another medical services provider:
    - Building must be designed to suit the market as well as ACME's needs—that is, it can't be so specific to ACME as to be less desirable to potential lessors.
- Reduce the Center footprint by collapsing to a reduced area and lease remaining floorspace after collapsing to a smaller footprint:
    - Building must be designed to allow for another tenant (i.e., include separate access and separate facilities).
    - Building must be capable of being modified/ refurbished in sections.
- Clean the site and sell the site or rebuild:
    - Building materials should suit a long life but also be amenable to easy removal in the event of building disposal, particularly given the tight space constraints around the Center location.

5.5 C1.2.3—Define Preliminary Validation Criteria

Once we have an idea of how the broad use of the system, an important top-level activity is the preliminary identification and definition of the criteria by which the customer will measure their satisfaction with the products of the acquisition phase before the system is placed into operational service. Key validation criteria (or measures) include, safety, reliability, supportability, maintainability, ease of use, and time and cost to train, as well as measures from each operational scenario.

Measures can be identified using both top-down and bottom-up approaches, described as follows:

- *Bottom-up:* In this approach we simply list all those things that are measurable without worrying about their level. We then analyze them and group the measures into a hierarchy.
- *Top-down:* In the top-down approach, we identify appropriate high-level measures and decompose/derive hierarchically from these lower level measures.

The two approaches are best illustrated by example. Sections 5.5.1 to 5.5.3 show how we might develop measures for the ACME Medical Center.

5.5.1 ACME Medical Center—Development of Measures

For the ACME Medical Center, we are interested in measures that allow us to verify that the goals and objectives have been achieved. We could start by using the bottom-up approach in which we write down as many measures as we can identify and then come back and group them into a hierarchical structure.

So, for the ACME Medical Center, our first list of measurable entities might include the following:

- Percentage profit;
- Numbers of patients;
- Rate of return of patients;
- Number of unused consultation times;
- Waiting time for each consultation;
- Length of waiting list/date booked out until;
- Proportion of consultations that did not start on time;
- Total (daily) working hours for each practitioner;
- Number of complaints received from patients (and staff);
- Maintenance rates for facilities and equipment;
- Staff retention (length of service)/turnover;
- Center prestige;
- Number of outside scripts filled by pharmacy;
- Number of requested pharmacy items that are not available;
- Staff security;
- Patient/customer security;
- Staff comfort;
- Patient/customer comfort.
- . . .

There is obviously some overlap in these measures and some differences in level. The next step is to group measures to provide a hierarchical set.

5.5.2 An Alternative Approach to the Development of Measures

As show in Section 5.5.1, a measure should be relevant—that is, it must be related to one or more requirements. One common way of doing this is to develop a matrix that directly relates measures to requirements. In fact, the use of relationship matrices is very common in systems engineering: we have traceability matrices that show inheritance of requirements, we have verification matrices to show how verification means relate to individual requirements or groups of requirements, and we could have matrices to relate measures to requirements.

The use of verification matrices is common because it is often perceived that the entities related by the matrix are orthogonal—that is, for example, there is a hierarchy of requirements, and there is a hierarchy of measures. But orthogonality in engineering and mathematics implies independence of variables, which is not always the case in the matrices we develop in systems engineering. In the case of requirements, we have developed performance statements and verification statements for each requirement. If we are looking for a measure of effectiveness or performance, we already have the basis in the information related to each requirement. Rather than developing the requirements database and then stepping sideways to develop independently the measures (which would then require the development of a matrix to relate them), it would make more sense to use the same functional/performance/verification requirements structure to develop the measures.

In doing so, it is common to develop a hierarchy of measures (see Figure 5.10), such as: critical strategic issues (CSIs) that relate to the measurement of system mission and goals, critical operational issues (COIs) that relate to the measurement of objectives, measures of effectiveness (MOEs) that relate to the next level of the requirements hierarchy, measures of performance (MOPs), and down to verification statements at the lowest level. This hierarchy is summarized as follows:

- *CSI:* A CSI is a measure of an aspect of the system that is of primary importance to business management and must be met before the system is allowed to proceed, either to the next development phase or into service. CSIs are commonly phrased as questions that must be answered in the affirmative.
- *COI:* A COI is a measure of an issue that must be examined to determine whether a system meets its mission; COIs are selected subsets of CSIs. COIs are also commonly phrased as questions that must be answered in the affirmative.
- *MOE:* An MOE is a measure (quantitative or qualitative) of the degree to which a system objective is met under specified conditions. MOEs address the effectiveness of the system in terms of both how well it accomplishes its mission and how well it integrates into the organization and operational environment (that is, how well it meets its interfaces). MOEs are supported by sibling measures of measures of suitability (MOSs), which measure subjective suitability of the system to the organization. The difference is often clouded, and MOE can

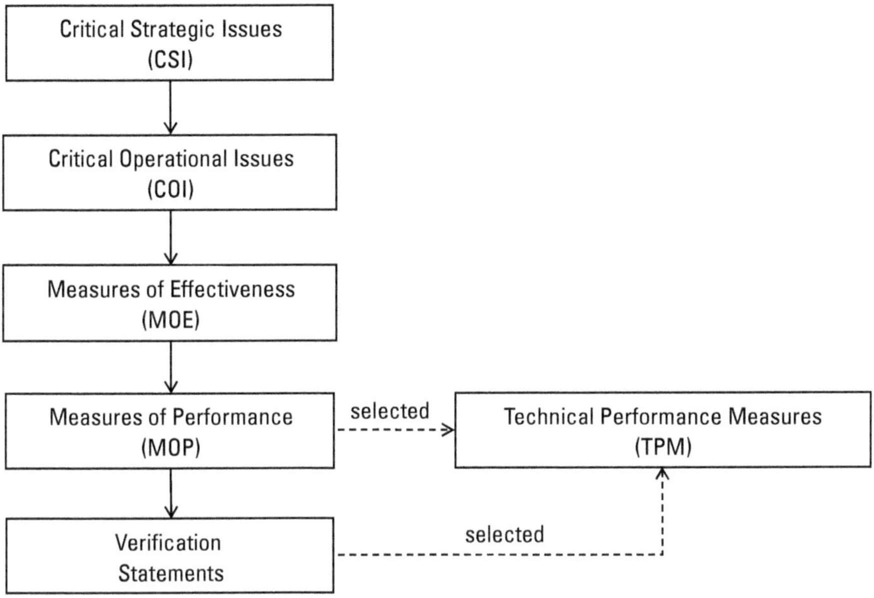

Figure 5.10 A typical measures hierarchy.

be qualitative, so here we avoid the complication, and include such measures as MOE.

- *MOP:* An MOP is a measure of performance that is one of a subset of measures that combine to form an MOE. Examples of MOPs include performance measures for range, velocity, frequency range, and all-up weight.
- *Verification statement:* It is good practice to ensure that, every time a requirement is articulated, a corresponding verification statement is made. While the common practice of attaching a letter to illustrate the verification methods is a useful indication of the customer's intent, it is much better practice to write a verification statement to make the verification methods more specific. That is particularly true for the analysis and test verification means, because the use of a single letter begs a number of important questions that really should be answered at the time the requirement is written.
- *TPM:* It is normally not possible for an acquirer and developer to jointly track each verification statement, or even each MOP. Therefore, selected MOPs and verification statements are nominated as TPMs that are used to assess conformance throughout the development process.

Commonly, TPMs are reported across the contractual boundary from the developer to the acquirer as a means of providing confidence to the acquirer that the contract is on track.

As discussed in Chapter 8, selected MOPs and verification statements are nominated as TPMs that are used to assess conformance in the development process.

Note that MOEs are related to their parent COI and grandparent CSI by a hierarchical system (just as objectives are related to goals, and so on). As illustrated in Figure 5.11, there then should be a direct relationship between the requirements hierarchy and the measures hierarchy.

At this stage in the process, the criteria being developed are upper level, preliminary criteria (probably at the level of CSI and COI) that will be refined further during the development of SNR (which will address MOE, MOP, verification criteria, and TPM).

In our ACME Medical Center example, we grouped the need elements for the project mission statement into five goals (1. Provide Medical_Center; 2. Deliver Medical_Services; 3. Manage Medical_Center; 4. Support Medical_Center; and 5. Attain Nominated_Profit). From these goals we can readily define the following five CSIs that indicate success in the project:

- CSI1. Was the Medical_Center provided?
- CSI2. Can the Medical_Services be delivered?

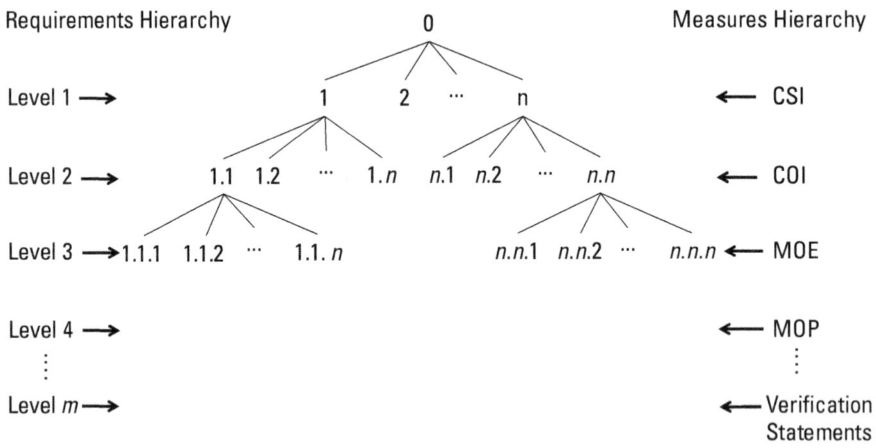

Figure 5.11 The relationship between the requirements hierarchy and the measures hierarchy.

- CSI3. Is the Medical_Center being managed?
- CSI4. Is the Medical_Center being supported?
- CSI5. Is the Nominated_Profit being attained?

Note that one advantage of a hierarchical decomposition of both requirements and measures is that we do not need a matrix to relate CSIs to goals because there is a direct one-to-one relationship that is implicit in the numbering system—that is CSI1 relates to Goal 1, and so on.

From each CSI we can derive COIs, from each of which we can define MOEs, and from each MOE we can derive MOPs—but we don't have to do that independently because the structure is already present in the project mission, goals, and objectives. For example, for CSI1. Was the Medical_Center_System provided? (which relates to Goal 1), we may use the subordinate objectives to identify the relevant COIs:

- COI1.1 Was the Medical_Center_System designed?
- COI1.2 Was the Site prepared?
- COI1.3 Was the Medical_Center built?
- COI1.4 Was the Medical_Center fitted out?
- COI1.5 Was the Medical_Center_System commissioned?

We can then continue to MOEs. For example, for COI1. Was the Medical Center designed? (which relates to Objective 1), we may use the subordinate objectives to identify the relevant MOEs:

- MOE1.1 Was the Building_Designer engaged?
- MOE1.2 Was the Building_Designer briefed?
- MOE1.3 Was the Building_Design conducted?
- MOE1.4 Was the Building_Design approved?

Note that MOEs are related to their parent COI and grandparent CSI by a hierarchical numbering system (just as objectives are related to goals and so on). As illustrated in Figure 5.11, there should be a direct relationship then between the requirements hierarchy and the measures hierarchy.

At this stage in the process, the criteria being developed are upper level, preliminary criteria (probably at the level of CSI and COI) that will be refined further during the development of SNR (which will address MOE, MOP, verification criteria, and TPM).

Note also that the example of CSI1 above is in support of the project mission statement; the elaborations of CSI2 to CSI5 would be in support of the system mission statement.

5.5.3 Measurement

The following issues that need to be considered when identifying measures:

- *Measurability:* The selected measure clearly must be measurable to be useful. That doesn't always mean that there must be some quantitative measurement possible—there may be a qualitative assessment. For example, in the ACME Medical Center, the number of complaints received is a quantitative measure, whereas staff comfort is qualitative. In both cases, what must be defined is the way in which the measurement is to be conducted. A quantitative measurement is normally straightforward to define (such as the measurement of length and weight, or the aggregation of events). A qualitative measure might require a questionnaire providing responses utilizing some scale (such as 1–10) to indicate degree.
- *Baseline:* For the measurement to be meaningful, it is often useful to be able to identify the change in the measurement, rather than the absolute value at any particular time. Most measurements therefore must be made at the start of the project and then at select times throughout the project (and the system's subsequent life) so that some change can be calculated or inferred—that is, we would prefer to derive measurements that reflect our progress as well as our current position.
- *Relevance:* A measure must be able to be related to at least one or more requirements. If that is not possible, the measure is seeking to observe a function that is not part of the design—in which case, either the measure is irrelevant, or a function is missing.

5.6 Summary

The C1.2—Define business needs activity involves defining the mission, goals and objectives for the system; defining the preliminary operational scenarios for its use; defining the upper level measures by which we will judge success; and defining the preliminary life cycle support concepts.

The definition of a system mission statement is a critical step in a project, because it provides a singular, complete, correct, unambiguous, feasible, justifiable, verifiable sentence describing a unique system. Since the project management scope and the subsequent system design use this sentence as their starting point, any endeavor that is not described by an adequate mission statement has a dramatically reduced chance of success. Once the mission statement is formulated, it is then expanded and qualified by short declarative statements of the system goals and objectives.

Operational scenarios are a very useful form of virtual prototyping that allow designers to "walk" through the system functions and performance with the business management and participants with the aim of garnering the attendant system-level requirements. Additionally, the use cases that are so helpful at the beginning of system design become test cases at the end of the acquisition phase—that is, if the system can perform as required within the operational scenarios, the users will be satisfied.

User satisfaction is described in the mechanisms by which the customer will measure their satisfaction with the products of the acquisition phase. Finally, business management must give some early guidance on the life-cycle concepts related to the development, production, testing, distribution, operation, support, training, and retirement of the system. While the conceptual design that follows will ensure a life cycle focus, it is important that business management focus on the major cost drivers that will have an impact on the supportability of the system.

Review Questions

1. Briefly describe the four main steps in the activity C1.2—Define business needs.
2. With the aid of a diagram, briefly describe the tasks associated with the step C1.2.1—Develop mission, goals, and objectives.
3. Briefly list and describe the eight desirable characteristics of a good system mission statement.
4. Briefly list and describe the six guidelines for the format of a good system mission statement.
5. Briefly describe the step C1.2.2—Define preliminary life cycle concepts including the role and likely content of the preliminary operational concepts.
6. Briefly describe the step C1.2.3—Define preliminary validation criteria.
7. Briefly describe the three tasks that designers may take into account when considering the retirement aspects of the system.

Endnotes

[1] Mager, R. F., *Preparing Instructional Objectives*, Belmont, CA: Pitman Learning, 1975.

[2] Bahill, A. T., and C. Briggs, "The Systems Engineering Started in the Middle Process: A Consensus of Systems Engineers and Project Managers," *Systems Engineering*, Vol. 4, No. 2, 2001.

[3] Grady, J. O., *System Requirements Analysis*, Burlington MA: Academic Press, 2006.

[4] IEEE-STD-1233, *IEEE Guide for Developing System Requirements Specifications*, New York: The Institute of Electrical and Electronic Engineers, 1998.

[5] Gause, D. C., and G. M. Weinberg, *Exploring Requirements: Quality Before Design*, New York: Dorset House Publishing, 1989.

[6] Buede, D., *The Engineering Design of Systems: Models and Methods*, Hoboken, NJ: John Wiley & Sons, 2009.

[7] Miller, G. A., "The Magical Number Seven, Plus or Minus Two: Some Limits on Our Capacity for Processing Information," *The Psychological Review*, 1956, pp. 81–97.

[8] Maier, E., and M. W. Rechtin, *The Art of System Architecting*, Boca Raton, FL: CRC Press, 2009.

[9] Van Gundy, A. B., *Idea Power: Techniques and Resources to Unleash Creativity in Your Organization*, New York: American Management Association, 1992.

[10] Project Management Institute (PMI), *Guide to the Project Management Body of Knowledge*, Newtown Square, PA: Project Management Institute (PMI), Version 6, 2017.

[11] ANSI/AIAA G-043A-2012, *Guide for the Preparation of Operational Concept Descriptions*, American National Standards Institute, American Institute of Aeronautics and Astronautics (sponsor), 2012.

[12] ISO/IEC/IEEE 29148, *Systems and Software Engineering—Lifecycle Processes—Requirements Engineering, 2018.*

[13] Specific detail on use-case modeling can be found in the following:

Bittner, K., and I. Spence, *Use Case Modelling*, Boston, MA: Addison-Wesley, 2003.

Cockburn, A., *Writing Effective Use Cases*, Boston, MA: Addison-Wesley, 2001.

Kulak, D., and E. Guiney, *Use Cases: Requirements in Context*, Boston, MA: Addison-Wesley, 2004.

Rinzler, B., *Telling Stories: A Short Path to Writing Better Software Requirements*, John Wiley & Sons, 2009.

[14] Ryan, M. J., "Design for Retirement," *Journal of Cleaner Production*, Vol. 70, 2014, pp. 203–210.

[15] Ryan, M. J., "Design for Retirement," *Journal of Cleaner Production*, Vol. 70, 2015, pp. 203–210.

6

Scope System

6.1 Introduction

Having identified the major stakeholders and constraints and then elicited the business needs, the next activities are to scope the system and to finalize the BNR.

6.2 C1.3—Scope System

The scope of the system is very important to systems engineers and requirements engineers, and perhaps even more so to project managers. Project scope management articulates the activities required to make sure that the project undertakes all of the work required, and no more than the work required, to complete the project. Scope management comprises initiation; scope planning; scope definition; scope verification; and scope change control [1]. To assist the project manager in scope management, this CDM activity, C1.3—Scope system, contributes to scope planning and definition by developing a context diagram, defining a system boundary, and then defining external interfaces. These activities, combined with the mission statement, LCs, and validation criteria, culminate in the endorsement of the business needs. Figure 6.1 shows the upper level steps involved in scoping the system.

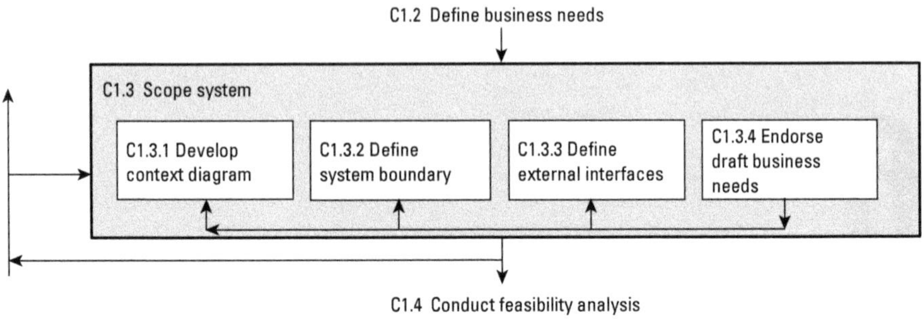

Figure 6.1 Steps involved in the activity C1.3—Scope system.

6.2.1 C1.3.1—Develop Context Diagram

To assist with the scoping activity, a tool called a context diagram may be used to illustrate the context within which the system exists—that is, the relevant regulatory environments, stakeholders, related external systems, interfaces, and the system boundary. There are a number of different forms of context diagrams in use as part of the various design methodologies; in CDM the context diagram is fairly free-form, with a format depending on the nature of the system under investigation.

As with the majority of the CDM artifacts, the context diagram can be developed bottom-up within a top-down construct, described as follows:

- The process begins with a simple listing all of the possible candidate elements for the context diagram. This can be produced as a textual list in a word processor, or as a list on a whiteboard, or perhaps as separate pieces of sticky paper on a board.
- Next, similar elements are grouped in accordance with their relationship with the SOI, their impact on it, or interface with it.
- The groupings are then drawn as goose eggs on a diagram in which two overlapping goose eggs represent an interface, and an arrow between two goose eggs illustrates an influence of one group over the other (in the direction of the arrow). The SOI is shown as a goose egg in the center of the diagram—nothing inside the system is shown because the system elements have not yet been defined (and will not be defined until subsystems and CIs are selected in preliminary design).
- Using this draft diagram, stakeholders move their focus around the boundary of the system (that is, around the outline of the goose egg that represents the system) verifying each relationship as it is depicted.

The diagram is adjusted as necessary, and the process continues until the diagram is complete.

If the context diagram becomes complicated, it may be useful to separate the components out into a number of context diagrams illustrating more than one view of the system. For example, there could be a context diagram for the operational view, one for the acquisition view, and one for the support/sustainment view. If the project is to have multiple phases during acquisition, there could even be one context diagram for each phase.

The context diagram is a very useful tool because, in this case, a picture is definitely worth a thousand words. In systems engineering there is a common heuristic that says that if you can't draw a picture of the system or you can't explain it in five minutes, it is either broken or you don't understand it [2]. The exercise of developing the context diagram is also very useful—it is not uncommon at the start of a requirements workshop to take a day of discussion to complete a suitable context diagram. Although the task seems simple enough at the outset to most stakeholders, it quickly becomes apparent that very few of them have a comprehensive overview of the system, its interfaces, or its boundary.

6.2.1.1 ACME Medical Center Example

In our ACME Medical Center example, we begin by simply listing all the possible candidate elements for the context diagram, as illustrated in Figure 6.2. Next, as illustrated in Figure 6.3, similar elements are grouped in accordance with their relationship with the system, their impact on it, or their interface with it. For example, the various medical regulators can be grouped, as can the local community, patients, and customers since they largely overlap as an individual grouping anyway. Similarly, specialists, other medical centers, and medical service providers can be grouped since they perform similar functions with regard to the medical center and will therefore have very similar interfaces.

This grouping process continues—the next stage is illustrated in Figure 6.4—until a balanced context diagram results. Further iteration of the process would result in refinement of the diagram. The process is not completed here, because we do not have sufficient detail in the background statement from the stakeholder. The next example illustrates a completed diagram (Figure 6.4).

Before moving on, however, even its immature state, Figure 6.4 is useful for a number of observations, discussed as follows:

- We noted earlier that it may be useful to separate the components out into a number of context diagrams illustrating more than one view

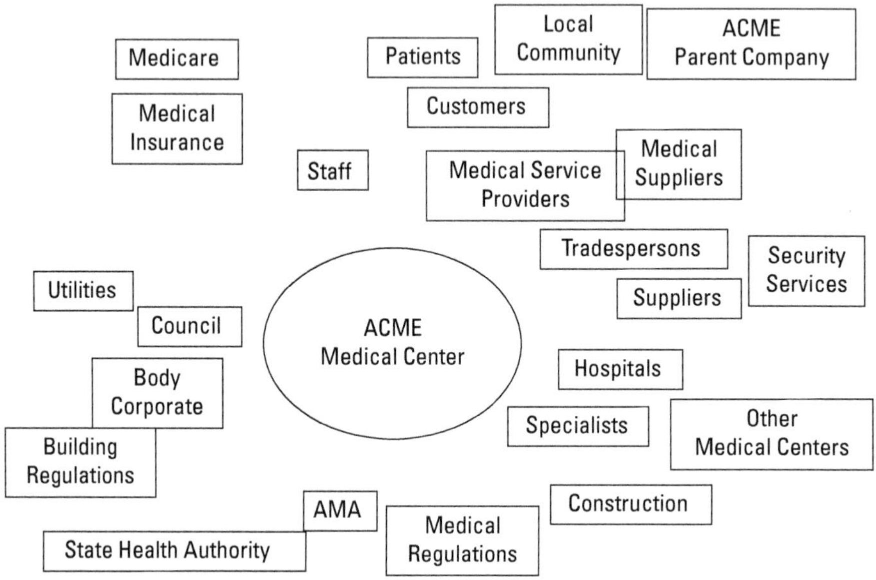

Figure 6.2 Identifying candidate elements of the context diagram for the ACME Medical Center.

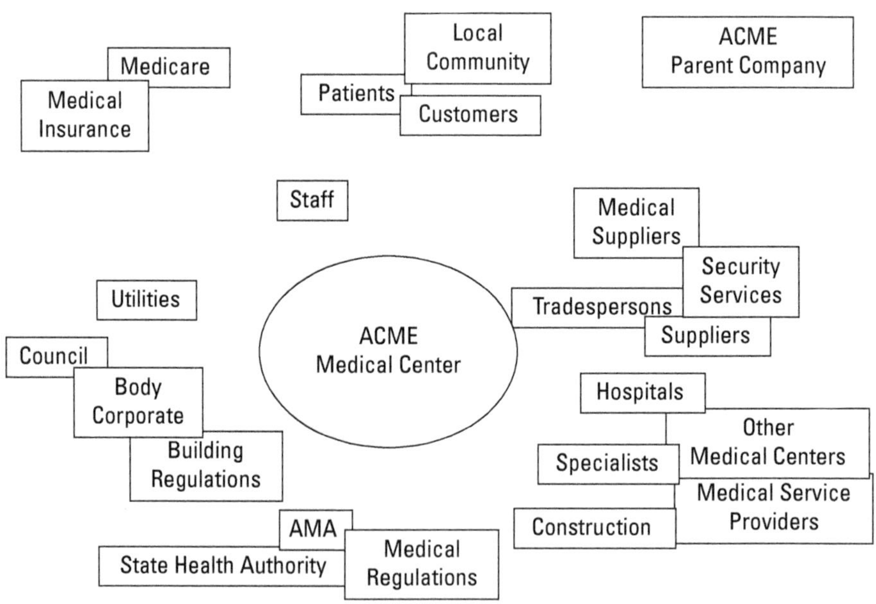

Figure 6.3 Grouping of candidate elements of the context diagram for the ACME Medical Center.

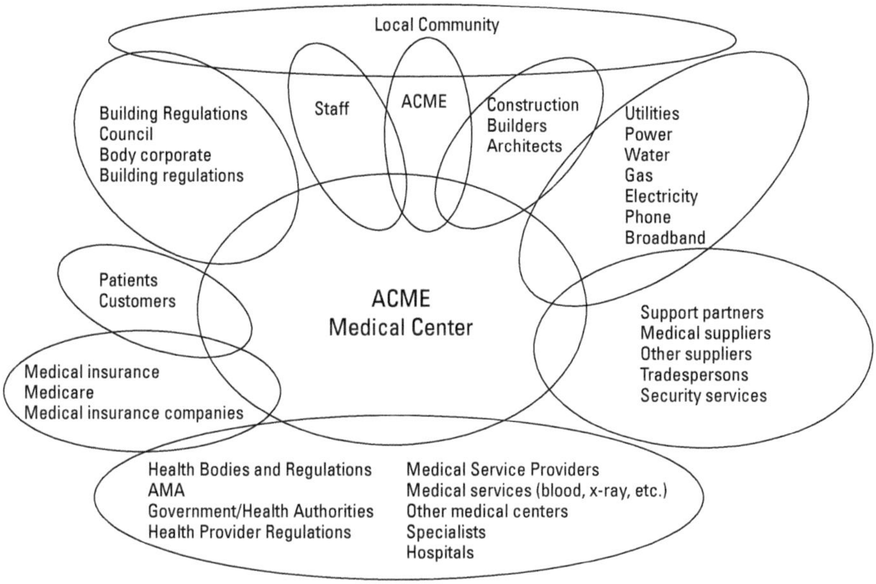

Figure 6.4 Further grouping of candidate elements of the context diagram for the ACME Medical Center.

of the system. In Figure 6.4, it is obvious that the Medical Center is better understood if we draw a context diagram for the operational view, one for the acquisition view, and one for the support/sustainment view. This is particularly useful since not all the entities on the diagram in Figure 6.4 will be present at the same time—for example, those involved in construction will be present in acquisition but not utilization, and patients will be present in utilization but not acquisition.

- Note also that the context diagram is a useful thinking tool in terms of understanding which entities are involved and how they interact with the Center. Then, the diagram can be used to make decisions. For example, how does ACME want the staff to be employed? Figure 6.4 shows that the staff have a relationship with ACME outside the Medical Center. Is that correct? Are the staff employed by ACME and work at the Center, or are they Center employees? In the latter case, the diagram needs to be redrawn. Once the design decisions have been made, the context diagram then becomes a useful communications tool that shows how the entities are interrelated. Section 6.2.1.3 discusses the use of the context diagram as a decision-making tool.

6.2.1.2 Domestic Security Alarm Example

We have not provided enough information to finish the context diagram for our Medical Center example. To illustrate a completed diagram in a narrower context to be manageable here, Figure 6.5 shows a context diagram for the operational mode of a Domestic Security Alarm (the maintenance and support modes would be shown on a separate diagram). In reality, of course, the diagram would be a little more detailed, but Figure 6.5 serves to illustrate the main elements of a useful diagram. Additionally, even this operational diagram would best be split to show the normal mode of operation (when the intruder is absent) and the exception case (when the resident is absent, but the intruder is present). The following considerations address the major elements of the context diagram in Figure 6.5:

- *Alarm:* It is obvious from the context diagram that the Alarm is entirely contained within the larger House system since one goose egg is a subset of the other. It is also obvious from the way the Alarm subset is drawn that the designers accept the prospect that the Alarm might not cover 100% of the area covered by the House system (since the boundaries of the two are not coincident). So, in this case, the context diagram shows a relationship between two elements—one is totally subsumed by the other.

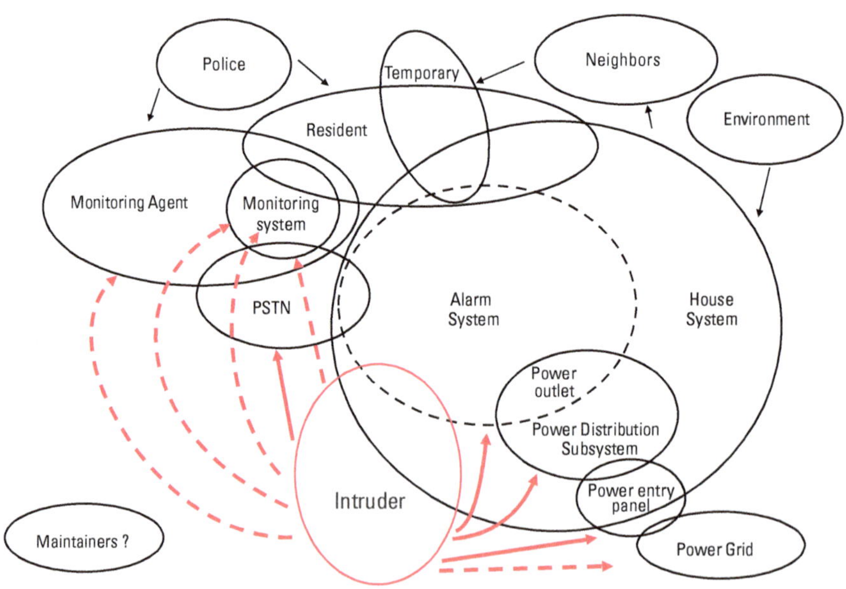

Figure 6.5 A simple context diagram for the operational mode of a Domestic Security Alarm.

- *Resident:* The Resident has a number of interfaces evident on the context diagram—shown as overlapping goose eggs. There is, of course, the physical interface in which the Resident first enters the House and then enters the sphere of influence of the Alarm. In this case, the context diagram shows an interface of the Resident with the House and the interface of the Resident with the Alarm. More than that, however, it shows that the Resident's interface with the Alarm occurs after their interface with the House—that is, the Resident has no relationship or interface with the Alarm until they are inside the House. Also evident is the interface between the Resident and the Monitoring Agent, which is shown as being independent of the Alarm system. Further, the diagram shows the notion that there might be temporary residents (guests of the residents, for example) who will interface with the alarm (in the same way as a resident) for the period during which there are guests.
- *Intruder:* The Intruder has a physical interface that is very similar to the Resident (both are humans that may enter the House and then the Alarm boundary by a number of means) so it can be concluded that the Alarm must be able to distinguish between the two in some manner. Note that, in addition to the physical interface, Figure 6.5 also shows a number of influences that the Intruder may have on the Alarm and House systems—the solid lines show the influences of most interest in this design (disabling Alarm sensor functions or their power source, disconnecting House power, and disconnecting the phone line); and the dashed lines show those influences that the threat analysis considered less likely for a domestic burglary (disabling the suburban power grid, logging across the phone line to mimic the Alarm, hacking into the monitoring system, paying an accomplice inside the Monitoring Agent to disable or ignore the Alarm, or destroying the Monitoring Agent).
- *Monitoring Agent:* Note that there is an expectation in this design that there is a Monitoring Agent (one does not necessarily have to exist in an Alarm system) and that the mandated interface is via the Public Switched Telephone Network (PSTN) (rather than a more modern version of alarm where a cellular data service might be incorporated). In this case the context diagram is assuming the presence of an element (the Monitoring Agent) and the use of a technology (the PSTN)—both assumptions are clearly obvious in Figure 6.5, which makes the context diagram a very useful communications device.
- *Power:* The diagram in Figure 6.5 shows the expected interface of the alarm to the House power subsystem—note that since a power outlet is identified explicitly, we can infer that the alarm is not hard-wired

into the power subsystem (perhaps by a licensed electrician) but instead is able to be plugged in by the resident so that the alarm can be transported between houses.

- *Police:* The context diagram illustrates that the Police do not have any specific role to play in this system—that is, for example, they are not expected to respond when the Alarm is triggered but would respond to any incident as a function of their normal policing duties.
- *Neighbors:* Again, Neighbors are shown to have no particular role in the operation of the Alarm system but are affected by it (perhaps by the noise it may make) and may influence the Resident (perhaps to complain about the noise).
- *Maintainers:* Maintainers are not shown on this operational view but would be more likely to be captured in a separate support context diagram, since operations and maintenance are unlikely to be undertaken at the same time.

6.2.1.3 The Context Diagram and Design Decisions (Trade Studies)

The context diagram helps capture stakeholder needs and requirements, but it must be remembered that it has been drawn by capturing the current understanding of stakeholders who have not necessarily considered how their current business could look now that modern technologies offer alternative solutions that were not available previously. To accommodate this, having drawn the diagram, we move around it asking stakeholders why they were happy to represent the relationships between entities (particularly the interfaces) in the way that they did.

Consequently, the context diagram is also useful to assist in performing the necessary feasibility analyses and upper level trade-offs at the business level. That is, the context diagram is a useful visual design artifact that allows stakeholders to visualize the system at an abstract level to allow them to make decisions at that level that will then drive (constrain) the subsequent considerations.

The several upper level trade-offs that can be conducted for our security alarm example are discussed as follows.

- We noted in Figure 6.5 that the Intruder may have a number of influences on the Alarm and House systems—the solid lines show the influences of most interest in this design (disabling Alarm sensor functions or their power source, disconnecting House power, disconnecting the phone line); and those that the threat analysis considered less likely for a domestic burglary (disabling the suburban power grid, logging across the phone line to mimic the Alarm, hacking into the monitoring system, paying an accomplice inside the Monitoring Agent to disable or ignore the Alarm, or destroying the Monitoring Agent). These latter potential influences (dotted lines) are less likely means of interference and, after some analysis, would most likely be considered out-of-scope.

- When developing the context diagram, the stakeholders included a monitoring agent connected to the alarm via the wired telephone system (the PSTN). Before progressing and transforming those implied needs into system requirements, there are some decisions to be made. For example, is the PSTN still relevant—after all, how many households have a fixed phone line these days? Wouldn't a mobile phone interface be more relevant in the modern world? In order to address this issue, a trade study would be established to decide whether the alarm has a phone line or mobile phone connection to the monitoring agent (or perhaps both types of connection). Of course, one of the advantages of the mobile phone (cellular) connection is that the associated SIM card would be installed inside the alarm controller and hence would be protected by the Alarm itself—that is, the Intruder would have to set off the Alarm in attempting to get to the SIM card and affect the external communications. An additional advantage is that the mobile phone can also be used for access authentication, as well as reporting.
- Similarly, why is there a monitoring agent? Although the business originally requested a back-to-base option, that was essential in the past but in a modern system the alarm could report directly to the resident when they are anywhere in the world (either across the internet, or by message via mobile phone, or both). Would the business like to reconsider its original need now that there is more detail available? Again, a trade study would be useful to determine whether the Alarm reports directly to the Resident, to the Monitoring Agent, or perhaps to both.

Figure 6.6 shows a modified context diagram, reflecting the design decisions to ignore the less likely Intruder interfaces, replace the PSTN with a SIM card, and remove the monitoring agent.

6.2.2 C1.3.2—Define System Boundary

Definition of the system boundaries is also critical to the success of the fielded system. It is essential that these boundaries are defined early in the acquisition phase so that it is clear which system elements are under the design control of the project and which are outside the control. This activity also identifies which aspects are inside the systems acquisition and which are not, which is

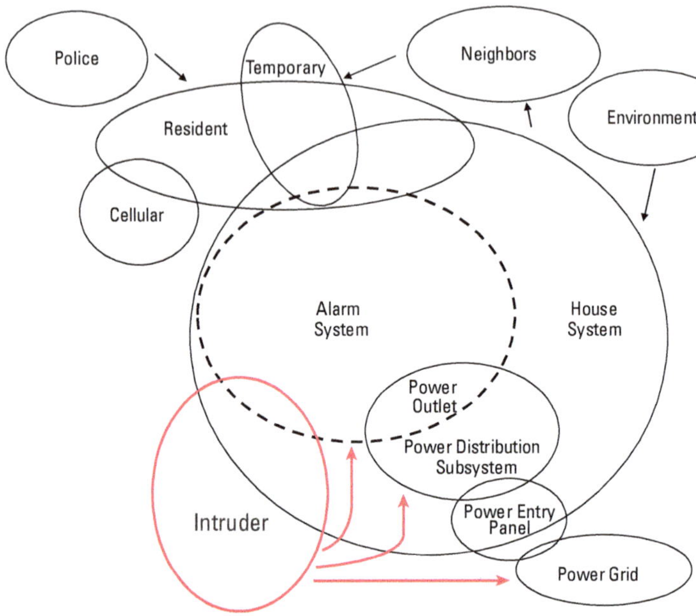

Figure 6.6 A simple context diagram for the operational mode of a Domestic Security Alarm reflecting design decisions made through trade studies (i.e., ignore the less likely Intruder influences, replace the PSTN with a SIM card, and remove the monitoring agent).

also particularly important to the project manager who is principally responsible for defining project scope—that is, what is to be included in the project as well as what is to be excluded.

The boundary of a system is normally straightforward to describe in physical terms (such as a fence line or external building walls), but it is often necessary to describe the boundary in conceptual or logical terms as well. Additionally, although we traditionally describe the boundary in terms of what is inside the boundary—that is, what is included inside the system—it is often useful to describe what isn't inside the boundary. This is particularly important when there are a number of elements that may commonly be assumed to be part of the system but are not in this particular case.

In Figure 6.5, the boundary of the alarm is shown by the dotted line. Note that, in this example, the boundary shows that the Alarm is totally contained within the House. Additionally, the two boundaries are drawn such that there does not seem to be any intention for the alarm to cover every part of the house—the extent of coverage will be defined as designers proceed with their analysis.

The boundary is also important in the next step in the scoping activity, which is to define the external interfaces—that is, those interfaces between the system of interest and those other external systems to which it is interconnected.

6.2.3 C1.3.3—Define External Interfaces

An interface is a place where an interaction occurs between two (or more) systems—often a boundary or a point that is common to the systems. At this stage in system design we are interested in the external interfaces of a system—that is, the places where there will be interaction between the system being designed and other systems. Later in the design process (in preliminary design), internal interfaces will be designed for the interaction between system elements.

6.2.3.1 Interface Definition

The definition of an interface requires considerably more detail than simply identifying and naming the interface. Broadly there are three main steps in interface definition (each of which is described in more detail in this section):

- *Interface description:* The interface is given a name, a short title, and an identifier. The nature of the interface is described in terms of who, what, when, where, why, and how.
- *Interface impact analysis:* The interface is analyzed in terms of its impact on the system. In particular, any constraints imposed by the system are identified. Additionally, a risk analysis is conducted to determine the impact of the interface on the operation and design of the system.
- *Interface control analysis:* Each external interface must be analyzed to determine the extent to which it can be controlled so that designers and operators of the system are not at the mercy of its external interfaces (noting the requirement to obtain agreement from the owners of systems on the other sides of external interfaces).

6.2.3.1.1 Interface Description

Interface description starts with ascribing the following attributes to each interface:

- *Identifier:* The interface is given a short identifier in a standard format. For example, external interfaces will begin with the letter "E" followed by a number (such as "E01"). If the interface can be broken down further, a hierarchical numbering scheme is used (such as "E01.1" and "E01.2").

- *Name:* The interface is given a name that, if possible, is also descriptive of its nature. So, for example the interface between system A and system B might be known as "E01. System A—System B Interface." If that interface could be broken down into an electrical and hydraulic interface, for example, those external interfaces would then be labeled "E01.1. System A—System B Hydraulic Interface" and "E01.2. System A—System B Electrical Interface."
- *Short title:* Although not always necessary, a short title is a useful way of abbreviating what can sometimes be long names for external interfaces. Short titles are easier for humans to use since they are more descriptive than a unique ID number but are easier to say than the full title. Because all external interfaces for System A will be between System A and some other system the "System A" words can be dropped from the interface name—for example, "E01.1. System A—System B Hydraulic" would become "E01.1. System B Hydraulic Interface" or, if there is only one hydraulic interface, the short title could be "E01.1. Hydraulic Interface."
- *Interface statement:* The nature of the interface should be described in a single short sentence, in a manner similar to that of a need statement.

The nature of the interface is then further elaborated in terms of the following:

- *Who?* Who is associated with the interface? Who crosses it? Who defines it? Who owns it?
- *What?* What crosses the interface (data, physical material, spares)?
- *When?* When is the interface in operation? When does each person or object cross the interface? Is the nature of the interface dynamic in some way (do some or all of its properties change with time)?
- *Where?* Where does the interface physically exist, or is it a functional interface? Is the location of the interface dynamic—if so, on what does the location depend?
- *Why?* Why is the interface in operation? Why does this individual person of object cross the interface?
- *How?* How does the person or object cross the boundary? For example, how is the person dressed, in what type of vehicle are they conveyed, and what might they be carrying? For objects, we might determine how it is packed, what state it is in (for example, is it turned on or off), in what vehicle is it conveyed, and what else might accompany it.

In the case of the domestic burglar alarm in Figure 6.5, the interfaces may be identified and labeled on the context diagram as in Figure 6.7. The nomenclature will be defined by the organization, but the example shows one simple method of listing each type of interface, from E01 to E06. Note that it is important to complete the context diagram before analyzing interfaces—that is, it is important for the business to decide on issues such as back-to-base monitoring, external communications link, and so on because those interfaces may or may not be present depending on decisions made (note the difference in the diagrams in Figure 6.5 and Figure 6.6). To allow more interfaces, Figure 6.7 shows the interfaces should the design decisions in Figure 6.5 not have been made.

6.2.3.1.2 Interface Impact Analysis

Each external interface is then analyzed in terms of its impact on the system. In particular, any constraints imposed by the interface on the system are identified, as well as any constraints that the system may impose on the interface. Many of these constraints associated with interfaces may already have been

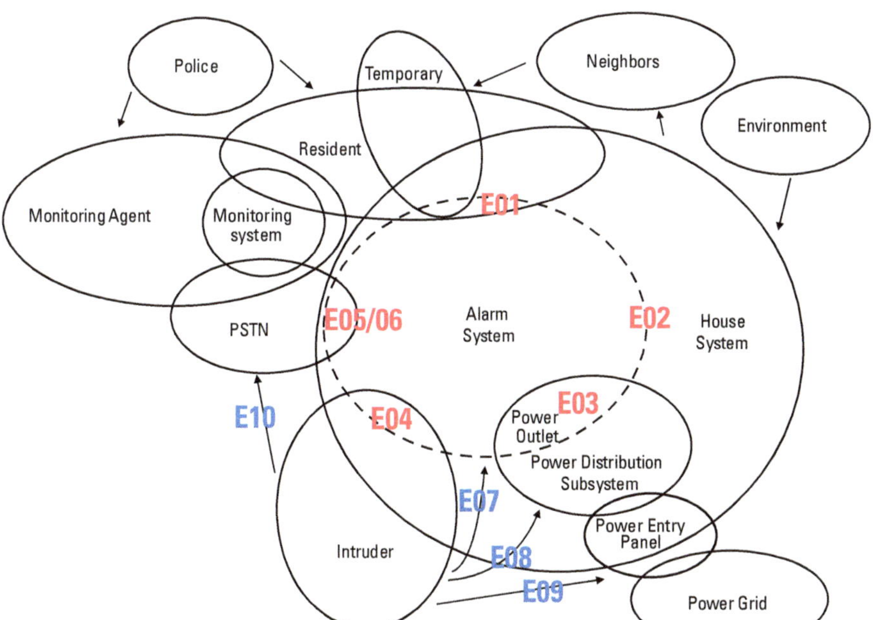

Figure 6.7 An example of the labeling of interfaces on the context diagram for a Domestic Security Alarm.

considered when conducting the step C1.1.3—Identify external constraints (as discussed in Chapter 4), since identification of external constraints will relate to interfaces and vice versa.

Additionally, a risk analysis is conducted to determine the impact of the interface on the operation and design of the system. Some of the impacts identified will remain isolated to an interface specification—that is, the system must produce reports in a specified format. Other impacts may flow across the system boundary and affect the way in which the system is designed internally.

It should not be assumed, however, that everything about the external interface is predestined. There will certainly be some aspects that are inviolate—for example, a new aircraft system will be useless unless it conforms to the interface requirements to allow it to be refueled from standardized refueling facilities. There may, however, be times when the interface is adjusted to suit the new system, rather than an existing one.

For example, if there was an interface already in existence between the existing System A and existing System B, it would make sense for the new System C to be designed to conform to that interface if it is to be connected to one or both of the existing systems. So, for example, it might make sense to specify that System C conforms to an existing 10-Mbps Ethernet interface because that is what is used by the existing System A and System B. If System C would prefer to use a 1-Gbps interface, however, it might be a useful time to reconsider the A-B interface to determine whether it should be upgraded, rather than develop System C to interface to an old standard. The issue is not always clear-cut and may not just involve the network interface cards—a move from 10-Mbps to 1-Gbps Ethernet may also require replacement of the underlying network cabling and hardware (such as hubs and switches). It is obvious, even from this simple example, how important interface analysis is to the design of the SOI as well as to the organization at large. Since interface definition has such a wide-ranging impact, it must be carefully considered early in the system development.

6.2.3.1.3 Interface Control Analysis

Unless it is unavoidable, the designers and operators of a system should not be at the mercy of that system's external interfaces. Each external interface must be analyzed to determine the extent to which it can be controlled—"own your interfaces" is a worthy catchphrase for system designers. As Meier and Rechtin point out, "The greatest leverage in architecting is at the interfaces . . . the greatest dangers are also at the interfaces" [3].

For example, if a social club is dependent on the adjacent two parking lot for interface to its members and guests, the club's livelihood is at risk should the town sell the parking lot for the development of an office building. When this interface is considered as part of the design process, conceptual design should examine mechanisms whereby this risk is mitigated, or perhaps even removed. For example, it may be removed if the local government can be persuaded to give the land to the club for use in perpetuity as a parking lot, or it may be mitigated by lobbying the town to pass legislation to zone the land to prevent any other use. Business owners must recognize that progressing without mitigating interface risk at this early stage may place the system at significant risk in the future.

6.2.3.1.4 Brief Example of Interface Analysis

By way of brief example, here are interface analyses for the Domestic Security Alarm:

- *E01 Resident:*
  - *Description:* Human/authorized/could have hands full/may be in groups with young children, pets, and so on. May be under duress to open house against their will.
  - *Impact:* Need to allow (and facilitate) entry. For the regions of interest, duress is not a great concern.
  - *Analysis:* Would like to make entry seamless and effortless but will need to ask to authenticate so the resident can be identified as authorized to enter. Duress alarm not included.
- *E02 House environment:*
  - *Description:* Describe ranges of temperature, pressure and humidity expected for regions in which the alarm is to be installed.
  - *Impact:* Need to know the ranges or the alarm will not work in low and high temperatures and will not work in regions of high humidity.
  - *Analysis:* Need to identify business needs for the environment of regions into which the business expects to sell the alarm. Then a trade study will be required to identify the price break point associated with meeting extremes. Business management may then wish to limit their expectations for markets by reducing one or more of the ranges to keep the alarm within its design-to cost ratio.
- *E03 Power interface:*
  - *Description:* Describe in terms of voltage (100/110/120/220/230/240V), frequency (50/60 Hz), and plug. May split into two interfaces: electrical (voltage, frequency, current) and physical (plug).
  - *Impact:* Need to know or the alarm will not work with the domestic power supply in the countries the business intends to market it.

- *Analysis:* Need to identify business needs for countries into which the business expects to sell the alarm. Then a trade study is required to identify the price break point associated with meeting all possible domestic power supplies. There are four to six distinct voltage/frequency combinations in practical use so a multipurpose (universal) power supply will be needed. Additionally, there are 15 recognized plug types worldwide designated by letters A through N. These plugs need to be separated from the power supply by a fly lead, or else the alarm can be supplied without a plug, and a local electrician can add one before installation. While these seem like engineering/technical design decisions that belong in preliminary design or detailed design and/or development, they really are business decisions based on marketing advice in terms of whether they include multiple fly leads with each alarm, specify a plug than can be adjusted by clipping on region-specific prongs, sell alarms by region (containing the appropriate fly lead), or sell alarms with bare wires and no plugs.
- *E04 Intruder:*
  - *Description:* Human/unauthorized/limited tools/mostly opportunistic/looking for cash or car keys.
  - *Impact:* Need to deter, detect, and report entry and need to deter from remaining in the house if entry is successful.
  - *Analysis:* Deter with blue light at the front of the house, stickers on the windows, and sensors located so that they are visible to anyone looking into the house. The intruder can't be identified by the alarm as unauthorized because the human intruder looks like a human resident from the alarm's perspective. Consequently, we need to derive a requirement to include authentication for the resident (since we cannot implement it for the intruder).
- *E05 Phone line (PSTN) interface:*
  - Description: Voltage, current, number of wires, plug. Need to be approved to connect to any phone line and phone service in the world potentially (although reduced by any decisions management make from previous deliberations).
  - *Impact:* Need to identify all possible interconnections; need to get alarm approved to standards such as the European CE standard.
  - *Analysis:* Business may again need to limit aspirations depending on the range of approvals required. Of course, if PSTN is replaced by SIM card, all these issues will be much simpler to address.
- *E06 Monitoring agent protocol interface:*

- *Description:* Alarm must conform to the back-to-base monitoring protocol so that the alarm can be monitored.
- *Impact:* Need to confirm whether this interface is needed, or the back-to-base monitoring system won't be able to connect to the alarm.
- *Analysis:* If required, need to identify which protocols exist. Is there a single standard protocol, or a number? Are protocols bespoke to each monitoring agent? Has ACME already chosen the partner, meaning that we can identify the protocol now? Again, while these sound like detailed design issues, they really are business issues because they need to be balanced (traded off) in a business sense before passing the desired solution to the technical specialists.

Note that the analyses above have been conducted briefly by way of demonstration of the methodology and the sorts of considerations that would need to be made. In reality, considerably more detail would be required.

6.2.3.1.5 Secondary (Referred) Interfaces

Figure 6.7 also shows examples of what can be called secondary interfaces (or referred interfaces)—E07, E08, E09, and E10. These are not explicitly interfaces to the alarm because they are outside the alarm's boundary. For example, the alarm cannot be aware if the intruder tries to cut the phone line—the alarm is not cognizant of either the phone line or the intruder. However, it is essential that the alarm is aware of the actions of the intruder in interfering with the phone line (and hence affecting E05 and E06) or the external power connection (and hence affecting E03).

These interfaces are called "referred," because they are not actually a direct interface with the alarm, but they are something the alarm has to consider as an interface. Therefore, we refer from an interface which the alarm does not recognize to one that it does.

- *E07 Intruder interference with alarm sensors:*
  - *Description:* The intruder can affect the sensors at the boundary of the system. For example, they could cover the sensor with a cloth to blind it.
  - *Impact:* The alarm would not be of much use if the sensors were so easily defeated.
  - *Analysis:* The alarm can't be aware of the intruder's actions outside its boundary, but those actions can be referred to something the alarm can be aware of. For example, to avoid tampering, the sensors

could have a movement detector that could alert the alarm if a sensor is interfered with and perhaps a tamper alarm to alert the alarm if the cover is removed from a sensor. Additionally, the alarm could continuously poll each sensor and then alert if it does not get a response. E07 would therefore be referred to an internal alarm interface, with the derived requirements to monitor movement of the sensor and to continuously poll each sensor.

- *E08/E09 Intruder interference with power supply internal to the house/at power entry panel:*
  - *Description:* The intruder can also disconnect the power source to the alarm either internally via the house power distribution subsystem or through simply turning off the power at the power entry panel. It would be noted that this intentional interference is also accompanied by a number of sources of unintentional power failures during blackouts.
  - *Impact:* The alarm cannot work without power.
  - *Analysis:* There is a derived requirement for the alarm to have a backup power supply (some form of battery) to allow the alarm to continue working if the mains power is disrupted for any reason. Business management will need to decide how long the battery should last. Presumably there will also be another derived requirement for the alarm to alert as soon as battery backup power is activated. The E08/E09 interfaces would then be referred to the E03 Power interface to perform the necessary monitoring.
- *E10 Intruder interference with phone line disrupting monitoring link:*
  - *Description:* The intruder could cut the phone line or simply disconnect it at the entry point to the house.
  - *Impact:* Back-to-base monitoring will not work if the phone line is not available.
  - *Analysis:* The interface analysis would refer E10 to E05 and E06—if the phone line is unavailable (loss of E05 or E06), the alarm would alert.

6.2.3.2 Interface Management

Once it has been defined, each interface has to be documented and managed. Interface management is very important because systems (and the projects that deliver them, for that matter) often live or die by their interfaces. This is even more evident in modern systems where the sheer number of interfaces and their complexity are significant sources of risk in system development.

The definition of a system's external interfaces assists in defining the system's scope—interface management is therefore an important part of the scope management activities undertaken by the project manager. It is highly likely that the scope of a system would be affected should there be a change to any aspect of a system's external interfaces throughout its development.

Interface management for external interfaces has a number of steps, described as follows:

- *Interface documentation:* Once defined, all external interfaces are documented in a common location to provide a single reference point for all external interfaces. Commonly, external interfaces are documented as requirements in the SyRS, although there may be a separate interface control document that is more descriptive.
- *Interface allocation:* Once control of the interface has been analyzed, the interface must be allocated to a system/subsystem and then to a group of designers to design and manage. The interface cannot be left without an owner responsible for its implementation and verification. Later, in preliminary design, the interface will be allocated to a single configuration item so that the interface can be implemented. External interfaces that exist will already be allocated to an existing system. A new interface (perhaps between two new systems) will need to be allocated to one system or the other.
- *Interface verification:* Each external interface must be tested once the system has been developed. The means of that verification should be identified and described and the verification activity allocated (normally to the same parties to whom the interface itself has been allocated).
- *Interface review:* Analysis of interface impact and control must be continually reviewed throughout the system development to ensure that nothing changes that can affect the definition of the interface.

Where there are a number of systems involved in a single interface definition, the organization may create an interface control working group (ICWG) to assist in the management of interfaces. The requirements that flow from interfaces are placed under configuration management through the organization's requirements management processes (see Chapter 2).

6.2.3.3 Example—ACME Medical Center

The external interfaces for the ACME Medical Center might include the following:

- ACME.
- AMA and other regulatory bodies.
- Medicare and medical insurance companies.
- Utilities—power, water, gas, electricity, phone, and broadband.
- Waste disposal—general and medical.
- Local government.
- Body corporate.
- Shopping center management.
- Medical suppliers.
- General suppliers.
- Medical service providers (e.g., of pathology and x-rays).
- Patients (including disabled access, etc.).
- Nonpatient customers (for pharmacy).
- Staff (including temporary staff).
- Tradespeople (maintenance) and suppliers.
- Other medical centers (transfer of medical records).
- Specialists (for referrals and reports).
- Ambulances (for taking patients to hospital, if needed).
- Hospitals (for transfer of records, reports, etc.).
- Financial institutions.
- Local community.
- ICT/network/internet interface.
- ICT/network support contractor.
- Maintenance contractors.
- Cleaning contractors.
- Marketing.
- Parking for staff and patients in parking lot for shopping center.
- Public transport.
- Road?
- Footpath access?
- Security (patrols and monitoring).
- Emergency services.

Each of these interfaces would need to be analyzed, as a number of needs and requirements will be derived by the need to meet them.

6.2.3.4 Types of Interface

Before leaving our discussion of interfaces, it is often useful to identify interfaces by focusing on the various categories or types of interface that exist. Types of interfaces include the following:

- *Physical interfaces:* Physical or mechanical interfaces may be pipes through which a fluid or gas flows or they may be a fiber-optic cable passing digital information between two computer systems. In defining the physical interface, factors such as mechanical connection type and specification, physical size of the interface (such as diameter of pipe), and layout of the interface are defined. Physical interfaces normally coexist with one of the other types of interfaces, such as electronic or electrical.
- *Electronic interfaces:* An electronic interface defines the flow of electronic signals (analog or digital) between two points. These signals may flow over a physical interface such as a cable or fiber, in which case the physical cable or fiber interface is also defined. Alternatively, the interface may make use of RF transmission such as microwave or satellite to pass the information. The characteristics of the data flow between the two points must be completely defined. The information requirements for this definition depend entirely on the nature of the interface but may include factors such as modulation method, frequency, data format, and data rates.
- *Electrical interfaces:* Electrical interfaces are normally associated with a physical interface of some sort such as a power cable or data cable. The type of information required in this sort of interface will include voltage levels, voltage type (for example, 110-V, 60-Hz), and any associated factors such as fault tolerance requirements. In the case of a data cable, pin assignments and descriptions may be required. If there are any electromagnetic compatibility issues relating to the interface, such as shielding and earthing, these requirements will also be defined.
- *Hydraulic/pneumatic interfaces:* Hydraulic and pneumatic interfaces are the mechanical equivalent of the electronic or electrical interfaces already described. A physical interface definition will coexist with this type of interface. Additional information such as flow rate, pressure, temperature, and fluid type will be required to define completely the hydraulic and pneumatic interfaces.
- *Software interfaces:* Software interfaces refer to the passing of data between different computer software components. This may occur within the one piece of hardware using internal hardware buses (as an internal interface) or it may be between two different pieces of hardware. If the latter is the case, a physical interface definition is also required to define the physical link between the two pieces of hardware. Software interfaces require information such as expected

inputs, generated outputs, format of the messages, and any protocols used during the transmission (such as error detection, for example).

- *Procedural interfaces:* Procedural interfaces relate to the conformance of the two systems being interfaces to a common set of procedures and/or processes. For example, procedural interfaces for a system may be required to interface to support and supply systems and to operational procedures such as those used for air-traffic control.
- *Environmental interfaces:* Environmental interfaces are sometimes not considered to be interfaces, but they form the interface between the system and its external environment. In some cases, the environment provides the system with important information such as air data in the case of an aircraft. The environment may also provide the system with challenging interfaces such as structural (vibration, shock, and acoustics), thermal (ranges of temperatures), magnetic (flux densities), and radiation (flux densities and types). All environmental interfaces need to be considered and documented to ensure that the most important integration of them all (the system with its operational environment) is successful.

Further, these interfaces may exist in a number of different ways. In particular, they may be static or dynamic; dynamic interfaces must be defined for the range of different states and modes of the system.

6.2.4 C1.3.4—Endorse Draft Business Needs

Before going on to develop formal requirements, business management must pause and endorse their business needs—since the next major activities are related to the transformation of those needs into formal requirements, it is important that the business is satisfied that their stated needs will result in satisfaction of the business case for the system. This endorsement is most likely undertaken through a review conducted by the business owner in the presence of business management and the major stakeholders from business operations.

6.3 C1.4—Conduct Feasibility Analysis

Once business management have defined their needs—the next major activities are related to the transformation of those needs into formal requirements. Figure 6.8 shows the steps associated with C1.4—Conduct feasibility analysis.

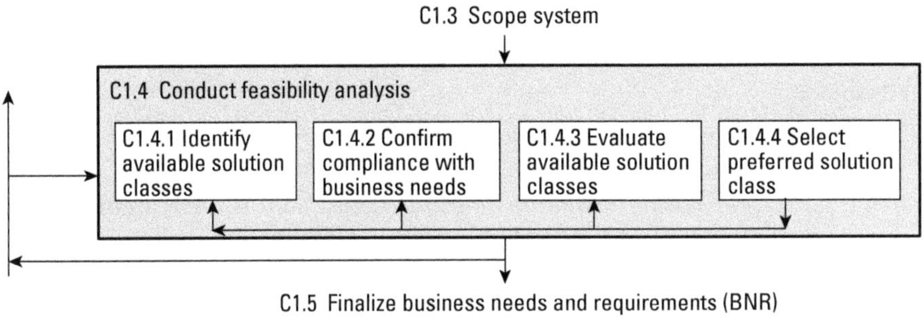

Figure 6.8 A summary of the steps involved in the activity C1.4—Conduct feasibility analysis.

At this stage of design, we have been careful to ensure that business management have stated their needs and requirements in terms of the problem domain (the business and operational environment)—that is, in logical terms. For any problem domain, however, there is normally a range of physical solution classes in the solution domain. Each alternative solution class may represent a completely different type of project—the feasibility analysis is therefore an essential step to narrowing the solution domain so that the subsequent project can be managed effectively. If we continue without narrowing down the design into one of those solution classes, we will have great difficulty coping with the breadth of the design. Consequently, we next undertake a system-level feasibility analysis, the aim of which is to select a desired solution class (which is then to be further elaborated on by the stakeholders at the business operations level).

Business requirements are generally couched in business or capability terms and are elicited first so that we can understand what is required by the business of the deployed system. Requirements analysis elaborates these business requirements into stakeholder requirements and then goes on to derive and decompose these stakeholder requirements into lower level system requirements. The idea of a feasibility analysis is to pause after identifying business requirements to determine what possible system-level alternatives there are at a system level before going on to conduct detailed requirements analysis and eventually to establish the system requirements. In other words, there are most likely a number of broad classes of system that might meet the upper level business requirements, and it is often very useful to choose one of these before proceeding to determine stakeholder needs and requirements, or else the SyRS that would result would be far too general to be useful (either for subsequent design or, more particularly, as a centerpiece of the statement of work for a contract). Not only that, but the choice among solution classes is invariably a business choice and should therefore be made before time is wasted in elaborating on classes of solution that the business does not wish to embrace.

The broad steps of the feasibility analysis are described as follows:

- *C1.4.1—Identify available solution classes.* Step C1.4.1 identifies the alternative system-level solution classes capable of satisfying the business needs. As identified earlier, the aim of identifying the available solution classes is to narrow considerations down into a useful solution space that can then be elaborated on further at the business operations level into stakeholder needs and requirements.
- *C1.4.2—Confirm compliance with business needs.* Step C1.4.2 confirms compliance of each solution class with the business needs and, if not completely compliant, the likely level of achievement is noted.
- *C1.4.3—Evaluate available solution classes.* Step C1.4.3 evaluates the alternative solution classes for their suitability in providing the desired capability—in terms of feasibility, performance, effectiveness, technical and project risk, and other selected measures of performance. In some cases, there may be only one suitable solution class, and business management may be comfortable with mandating a specific solution on the basis that any wider consideration would not present value for money. In other cases, there may be a small number of useful solution classes that may warrant expending the effort on a more exhaustive evaluation process. There may be an opportunity to examine the marketplace to identify existing systems that may be assessed to assist with feasibility analysis. Assessing operational systems may also assist with the refinement of the mission, goals, and objectives of the new system based on a realistic understanding of what is currently available. The feasibility of using existing systems to meet emerging needs may be assessed as a possible way of reducing the schedule and technical risk associated with the system development. The process is sometimes called concept demonstration or technical demonstration.
- *C1.4.4—Select preferred solution class.* Step C1.4.4 selects the best of the alternative solution classes (ensuring that the options are narrowed down as much as is practicable at this stage).

A great deal of care should be taken during this stage of the project, as the feasibility analysis sets the direction for the rest of the system life cycle. To that end, the feasibility analysis should address the entire life cycle, not just the operational aspects of the system.

6.3.1 Feasibility Analysis Examples

There are a wide range of feasibility analyses that can be conducted, and the number and depth are not prescriptive. The following examples are chosen to illustrate that range.

6.3.1.1 Feasibility Analysis—Solution Class for the Aircraft System

ACME wants to service selected short-to-medium domestic and international routes but doesn't have the airframes to do it. Based on this simple statement of business needs, the feasible alternative solution classes might be the following:

- Leasing or buying an existing aircraft;
- Leasing or buying and modifying an existing aircraft;
- Contracting to have a new aircraft developed;
- Outsourcing the operation to another airline;
- Not servicing these medium routes.

There is therefore a wide range of possible solution classes for ACME in replacing its existing aircraft. Each alternative is a completely different type of project with a markedly different set of stakeholders. Feasibility analysis is therefore an essential step to narrowing the solution domain so that we can manage the subsequent project. It is also an important step for the business to confirm for itself which type of solution it prefers.

Then, for each approach, there is still a range of subordinate alternative system solutions. It may be desirable to establish another feasibility analysis to further narrow down the solution space.

6.3.1.2 Feasibility Analysis—Family Vehicle(s)

Consider a family of four who are also avid hobby farmers on a small block of land at the edge of town. They have a family car, but they also need a way to transport significant amounts of waste to the rubbish dump as well as gardening supplies such as fertilizer, garden soil, sod, and pavers.

The following briefly addresses the steps of a feasibility analysis:

- *C1.4.1—Identify alternative system-level solution classes.* There are three broad solution classes listed as follows:
  - Fit a tow hitch on the family vehicle to tow a trailer (buy or rent when needed);
  - Buy or rent an additional utility vehicle when needed;
  - Replace the family car with a large utility vehicle that has a dual

cab to accommodate four people and a sufficient cargo bed to carry garden waste and supplies.

- *C1.4.2—Confirm compliance with BNR.* Each of the candidate solutions is suitable in terms of meeting the need for transporting humans and material. Note that we will need to consider all the BNR before we can make a choice—is there sufficient parking space available for the two-vehicle option, what is the total cost of ownership of each option through its life cycle (or perhaps compared over 10 years), are sufficient funds available for an outright purchase of a second vehicle or is a loan required, and many more. Note that the decision is a business decision—what is best for the family business (not which vehicle one family member might prefer to drive).
- *C1.4.3—Evaluate alternative solution classes.* Criteria would be established and each solution class evaluated against those criteria (noting that the first two solution classes have two ways of being implemented and may be considered as separate solution classes).
- *C1.4.4—Select the best of the alternative solution classes.* With the lack of additional parking space and a limited budget, the family may opt to simply purchase (or rent) a trailer and look to replace the family car with a utility vehicle when the car has reached the end of its economic life.

6.3.1.3 Feasibility Analysis—Long-Range Communications Capability

The business need for a small defense force is for a long-range communications capability to connect the national network to forces deployed up to 1,200 miles away for a limited time. Business management will begin by framing their needs in capability terms—that is, in terms of the capacity required, the range required, the mobility of the terminals at either end of the link, the networking requirements, security required, and noise and delay that can be accommodated. Before proceeding, however, there is considerable utility in identifying what system-level options are available. Such a process is described as follows.

- *C1.4.1—Identify alternative system-level solution classes.* There are broadly four classes of beyond-the-horizon communications systems: microwave radio relay, cable, high-frequency (HF)-based systems, and satellite-based systems.
- *C1.4.2—Confirm compliance with BNR.* Each alternative, described as follows, can be compared against the needs to determine compliance:

    - *Microwave radio relay:* Meets all needs but not be able to cope with mobile users very well. The cost would be significant as some 80 point-to-point radio-relay links would need to be established to cover 1,200 miles. Additionally, the time taken to establish and disestablish the network is prohibitive.
    - *Cable:* Meets all needs but not able to cope with mobile users very well. It would be costly to lay and recover 1,200 miles of cable to support each deployment. Additionally, the time taken to establish and disestablish the network is prohibitive.
    - *HF-based system:* Meets all needs, although capacity may be too small, and, while able to support deployed forces that are mobile, the end terminals can be slow to establish and disestablish.
    - *Satellite-based system:* Meets all needs.
- *C1.4.3—Evaluate alternative solution classes.* Radio relay and submarine cable solutions are suitable for most needs except they cannot meet the need for mobile users, and they are really permanent solutions to a short-term problem. HF-based systems do not have sufficient capacity, which means that the satellite-based system is the only viable solution class.
- *C1.4.4—Select the best of the alternative solution classes.* Satellite-based system is the best solution class. Note that decisions have yet to be made regarding the height of the orbit and whether the satellite is owned outright or a transponder is leased or any physical detail at all.

Note that the set of stakeholder and system-level requirements for each of the candidate systems is very different. For example, for the choice between HF-based and satellite-based solutions, we would need to develop very different stakeholder and system requirements for a system that uses refraction from the ionosphere, as opposed to one that utilizes retransmission through a space-based transponder. The narrowing of a feasibility analysis is therefore quite important; we need to make a design choice (or several design choices) at the business management level before progressing so that the subsequent requirements analysis activities result in an StRS and an SyRS that are sufficiently focused to be useful.

6.3.1.4 C1.4.5—Define Business Requirements

The feasibility analysis and selection of the resultant desired solution class may require the draft business needs to be revisited (Figure 6.8, step C1.4.5) in light of the more detailed investigations of the feasibility analysis. The hierarchical representation of business needs (mission, goals, objectives) is now further elaborated (decomposed and derived) and formalized into a balanced set of business requirements, which are recorded in the BRS.

A representative requirements framework in Figure 6.9 for the ACME Medical Center example and in Figure 6.10 for the Domestic Security Alarm. If desired, the three-word, verb-based elements can be expanded to form a requirement statement by simply adding the subject.

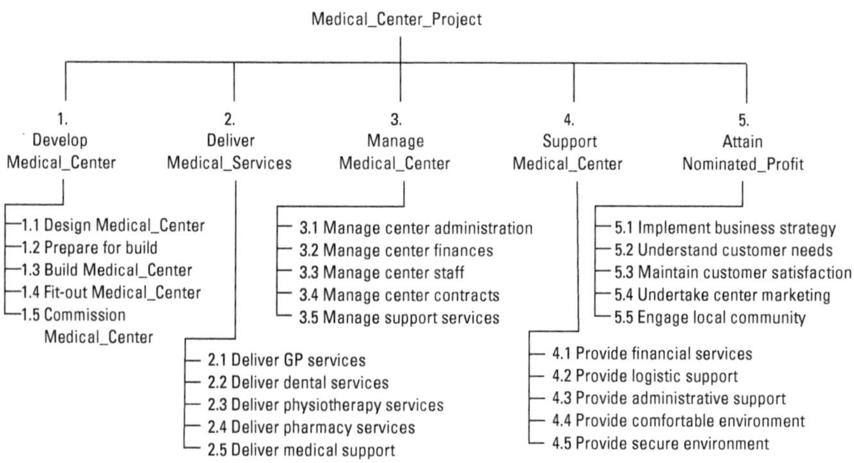

Figure 6.9 Example RBS for the ACME Medical Center.

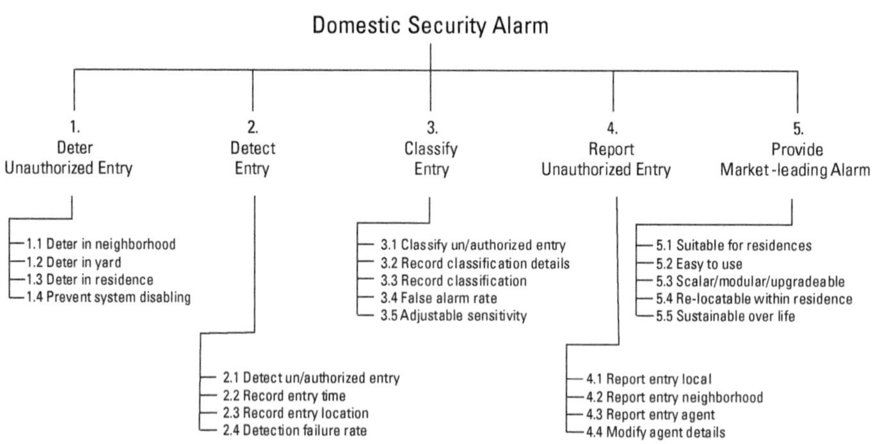

Figure 6.10 Example RBS for the Domestic Security Alarm.

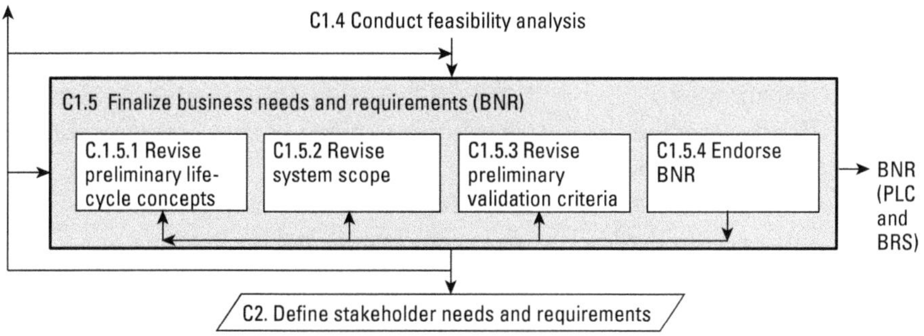

Figure 6.11 A summary of the steps involved in the activity C1.5—Finalize business needs and requirements.

6.4 C1.5—Finalize Business Needs and Requirements

At this stage, the business now understands the system much better than it did when it first drafted the mission statement—not only has much more detail been elicited, but a solution class has been identified, providing a much more concrete view of the system and its operation. Consequently, the BNR can be finalized by revisiting earlier artifacts and revising them to incorporate the increased understanding from subsequent steps.

The preliminary life-cycle concepts are revised (Figure 6.11, step C1.5.1) as are the solution scope in terms of the mission, goals, objectives, context diagram, boundary and external interfaces (Figure 6.11, step C1.5.2) and the preliminary validation criteria (Figure 6.11, step C1.5.3). Finally, the BNRs are endorsed (Figure 6.11, step C1.5.4) by business management (and perhaps enterprise management), most likely at a major review.

Review Questions

1. Briefly describe the four main steps in the CDM activity C1.3—Scope system.
2. With the aid of diagrams, illustrate what is involved in creating a context diagram for a system.
3. Draw a context diagram for a simple system (for example, for the building of your own home).
4. Briefly discuss the issues associated with the definition of the system boundary.

5. Briefly describe the three steps associated with interface definition: interface description, interface impact analysis, and interface control analysis.
6. Briefly describe the four attributes ascribed to each interface during interface definition.
7. Briefly describe what is involved in interface management.
8. List and briefly describe the types of external interface.
9. Briefly describe the five main steps in the CDM activity C1.4—Conduct feasibility analysis.
10. Briefly explain why it is important to conduct a feasibility analysis in conceptual design.
11. Briefly describe the five main steps in the CDM activity C1.5—Finalize business needs and requirements (BNR).

Endnotes

[1] *A Guide to the Project Management Body of Knowledge (PMBOK)*, Upper Darby, PA: Project Management Institute, Version 6, 2017.

[2] See, for example: Maier, M. W., and E. Rechtin, *The Art of System Architecting*, Boca Raton, FL: CRC Press, 2009.

[3] Maier, M. W., and E. Rechtin, *The Art of Systems Architecting*, Boca Raton, FL: CRC Press, 2009, p. 403.

7

Define Stakeholder Needs and Requirements

7.1 Introduction

The endorsement of the BNR signifies that business management are confident that they have defined the problem domain in sufficient detail to communicate their needs and requirements to the stakeholders at the business operations level. These stakeholders then go on to develop their needs and requirements (the SNR) within the context set by business management in the BNR. Figure 7.1 summarizes the activities involved in conducting C2—Define stakeholder needs and requirements resulting in the development of the SNR (as captured in the LC and StRS).

7.2 C2.1—Define Stakeholder Needs

Once the BNR have been confirmed for the preferred solution class, the nominated stakeholders at the operations level can begin to articulate their needs and requirements. As we noted when considering stakeholders in Chapter 5 and feasibility analyses in Chapter 6, the stakeholders in C1—Define business needs and requirements are mostly at the business management level. There will have been some representation from the operations level, but their interests will often be taken on by an appropriate manager from the business

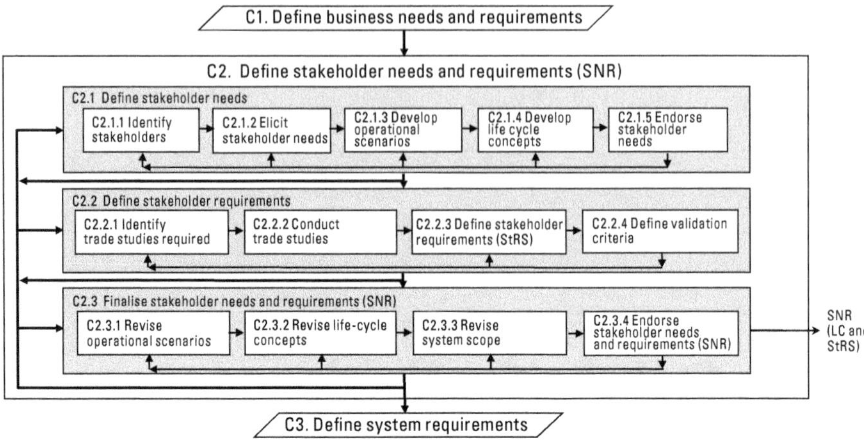

Figure 7.1 A summary of C2—Define stakeholder needs and requirements process.

management level who will act as their representative. Now that a feasibility analysis has selected an appropriate solution class, stakeholders can be identified at the operations level.

7.2.1 C2.1.1—Identify Stakeholders

Although the major stakeholders were nominated by business management during the definition of BNR, the first act of the process of defining SNR is to refine the stakeholders by naming individuals from the business management level that will remain on the stakeholder board as well as adding lower level stakeholders from the operations level.

Once the business has selected the appropriate solution class, the operations-level stakeholders can be identified and suitable representatives nominated to the stakeholder board. The same process is followed in identifying these operations level stakeholders as was followed for identifying business management-level stakeholders (see Section 4.2).

As an illustration of why we need to split stakeholders between the management and operations level, consider our ACME Aircraft System example. We saw in Section 6.3 that the feasible alternative solution classes for the replacement aircraft might be the following:

- Leasing or buying an existing aircraft;
- Leasing or buying and modifying an existing aircraft;
- Contracting to have a new aircraft developed;

- Outsourcing the operation to another airline;
- Not servicing these medium routes.

We noted that there is therefore a wide range of possible solution classes for the replacement aircraft, and each alternative is a completely different type of project with a markedly different set of stakeholders at the operations level.

We also discussed in Section 6.3 the feasibility analysis for the long-range communications system and noted that the two alternatives that were most feasible were a HF-based solution or a satellite-based solution. Although both solution classes have the same business management level stakeholders, each has a different set of operations level stakeholders. For a HF-based system the operations level stakeholders include representatives of the HF radio operators, the large strategic HF transmitter and receiver sites, and HF communications planners. The operations level stakeholders for a satellite-based solution would include satellite operators, satellite antenna owners, and satellite network operations centers. Clearly, the operations level stakeholders don't become evident until the feasibility analysis has been conducted.

Consider also the ACME Timber example in which the sawmill managers will be stakeholders at the operations level if the business is introducing the capability to fell more trees. If, however, the chainsaw is being introduced into service to cut down the same amount of wood more efficiently, the sawmill managers will not even know about the project and would not be included as operations—even stakeholders.

7.2.2 C2.1.2—Refine Mission, Goals, and Objectives

Within the context set by the BNR, stakeholders at the operations level articulate their needs through the refinement of the mission, goals, and objectives. In some cases, the mission may not change substantially from that which was developed during the development of the BNR. In other cases, the changes are subtle to reflect the chosen solution class. In cases such as the aircraft system example the mission statement will change significantly because there are significant differences among the solution classed.

7.2.3 C2.1.3—Develop Operational Scenarios

The preliminary operational scenarios outlined by business management are now defined in more detail, and other lower level scenarios (often called vignettes) are developed now that the solution class has been selected and those lower level use cases have become obvious.

Consider our Domestic Security Alarm example. The operation of the Alarm can be described in terms of four normal modes noting that Normal Case 2 is the reason why the Alarm exists:

- *Normal Case 1.* Resident is at home, and Alarm is not set.
- *Normal Case 2.* Resident leaves Residence, sets Alarm, and then returns sometime later to disable the Alarm and reenter Residence.
- *Normal Case 3.* Local maintenance is conducted on Alarm system.
- *Normal Case 4.* Remote maintenance is conducted on Alarm system.

The normal scenarios can be elaborated in terms of the exception conditions that would apply and would need to be accounted for in the design. For Normal Case 2, the exception cases would be the following:

- *Exception Case 2-1:* Resident doesn't disable the Alarm on re-entry.
- *Exception Case 2-2:* Unauthorized Entry (successful or unsuccessful) is attempted while Alarm set and Resident is absent.
- *Exception Case 2-3:* Unauthorized Entry (successful or unsuccessful) is attempted while Alarm is not set and Resident is absent.
- *Exception Case 2-4:* Unauthorized Entry (successful or unsuccessful) is attempted while Alarm is set and Resident is present.
- *Exception Case 2-5.* Unauthorized Entry (successful or unsuccessful) is attempted while Alarm is not set and Resident is present.
- *Exception Case 2-6:* Unauthorized Entry is detected while Intruder is not present while Alarm set, and Resident is absent (false alarm).
- *Exception Case 2-7:* Unauthorized Entry is detected while Intruder is not present while Alarm set, and Resident is present (false alarm).
- *Exception Case 2-8:* One of the components of the Alarm (sensor, processing, power, or PSTN) is interfered with by Intruder.
- *Exception Case 2-9:* One of the components of the Alarm is interfered with by accident.
- *Exception Case 2-10:* One of the components of the Alarm fails.

More detail can be added for each of the normal and exception cases. For example, for Normal Case 2, the following could be included.

- *Alarm is inactive:* Resident is inside. Resident sets Alarm and receives indication that Alarm will be active shortly. Resident has time to evacuate residence before Alarm is active. Resident receives indication that Alarm is active. Resident departs.

- Alarm system monitors residence for entry.
- Resident returns and enters residence.
- Alarm detects entry and provides warning that it will report an unauthorized entry unless the resident authenticates.
- Resident authenticates within required time, and Alarm returns to Inactive.

It is useful to avoid any of the flourishes normally associated with written English, making sure that no superfluous words are used. As with the RBS elements, it is useful to stay with the functional components of a sentence (subject, verb, object) to be as pithy and as succinct as possible.

After this first pass, the scenario can be tightened up using glossary items Entry: Entry into the Alarm's sensor range by a Resident (Authorized Entry) or an Intruder (Unauthorized Entry); Activitation_Time, Authenticate, Inactive, Active, Set, Activate, Monitor, Detect, Alert, Report, and so on. The scenario above can then be rewritten as follows:

- Alarm Inactive. Resident inside Residence. Resident Sets Alarm and receives Indication that Alarm will be Active shortly. Resident Evacuates Residence within Evacuation_Time before Alarm is Active. Resident receives indication that Alarm is Active. Resident Departs.
- Alarm Monitors Residence for Entry.
- Resident Returns and Enters Residence.
- Alarm Detects Entry and Alerts that it will Report an Unauthorized Entry unless Resident Authenticates.
- Resident Authenticates within required Authentication_Time and Alarm returns to Inactive.

The use of normal exception cases does not suit the development of all system operational scenarios but it is often a useful way to describe system operation, We demonstrate why this approach is particularly useful in Chapter 8.

7.2.4 C2.1.4—Develop Life Cycle Concepts

Similarly, the other PLCs are developed further and fleshed out with details from the various stakeholders to be more complete concepts in the LC. The feasibility analysis will have identified an appropriate solution class that will often have quite different (and certainly more detailed) LC than originally envisaged.

7.2.5 C2.1.5—Endorse Stakeholder Needs

Stakeholder needs are endorsed first by the stakeholders at the operations level and then by business management, and perhaps by enterprise management. Again, endorsement is most likely achieved through a major review.

7.3 C2.2—Define Stakeholder Requirements

Once the stakeholder needs have been endorsed, the process of transforming those needs into formal stakeholder requirements is undertaken. As noted earlier, the RBS is a useful framework for refining the BRS into the detailed requirements of the StRS while also assisting in the maintenance of traceability.

7.3.1 C2.2.1 and C2.2.2—Trade Studies

In a similar manner to that in which business management made use of a feasibility analysis to narrow down the alternative solution classes, stakeholders at the business operations level will have a number of alternative ways in which their needs can be met. The selection among various options is best undertaken by a formal trade (trade-off) study. The first activity, C2.2.1—Identify trade studies required lists those trade studies that are required, groups related trade studies for efficiency, prioritizes studies, and allocates appropriate resources. Those trade studies are then undertaken in C2.2.2—Conduct trade studies.

The progress of conceptual design is greatly assisted by the use of trade studies. If we are not careful, the requirements decomposition/derivation (hierarchical devolution) process will result in the development of an RBS that is far too broad for us to cope with. We therefore need to prune the RBS as the design progresses in order to keep the design under control. Design is about choice, and we should be careful to ensure that we make choices whenever we can, so that we limit the scope as well as limit the requirements hierarchy below us. Informed choices in design are made through trade studies.

7.3.1.1 Role of Trade Studies

The crucial role of trade studies in systems engineering and requirements engineering is reflected in the major lower level systems engineering standards. MIL-STD-499 states the following about trade-off studies: ". . .desirable and practical trade-offs among stakeholder requirements, technical objectives, design, program schedule, functional and performance requirements, and life cycle costs shall be identified and conducted. Trade-off studies shall be defined, conducted, and documented at the various levels of the functional or physical architecture in enough detail to support decision making. The level of detail of each study shall be commensurate with cost, schedule, performance, and risk impacts" [1].

Both MIL-STD-499 and IEEE-STD-1220 state that trade-off analyses are used for the following reasons:

- Used in requirements analysis to establish alternative performance and functional requirements to both resolve conflicts with and satisfy customer requirements.
- Used in functional analysis/allocation to support functional decomposition and allocation of performance requirements.
- Used in synthesis to support design decisions.

7.3.1.2 Types of Trade Study

IEEE-STD-1220 defines the following three types of trade studies [2]:

- *Judgmental:* Based on the judgment of the designer, where the consequences are not too important; one alternative is clearly superior; and/or time may not be available for a more formal approach. (Most trade-off analyses are of the judgmental type.)
- *Formal:* Formally established, conducted, and documented with results reviewed at technical reviews.
- *Informal:* Follows the same methodology of a formal trade-off analysis but is not documented as formally and is of less importance to the acquirer.

7.3.1.3 Trade Study Process

The process employed when conducting a trade-off analysis will vary with the individual needs of the decision-making process. The steps discussed in Sections 7.3.1.3.1 to 7.3.1.3.7 provide a guide to the general process that may be employed [3].

7.3.1.3.1 Step 1: Definition of Objective and Scope

The objective and scope of the problem to be investigated must be documented in precise terms. Systems engineering documentation, including specifications and design aids such as FFBD, is a good source of this information. Sources will vary as the development moves through the system life cycle. This step also defines processes and methodologies to be used in the trade study, including data and information collection requirements as well as a schedule of activities and a budget.

7.3.1.3.2 Step 2: Identification of Alternative Solutions

The trade-off analysis must identify the widest range of alternative solutions possible to maximize the chance of selecting the optimal solution to the problem. There are many sources of alternative solutions, including systems engineering synthesis activities, OTS equipment, and other existing equipment. Once the range of solutions has been identified, the possible solutions should be passed through a screening process to eliminate infeasible or unsuitable solutions from further consideration. This will avoid investing valuable time and effort developing solutions that ultimately cannot be implemented for one reason or another. The remaining (feasible) solutions should be described in sufficient detail to allow further detailed consideration at a later stage in the process.

7.3.1.3.3 Step 3: Nomination of Selection Criteria

Criteria must be formulated to provide measures of the effectiveness of the alternatives to meet the requirements of the problem. If possible, these measures should be quantitative in nature to avoid subjective assessment of the alternatives as much as possible. The different criteria should be independent to avoid assessing the same characteristics a number of times, and the criteria should be directly related to the defined problem.

7.3.1.3.4 Step 4: Determination of Criteria Weighting

Once the criteria have been identified, they should be weighted according to their relative importance. Some criteria may be considered equally important as others and be assigned the same weighting; however, in most cases, there will be a clear difference in the level of importance among the criteria. This should be reflected in weighting. If considered necessary, the weightings may be withheld from the individuals performing the evaluation to help ensure the objectivity of the assessment. Numerical assessment through the use of weightings will assist in the comparative assessment of the alternatives under consideration. By combining an assessment of how well an alternative meets a certain criterion with a measure of the importance of that criterion, the evaluation will provide an excellent indication of the alternative's performance.

Figure 7.2 illustrates how the weighting of criteria may be allocated for our aircraft example.

7.3.1.3.5 Step 5: Definition of Scoring Functions

It may be worthwhile to assign scoring functions to each criterion to ensure that the performance of each of the alternatives against the criteria is scored consistently. The scoring functions will be used by the scorers during the

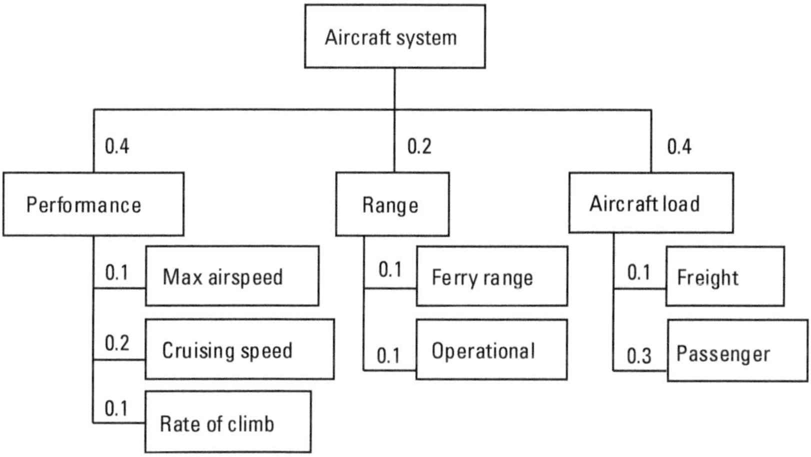

Figure 7.2 An illustration of weighting of criteria for our aircraft system.

assessment of each alternative—scoring functions can be linear or nonlinear, depending on the nature of the criterion. A range of scores should be selected (for example, 0–10) and this range applied to all criteria. By applying different scoring ranges to different criteria, inadvertent weighting will be introduced. Having said that, the score for each criterion should be described in such a way that any person scoring the criterion would allocate the same score.

Using the aircraft example, the scoring function in Figure 7.3 may be applicable to the cruising speed criterion.

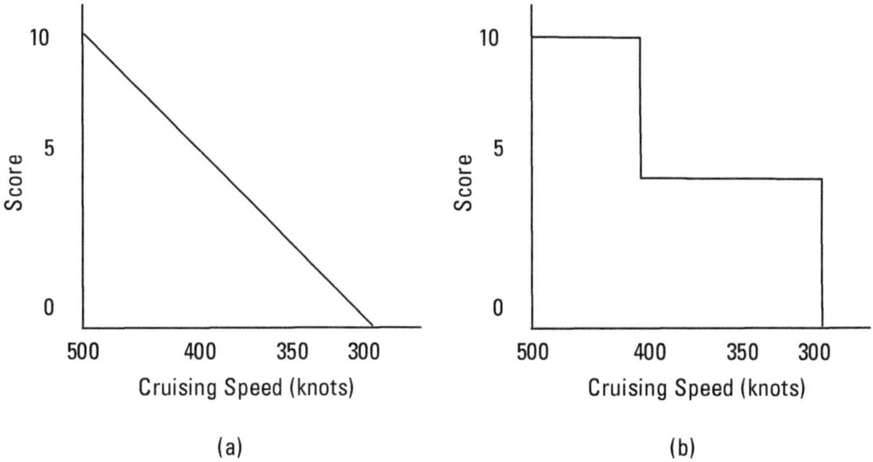

Figure 7.3 Illustration of a scoring function for the cruising speed criterion for our aircraft system: (a) nonlinear and (b) linear.

In both examples, candidate aircraft that have a cruising speed greater or equal to 500 knots will score 10 points. Speeds of less than 300 knots will be given a score of 0. In the linear example, however, 5 points is given for speeds between 300 and 400 knots whereas, in the nonlinear example, a higher score is available as speeds increase from 300 to 400 knots. Both functions then provide higher scores for greater speeds between 400 and 500 knots.

7.3.1.3.6 Step 6: Evaluation of Alternatives

Once all the weights and scoring functions have been produced, the alternatives can be assessed. Normally, the scores and weightings will be recorded in a tabulated form, and the total score for each alternative will be calculated using the following (or similar) expression:

$$\text{Weighted score} = \sum_{i=1}^{N} w_i s_i$$

where

N = number of criteria

w_i = weighting of ith criteria

s_i = score of ith criteria

Provided that the criteria, weightings, and scoring functions have been correctly compiled, the weighted scores will then provide an indication of the relative order of merit of each of the alternatives.

7.3.1.3.7 Step 7: Sensitivity Study

A sensitivity study, or sensitivity analysis, should be performed to confirm the validity of the results of the overall trade-off analysis. The sensitivity check will concentrate on the allocation of weights to each criterion and the range of possible scores that could have been achieved by each alternative against each criterion (given that the score assessment may have been subjective to an extent). The allocated weights and the range of scores will then translate into a range of weighted scores for each alternative. Sensitivity studies attempt to avoid the situation where the incorrect alternative is selected due to poorly informed assessment by investigating the situation where even a small change in weights or in a single score can have significant effect on the overall weighted score. If the score ranges still clearly identify the preferred alternative, the sensitivity study will have reinforced the validity of that alternative. If, on the other hand, the score ranges overlap, the decision makers need to consider the following courses of action:

- Obtaining additional information for each alternative to further differentiate between them;
- Reviewing the criteria, weightings, and scoring functions to ensure that they are correct;
- Ensuring that risk-management planning is in place to handle the possibilities arising from potential incorrect selection.

7.3.1.4 Example—Trade Studies for the ACME Medical Center

By way of example, the trade studies that we might conduct early on for the ACME Medical Center could include consideration of options for the following:

- Safe access by patients (of all classes);
- Factors such as provision of services, layout of building, and number of floors;
- Other services provided (e.g., x-rays and day surgeries);
- The relationship between profit level and services (e.g., consultation times, numbers of staff, and opening hours);
- Types of patients (in terms of services to be provided);
- Marketing (attraction of patients);
- Waiting arrangements;
- Central reception versus individual receptions;
- Lease versus purchase of equipment;
- Staffing arrangements (e.g., salary and contract);
- Professional indemnity insurance;
- File management (manual versus automated, centralized versus distributed);
- Billing and accounts processes;
- ICT/network provision;
- Provision of cleaning services;
- Provision of ad hoc and planned maintenance.

7.3.1.5 Example—Trade Studies for the Domestic Security Alarm

By way of example, the trade studies that we might conduct early on for the Domestic Security Alarm could include consideration of options for the following:

- Features versus target market price versus market-leading;
- Choice for feature bundling (numbers of sensors included, expansion packs, and after-market add-ons);
- Choice of sensor type;

- Type of sensor connection (wired/wireless/both);
- Choice of processor;
- Authentication means;
- Stand-alone versus back-to-base (or both);
- Remote monitoring options (agent/customer);
- Back-to-base communications options (PSTN/internet/cellular);
- Installation options (particularly for user installation);
- Support options.

7.3.2 C2.2.3—Define Stakeholder Requirements (StRS)

The development of the StRS, an essential first step toward successful system development, is given a number of titles, including the system design document (SDD) or simply stakeholder requirements or user requirements. A reader of the StRS should be able to understand completely the likely applications or missions for which the system is intended; the major operational characteristics to be exhibited by the system; the operational constraints that limit the design and development of the system; the external systems and interfaces with which the system under development must operate; the operational and support environment within which the system must exist; and the support concept to be employed to support the system and enable it to continue performing in accordance with customer expectations.

Although the StRS is a formal requirements document, it must be written in the language of the stakeholders to ensure that the document is readable by all stakeholders. However, since it is critical that the StRS is comprehensive, readable, unambiguous, and subsequently easily traceable, the language must still be formal. In many cases the difference between the StRS and the SyRS is marked because the users are operationally focused and use very little technical language. In other cases, such as in an aircraft system, the stakeholders are such groups as air crew, ground crew, maintainers, and engineers, all of whom will use relatively technical language in their StRS statements. The difference is always, however, that StRS statements are high-level and made in more natural language, which is generally much more easily read and understood than the language of formal specifications.

The potential audience for the StRS includes everyone who has an interest in the delivered system, from the eventual users, operators, and support personnel to the systems engineers and designers responsible for designing, developing, and manufacturing the system as well as the test personnel responsible for confirming that the system is fit for its intended purpose. Each audience set provides a different view of the requirements.

While there may be a large number of stakeholders who have an interest in the development of the StRS, the population of the StRS document is generally best conducted by a small team of experienced stakeholder representatives. It is important that all stakeholders contribute to the preparation of the document, however, and elicitation/generation techniques such as structured workshops, surveys, interviews, and prototyping are necessary to ensure that all relevant information is considered.

7.3.3 C2.2.4—Define Validation Criteria

Each requirement in the StRS must be accompanied by a validation statement—that is, the stakeholders must not only say what it is that they require but also how it is they will know that the system has met that requirement. These validation criteria are used during OT&E to ensure that the system meets the stakeholder requirements before it is accepted into service. In Step 1.2.3 (see Chapter 5), upper level, preliminary validation criteria (probably at the level of CI and COI) were developed and now need to be refined further to develop MOE, MOP, verification criteria, and TPM.

7.4 C2.3—Finalize SNR

At this stage, the stakeholders now understand the system much better than they did at the beginning of this process—the broad guidance of the BNR has been significantly refined, and much more detail has been added as a result of the additional needs elicited from stakeholders and the decisions made as a result of a number of trade studies. Consequently, the SNR can be finalized by revisiting earlier artifacts and revising them to incorporate the increased understanding. The operational scenarios are revised (Figure 7.1, step C2.3.1) as are the other LCD (Figure 7.1, step C2.3.2) and the solution scope in terms of the mission, goals, and objectives; context diagram and boundary; and external interfaces (Figure 7.1, step C2.3.3).

Finally, the SNR are endorsed (Figure 7.1, step C2.3.4) by stakeholders and by business management, most likely at a major review. Endorsement of the SNR is a critical step in developing the system in order to provide confidence to the system designers that all of the fundamental stakeholder requirements have been captured. Consequently, endorsement of the StRS is not to be taken lightly. For requirements engineers, the StRS is the basis of further development of the requirements to populate the SyRS that forms the centerpiece of the FBL of the system. From a project management perspective, the StRS assists in bounding the scope of the project from which cost and schedule estimates can be developed.

7.4.1 Capability System

As discussed in Chapter 1, because a system is often seen as an operational capability, it is commonly referred to as a capability system. Recall also that a capability system comprises constituent elements such as the major equipment solution (hardware and software), organization, personnel, collective training, facilities, data, support system (including supplies), and operating procedures and organizational policies. At this point in the design, the business owners and stakeholders must agree on how the elements of the capability will be acquired. There could be a single SyRS developed for the whole capability system or, as illustrated in Figure 7.4, there could be one SyRS developed for each element of the capability system or for any combination of elements.

As outlined in Chapter 1, we have focused here, for simplicity, on the acquisition of the major equipment element (often called the material system) of the capability system. We must always keep in mind, however, that all the elements need to be brought back together prior to introduction into service in order to field an operational capability.

7.4.2 Traceability

The StRS is the centerpiece of the set of BNR and SNR developed so far during conceptual design. During the subsequent synthesis and analysis, requirements engineers must ensure that each requirement in the StRS is traceable back to at least one requirement in the BRS and forward to at least one requirement

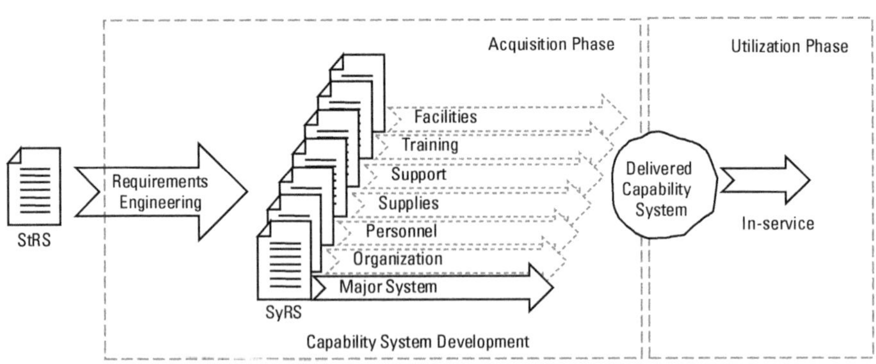

Figure 7.4 An example in which one SyRS is developed for each of the elements of a capability system.

in the SyRS (and thence from the SyRS to lower level specifications). In doing so, business management can be confident that all BRS requirements have been addressed in the StRS and then that all StRS requirements have been addressed in the SyRS. We saw in Chapter 2 that traceability from the BRS to the StRS to the SyRS (and onward) is called forward traceability; it ensures that the design process has appropriately accommodated each requirement at the next level of design. There must also be backward traceability from SyRS back to the StRS and thence back to the BRS. Such traceability ensures that additional (not formally endorsed) requirements have not crept into the specifications during conceptual design.

Requirements creep must be guarded against as additional, unendorsed requirements ultimately compete with endorsed requirements for valuable design space and may end up reducing the performance of the system against the endorsed requirements. Unendorsed requirements may also add significantly to the cost and schedule associated with the system development, without adding to the system's ability to solve the original business and stakeholder needs.

Review Questions

1. Briefly describe the five main steps in the CDM activity of C2.1—Define stakeholder needs.
2. Briefly describe the four main steps in the CDM activity of C2.2—Define stakeholder requirements.
3. Briefly describe the four main steps in the CDM activity of C2.3—Finalize SNR.
4. Briefly describe the role of trade studies in scoping a system.
5. Briefly describe the three types of trade studies, identifying when each might be conducted.
6. Briefly describe the seven steps involved in the trade study process.

Endnotes

[1] MIL-STD-499B, *Military Standard—Systems Engineering—Draft*, Washington DC: US Department of Defense, 1994.

[2] IEEE-STD-1220, *IEEE Standard for Application and Management of the Systems Engineering Process*, New York: IEEE Computer Society, 2005.

[3] Faulconbridge, R. I., and M. J. Ryan, *Applied Systems Engineering*, Norwood, MA: Artech House, 2026.

8

Define System Requirements

8.1 Introduction

The next process in conceptual design is the definition of system requirements, which starts with the endorsed SNR and decomposes/derives them into system requirements in the SyRS. Figure 8.1 summarizes the activities involved in conducting C3—Define system requirements, which is based on requirements analysis. The aim of system requirements definition is to determine what the system must do to meet the stakeholder needs and requirements—that is, to describe the requirements at the system level and be able to relate the logical design back to the SNR, particularly the StRS. System requirements definition therefore uses elaboration to bridge from stakeholder requirements to system requirements.

During system requirements definition, it is necessary to perform the following tasks:

- Establish an appropriate requirements framework;
- Perform requirements analysis and allocation—define the major requirements (with associated performance and verification) required of the new system (including maintenance, support, operational, and any other necessary functions);
- Assign a rationale to each of the requirements;
- Determine the TPMs to be applied during the systems engineering effort;

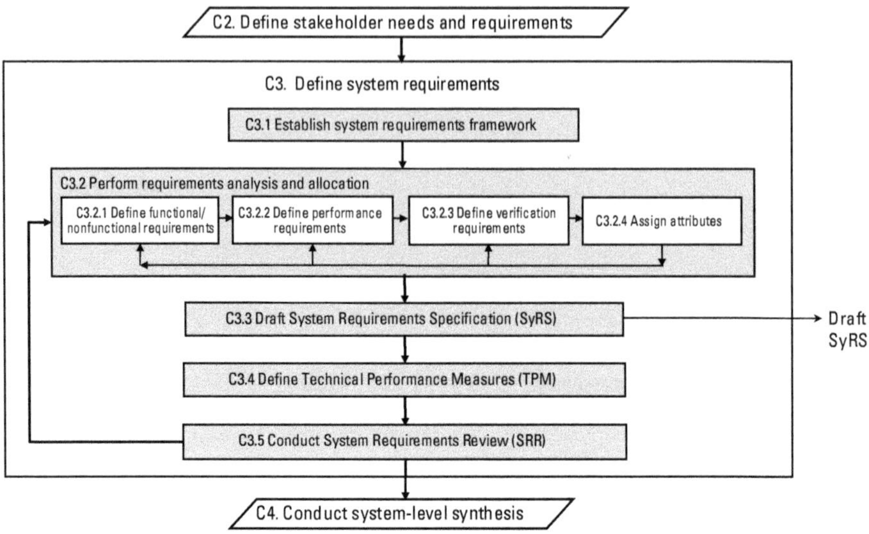

Figure 8.1 A summary of C3—Define system requirements.

- Produce the draft SyRS;
- Conduct SRRs.

These activities are described in more detail in the following sections.

8.2 C3.1—Establishing the Requirements Framework

The first step in the C3;Define system requirements activity is to establish a requirements framework (Figure 8.1, activity C3.1) around which the logical design of the system is to be based. Here we use the RBS. Remember that the RBS is grouped by function, the WBS is structured by physical project elements and contains other project-related information. Since we have started with a hierarchical decomposition of mission, goals, and objectives, we can use that information to begin our development of a logical hierarchy, which can be captured in an RBS.

Recall the RBS for the Medical Center and the domestic security alarm shown in Chapters 6 and 7—in this step, the hierarchical elaboration (decomposition/derivation) proceeds to successive lower levels. Once again, the RBS provides the structure for the decomposition and derivation, and the resultant three-word, verb-based terms can be quickly turned into sentences.

8.3 C3.2—Perform Requirements Analysis and Allocation

The requirements-analysis activity involves the elaboration of major requirements for the system (with associated performance, verification, and rationale) including operations, maintenance, support, and other necessary functions. EIA-632 defines requirements analysis as:

> "The determination of system-specific performance and functional characteristics based on analyses of customer needs, requirements and objectives; missions/operations; projected utilization environments for people, products and processes; constraints and measures of effectiveness. The bridge between customer requirements and system specific requirements from which solutions can be generated for the primary system functions" [1].

8.3.1 C3.2.1—Define Functional/Nonfunctional Requirements

Defining functional and nonfunctional requirements (Figure 8.1, step C3.2.1) is an important part of the overall system requirements analysis that, as the name suggests, identifies the many different functions that the system needs to perform in order to meet the endorsed mission, goals, and objectives contained in the BNR. The analysis identifies all functions including maintenance functions that are necessary to maintain or return the system to operational use. At this stage in the development, the concentration is still on what needs to be achieved, not how it should be done.

Operational requirements are normally given the highest priority during the early stages of a project, because they tend to be the most tangible and most easily understood. The life cycle approach emphasized in systems engineering also acknowledges that design decisions made early in the acquisition phase can have a large impact on the maintenance and support of the system. To that end, maintenance and support requirements need to be determined and specified during conceptual design. In this way, the design of the system ensures efficient and effective support during the utilization phase. Human factor considerations also produce a number of additional requirements. Functional requirements for the system are obtained from an analysis of such aspects as design space limitations, climatic limits, eye movement, operator reach, occupational health and safety issues, ergonomics, cognitive limits, and usability. The assessed skill levels of known users also constrains the design.

8.3.2 FFBD Examples

This section illustrates requirements analysis using FFBD for a couple of our examples.

8.3.2.1 FFBD Example—ACME Medical Center

As discussed in Chapter 2, although the RBS is very useful to show a hierarchical decomposition of the requirements, FFBDs are useful in showing the interrelationships of the functions to be performed by the system in accomplishing its mission. Some functions must be performed sequentially; some may be performed in parallel; sometimes alternative paths may be taken based on certain conditions—the FFBDs can show these relationships much more easily than the RBS.

Figure 8.2 is our first cut of the level-1 FFBD for a patient using our Medical Center example. Having completed the initial diagram, we need to test it. Our first major consideration should be the centerpiece of the FFBD, which is the flow of patients through reception to one of the practitioners and then back through reception. We must ask ourselves whether flow is appropriate. In this case, we may consider that it is because we do not want patients accessing practitioners without coming through reception, and we do not want patients to wander away after their appointment—at least we should get their payment first!

Note that later, when we derive subsequent functionality, we will want to make some functional statements to ensure that any patient visiting the center must pass through reception before and after their appointments.

As illustrated in Figure 8.3, in examining the parallel activities of a patient when being called forward from waiting (to go to see a GP, dentist, or physical therapist), we note that we have not accounted for the possibility of a patient being called away (perhaps by an emergency at home) before seeing the medical professional; we need, therefore, to have a null path through the

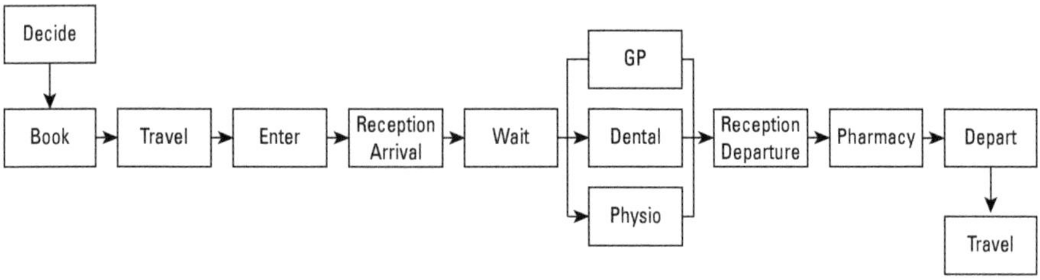

Figure 8.2 An initial attempt at an FFBD to describe the patient flow for the ACME Medical Center.

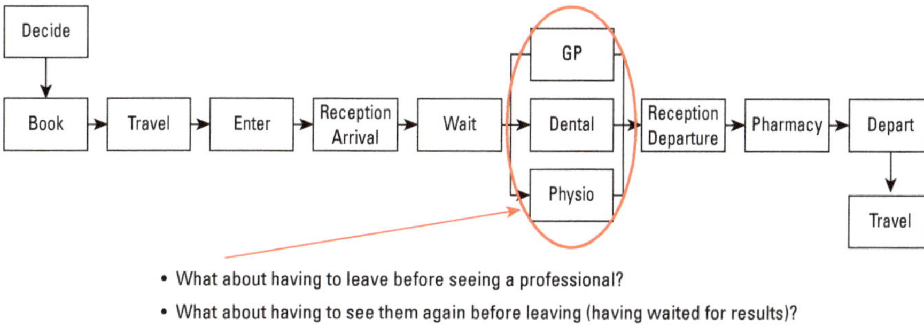

Figure 8.3 Investigation of the FFBD to account for better waiting arrangements.

system so that they can decide to leave. Note that, once again, it is hoped that all patients check out with reception before they leave the building.

Note also that we need to accommodate the case where a patient is seen by one of the professionals and then needs to wait (perhaps for some test results) before being admitted to the professional again.

Having augmented the diagram in Figure 8.4 to accommodate better the flow required to handle patients waiting for an appointment, we can now look at the flow associated with booking an appointment. Our current flow has the patient only able to book before coming to the Medical Center. Often, however, a patient may (perhaps while on other business at the shopping center) drop in to book an appointment, either in the hope that one might be available at the time, or to make an appointment for a future date and time.

As illustrated in Figure 8.5, we need to adjust our FFBD to accommodate these two booking mechanisms. Having made those adjustments, we are now comfortable that the booking arrangements make more sense.

We look further and examine the flow through the pharmacy, illustrated in Figure 8.5. In our current flow, we are not handling well the needs

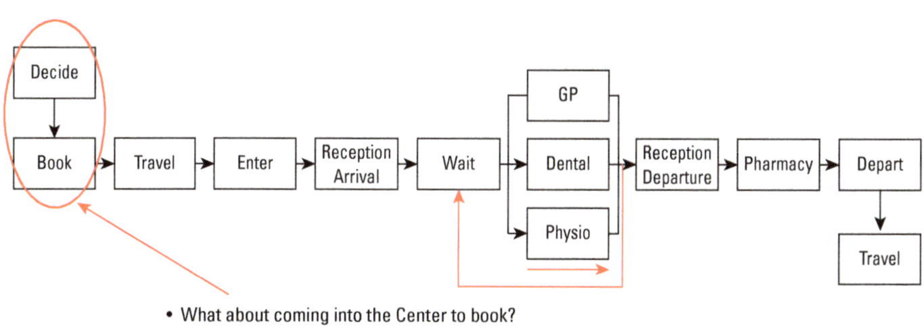

Figure 8.4 Modification of the FFBD to account for better waiting arrangements and to investigate booking arrangements.

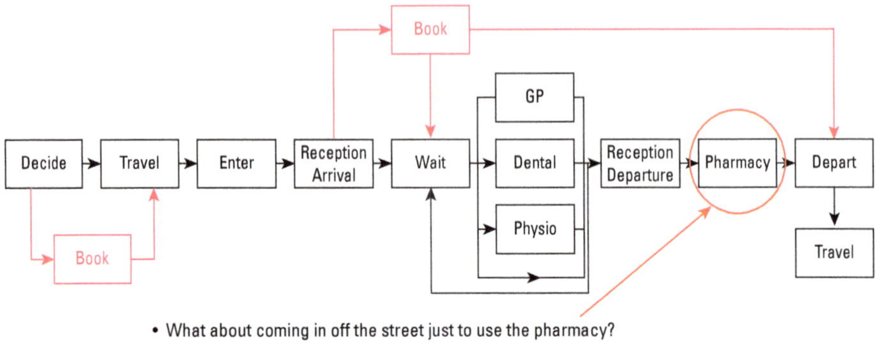

Figure 8.5 Investigation of the FFBD to account for better use of the pharmacy.

of customers who come in to just use the pharmacy (that is, they want to visit the pharmacy without having had an appointment first). These people should be able to enter the Center and use the pharmacy without passing through reception. A further issue is that the FFBD shows that patients must visit the pharmacy before they depart, whether they need to or not.

In Figure 8.6, we have adjusted the FFBD to reflect the that patients can depart without visiting the pharmacy and that customers can access the pharmacy without coming through reception. Note, however, that we still have the customer entering the Center to access the pharmacy. Here, we would make a note to discuss this issue with the stakeholders since we have the option (as illustrated by the dotted line) to provide access to the pharmacy from outside the Center so that only patients enter. The physical impact of exercising this option (the need for another external door to the Center) would most probably be the subject of a trade study.

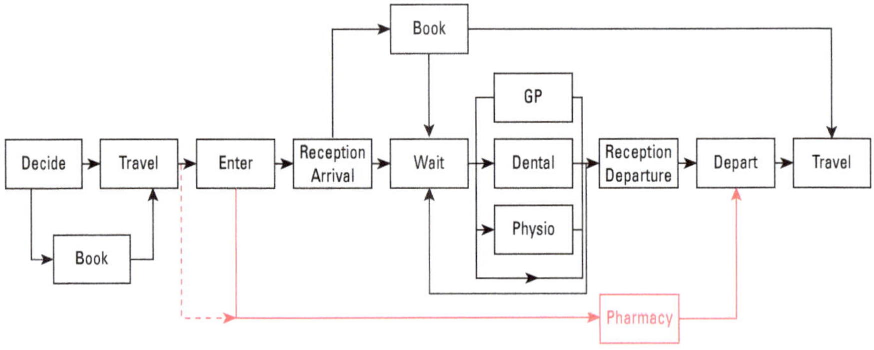

Figure 8.6 Modification of the FFBD to account for better use of the pharmacy.

Once the FFBDs have been completed to the required level, all major system functions and their interrelationships should have been identified. Some functional allocation may be undertaken where the functions are grouped together in logical groups and assigned to one of the headings in the RBS. It is not necessary to have an FFBD that describes every requirement, but every requirement must be included in the RBS.

The system-level functions in the FFBD are assigned numbers, and those numbers are continued through to the lower level functions. In this way, even the lowest level functions and requirements can be traced back to the original system-level function. Traceability has already been introduced as a cornerstone in systems engineering, and it becomes increasingly important as the design matures and becomes more complex.

8.3.2.2 FFBD Example—Domestic Security Alarm

FFBD may be used as a supporting thinking tool. For example, if we are not clear how the Domestic Security Alarm will work, we could augment our scenarios with the FFBD in Figure 8.7 for Normal Case 2 and for two of the exception cases.

Having fleshed out our first thoughts in the FFBD in Figure 8.7, we can note that there are quite a few elements of each FFBD that are common in each (Resident sets alarm, Alarm activated, Alarm provides alert, No authentication, Alarm triggers). Since that is the case, we can consolidate the three diagrams from Figure 8.7 into the one diagram in Figure 8.8.

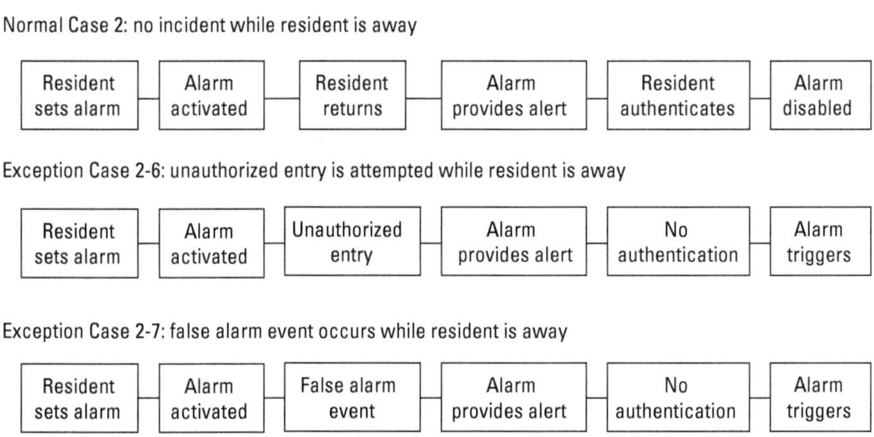

Figure 8.7 Using FFBD to illustrate Normal Case 2 and two exception cases for the operation of the Domestic Security Alarm.

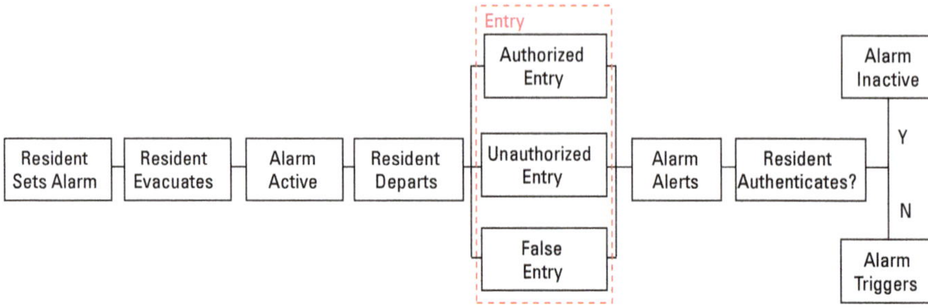

Figure 8.8 Using FFBD to consolidate Normal Case 2 and two exception cases for the operation of the Domestic Security Alarm.

Note that in Figure 8.8, we have taken the opportunity to create a class of Event called Entry which has the type of entry (Authorized Entry, Unauthorized Entry, and False Entry). This is example of a very important aspect of system design. System designers must be cognizant of what is going to happen with their requirements. In this case they know that a software developer will be coding controller instructions, probably using an object-oriented or class-oriented software language. If system designers can couch the system requirements in what is effectively pseudocode, stakeholders will still understand the requirements, and designers will find them much easier to implement, because they have system requirements that have already been predigested on their behalf.

8.3.3 Phases, Modes, and States

Note that the system operations may be defined in terms of modes and states. Although most system development methodologies would agree on the utility of using states and modes to describe a system, there is little guidance in the literature as to what constitutes a state or a mode, which is made worse by poor usage of the terms such that it is not commonly agreed-on as to whether states include modes or vice versa [2]. Further, some add phases to modes and states [3].

There are many definitions of state and mode. One of MIL-STD-498 DIDs [4] even states that the distinction between states and modes is arbitrary and that a system may be described in terms of states only, modes only, states within modes, modes within states, or any other scheme that is useful. Perhaps the most useful view is that a state is a static condition of the system and a mode is a particular set of sequences as the system moves between states.

For our aircraft example, there may be a Parked_State in which the aircraft is fully loaded and stationary, so that a requirement might state: "When in the Parked_State, the Aircraft shall exert no more than Y kN/m^2 of pressure on the tarmac when the aircraft is stationary and at maximum all-up weight." That Parked_State might then be part of the "Takeoff" mode as well as part of the "Landing" mode.

Describing the system in states and modes doesn't always make sense, but it is often useful—see the state transition diagram for our Domestic Security Alarm example in Section 8.3.4.1.

8.3.4 SNR Versus SyRS Descriptions and Requirements

The language utilized in the SNR is intended for a wide audience and is easy reading compared to the specification language, which tends to be more precise. For example, the StRS requirement statement: "The Aircraft shall operate from Class X airfields," will spawn, among many other requirements, the more detailed SyRS statement: "The Aircraft shall exert less than Y kN/m^2 of pressure on the tarmac when the aircraft is stationary and at maximum all-up weight."

To illustrate the differences between SNR and system descriptions and requirements, let us look at two of our examples: the Domestic Security Alarm and the CVR.

8.3.4.1 SNR versus SyRS: Domestic Security Alarm Example

Remember the stakeholders' operational scenario for Normal Case 2, which (quite rightly) is written from a stakeholder (in this case the Resident's) perspective. This stakeholder scenario can't simply be rewritten as a system scenario, because the alarm system knows nothing of the Resident or the Intruder—all the alarm can do is monitor in order to detect movement. Consequently, the Resident and Residence must be removed from the system perspective since the alarm cannot know anything of them. Revisiting the Normal Case 2 scenario and highlighting in bold the entities that are irrelevant to the alarm yields the following:

- Alarm Inactive. **Resident** inside **Residence**. **Resident** Sets Alarm and receives Indication that Alarm will be Active shortly. **Resident** Evacuates **Residence** within Evacuation_Time before Alarm is Active. **Resident** receives indication that Alarm is Active. **Resident** Departs.
- Alarm Monitors **Residence** for Entry.
- **Resident** Returns and Enters **Residence**.

- Alarm Detects Entry and Alerts that it will Report an Unauthorized Entry unless **Resident** Authenticates.
- **Resident** Authenticates within required Authentication_Time and Alarm returns to Inactive.

If we remove those irrelevant entries we can then translate the stakeholder-focused scenario into a system-focused scenario as follows:

- Alarm is On.
- Alarm is Set.
- Alarm waits Evacuation_Time before Alarm is Active.
- Alarm Monitors until Entry.
- Alarm Detects Entry is Alert.
- Alarm waits Authentication_Time and Alarms if Authentication is received returns to On, else remains in Alarm until Time-out after which returns to Active.

Note that the system descriptions are now independent of the surrounding world and just describe what the system has to do, not how the actors act (as would be described in the SNR). Consequently, the system requirements in the SyRS must be written to implement the system scenarios, not the stakeholder scenarios without having been transformed to reflect just the system view.

Note also that the system scenarios immediately imply a sense of states and modes. For example, Figure 8.9 illustrates a state transition diagram for our Domestic Security Alarm. You can see that in such systems the state transition diagram is a useful way to communicate the system's operation.

8.3.4.2 SNR Versus SyRS: CVR Example

By way of further illustrating the difference between SNR and SyRS descriptions and requirements, consider the CVR example. In the SNR, the stakeholders will address, among many other issues, needs and requirements such as that the CVR is not to interfere with any aircraft equipment, and the aircraft equipment is not to interfere with the CVR. These might be reasonable as stakeholder needs, but they cannot simply be repeated as system requirements.

Designers have to transform from the stakeholder perspective to the system perspective. All system requirements must be independent of the stakeholder and any actors—they must be solely about what the system does. A context diagram such as that in Figure 8.10 must be drawn to illustrate how the system works. So, for example the stakeholder need to not interfere with aircraft equipment must be transformed into a number of system requirements

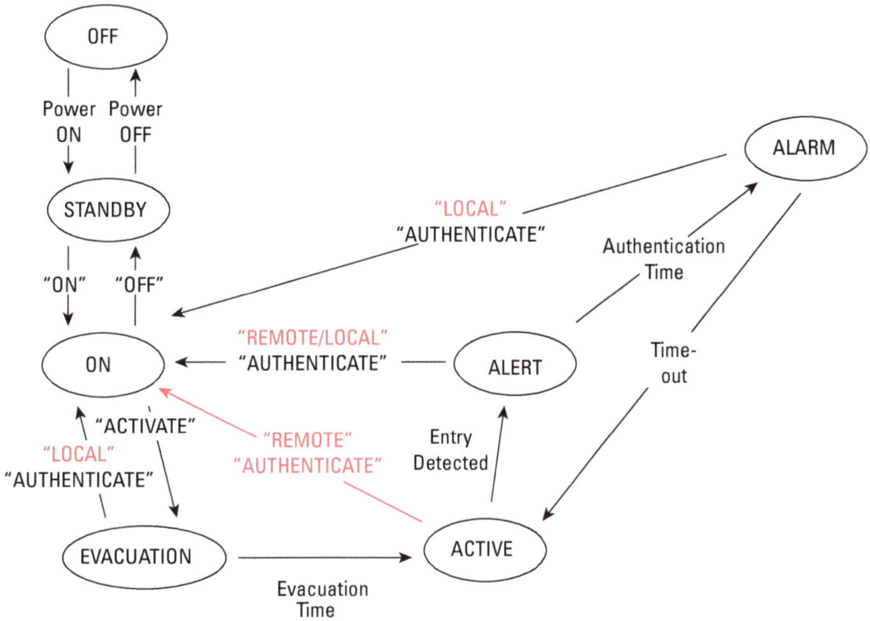

Figure 8.9 State transition diagram for the Domestic Security Alarm.

articulating what power is transmitted by the CVR across the electromagnetic spectrum (including light, infrared, and radio frequency). To do that, designers must understand what power levels will affect aircraft equipment and then work that back to a transmitted power level at the CVR (aided of course by a number of standards for devices to be fitted into commercial and military aircraft).

By way of further example, consider the stakeholder need to be crashworthy. That is, the CVR needs to be recoverable after a crash on land or at sea. The designers must transform this need to system requirements to survive a specified impact (chosen to be the maximum impact of a rotary wing aircraft); to survive a maximum temperature (the maximum temperature of an aircraft crash); to survive a minimum temperature (the lowest of the temperature at the maximum altitude of the aircraft and the temperature at the deepest part of the ocean); to survive in water with a maximum salinity (chosen from the worst ocean scenario); to survive a maximum pressure (of the deepest part of the ocean); and so on. That is, the system specification need make no mention of a crash, specifying instead the maximum/minimum environment in which the CVR must survive during normal operation in the cockpit and during and after a crash.

Note that, as discussed in Chapter 1 and illustrated for the CVR in Figure 8.10, we must be able to excise the system from its environment and then describe it in sufficient detail to be able to develop the system, verify its performance, place it back into its environment, and then validate its performance. Verification of the CVR cannot include taking the aircraft up to various heights and crashing into the ground at various angles and speeds to determine if the CVR can survive a crash. Rather, verification of the system in Figure 8.10 is in terms of how the CVR survives an impact, maximum temperature, maximum pressure, and so on; how it performs in the presence of external radiation of a range of frequencies and powers; how much radiation it produces in terms of frequency and power; what spurious signals it produces on its power input; and so on. All of those parameters can be verified in test laboratories.

8.3.5 C3.2.2—Define Performance Requirements

Once the functions have been identified and grouped according to the agreed RBS, design focuses on the performance-related parameters (Figure 8.1, step C3.2.2) that the new system must achieve. Having decided what the system must do, the designer must now determine how well the system is required to perform each of those functional requirements. A good discipline is to ensure that, every time a functional requirement is articulated, at least one corresponding performance statement is made. (See Chapter 9 for more detail.)

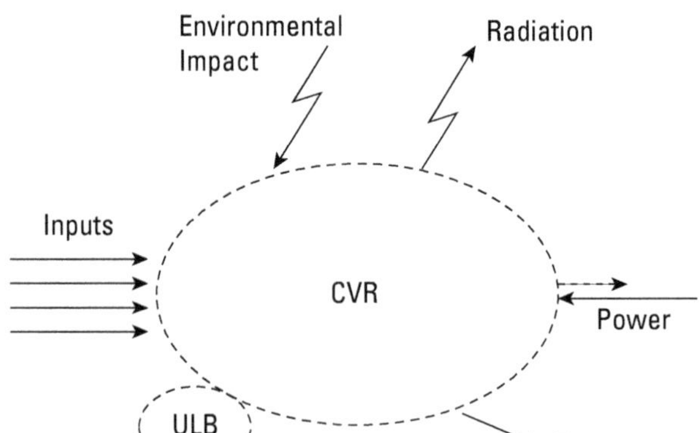

CVR: Cockpit voice recorder
ULB: Underwater location beacon

Figure 8.10 System context diagram for the CVR.

Consider that a CVR requires that the pilot's voice, the copilot's voice, and ambient cockpit sounds to be recorded during an aircraft's flight to assist in incident investigation.

One of the requirement statements we will write is, therefore, (giving it an arbitrary reference number): 1.1 The CVR shall record pilot audio.

Immediately, however, we realize that there are several performance requirements that must be articulated. Let us just focus on two: within what frequency range and with what duration? We can (at least) therefore attach the following two performance requirements, 1.1P1 and 1.1P2. (There will be other performance requirements, but we limit them to the following two here, by way of example.)

- 1.1 The CVR shall record pilot audio.
- 1.1P1 The CVR shall record pilot audio frequencies greater than 300 Hz AND less than 3.4 kHz.
- 1.1P2 The CVR shall record pilot audio for 60±2 minutes.

Note that the performance requirement uses the same stem as the functional requirement (in this case, "The CVR shall record pilot audio") and then appends the performance clause.

Most of the operational functions will have obvious performance parameters associated with them such as speed, accuracy, endurance, and acceleration. Support and other functions also require parametric definition to define completely the requirement. Examples of these parametric requirements are listed in the following nonexhaustive list:

- Environmental considerations, including temperature ranges for both operation and support, humidity tolerances, and operation under conditions of obscuration such as rain, snow, or fog. Operation under mechanical stresses such as noise, vibration, and shock should also be considered, as should chemical, nuclear, and biological contexts if applicable.
- Quality factors, such as availability requirements, reliability and maintainability limits, and manufacturing tolerances.
- Support requirements, including storage capacity, spares allowances, personnel establishments, training goals, and maintenance targets. The intended levels of maintenance employed to support the system once it becomes operational should also be included and related to the overall logistics support concept described or referenced in the life-cycle concepts.

- Utilization requirements, including duty cycles, hours of operation per day, and operational days per year.

The preceding example illustrates how two performance requirements (recording bandwidth and duration) might be described. An alternative approach might be to define Pilot_Audio in the glossary and write the requirement as follows:

- 1.1 The CVR shall record Pilot_Audio.

In the glossary, Pilot_Audio could be defined as having bandwidth of frequencies greater than 300 Hz and less than 3.4 kHz, a recording duration of 60 minutes, and so on. While this latter approach may be expedient, it does have the disadvantage that the performance requirements are aggregated in the glossary entry and may not be so readily edited if one individual performance requirement is to be modified.

8.3.6 C3.2.3—Define Verification Requirements

The definition of functional and performance requirements is not complete until verification requirements (Figure 8.1, step C3.2.3) are also included. Every time a functional requirement is articulated [with appropriate performance statement(s)], it is good practice for a corresponding verification statement to be made (see Chapter 9 for more detail). It is often difficult to write verification requirements for system-level functions, but the discipline of doing so is important since there is little point in stating a requirement for a function without consideration of how that function is to be tested. An additional benefit of adhering to the discipline at this stage is that test plans become much easier to write because the tests required at any level can be considered within a framework provided by an aggregation of the verification requirements at that level.

Unfortunately, it is not common to include verification statements in the SyRS. In many instances, the functional and performance statements are included in the SyRS, and the verification requirements are included in a separate document—such as a test concept document, or through a verification cross reference matrix (VCRM). This represents the view of the older style of specification. A more modern standard, EIA-632, defines a specification as: "A document that contains specified requirements for a product and the means to be used to determine that the product satisfies these requirements." Note, therefore, that the two parts are necessary—that is, the specification should include functional and performance requirements as well as the means to test them (the verification requirements).

Although the inclusion of verification statements is not common, it is reasonably common to attach a letter after each functional requirement to show the method that is proposed to be utilized to accomplish verification ([A], [D], [I], and [T]) as described in the following:

- *Analysis [A]:* Analysis of data through simulation or modeling to verify functionality;
- *Demonstration [D]:* Operation of a system or system element under defined conditions to demonstrate that a requirement can be achieved by, or is possessed by, the system or system element;
- *Inspection [I]:* Visual examination of the system or system element, normally used to verify physical design features;
- *Test [T]:* The application of scientific method is used to obtain quantitative or qualitative data relating to the desired functionality.

While the attachment of a letter to illustrate the verification methods is a useful indication of the customer's intent, it is much better practice to write a verification statement to make the verification methods more specific. That is particularly true for the analysis [A] and test [T] verification method because the use of a single letter begs a number of important questions that really should be answered at the time the requirement is written.

Note that each of the performance requirements will have a different verification requirement: for example, a spectrum analyzer to verify the recorded frequency range, and a stopwatch (or similar device) to verify the recorded time. Even so, there are a number of ways of verifying each performance requirement, each of which can have a significantly different impact on the project's cost and schedule.

For the first performance requirement, for example, when considering the verification of the frequency range, we need to consider how we are going to be satisfied that the delivered system meets that requirement. There are, of course, a number of options, described as follows:

- We could very simply have a test subject speak into the pilot's microphone and then determine that the voice has been recorded by playing it back through speakers.
- We could generate a test tone and sweep it through the desired frequency range, record each tone, and then play it back to confirm that the tone has been recorded without any loss.

- We could use a test voice sequence that is recorded and then compared quantitatively with the original.

Note that each of the above verification methods has a different requirement for resources, personnel, and time. Therefore, the customer cannot simply place [T] at the end of a requirement in the draft SyRS, because, although it gives the tenderers some clue as to an appropriate verification method (test), it gives no clue as to what the customer is thinking with regard to the extent of that method. Since the tender conditions will require tenderers to declare (and cost and schedule) a verification method, tenderers can really only choose the cheapest fastest method or else their bid will have greater cost and take longer but be no more compliant. Since they have to win the tender with the lowest compliant bid, tenderers cannot afford to add any cost or time for verification for fear they may not then be the best value for money in the tender evaluation process.

So, is it sufficient to just speak into a microphone and hear the tester's voice played back? If a test tone is to be generated, at which frequency increment is it to be swept across the range? In either case, how many times is the test conducted? For how many devices? Each of them at the end of production? Only the first item? Or for an agreed-on representative sample? These questions can't be left hanging because they have considerable impact on contract price and schedule. The consequence is that T&E is almost always undervalued in a contract, because the extent of verification is not properly agreed on at the start of the contract.

8.3.7 C3.2.4—Assign Attributes

As outlined in Section 2.2.2, an attribute is additional information associated with a need or requirement that is used to aid in its definition and management. Attributes also aid in achieving many of the characteristics of needs and requirements defined in the INCOSE *Guide to Writing Requirements* [5]. Need and requirement statements are therefore supported by associated attributes (Figure 8.1, step C3.2.4) that aid in the definition and management of needs and requirements and of sets of need and requirements [6].

One of the attributes, the rationale, is the reason why the requirement exists. The rationale may record the design decision(s) leading to the requirements, the assumptions behind it, or something as simple as the logic behind the selection of the numbers in the statement (for example, why they have been rounded to two decimal places). A requirement should not be accepted unless the rationale is well-understood. The recording of a rationale for each requirement has a number of advantages, discussed as follows:

- *Justification:* The need to articulate the rationale for each requirement forces the source of the requirement to be able to justify why the requirement should be included—just "because . . ." is not sufficient. The additional consideration may force the stakeholder to surface unstated assumptions, a number of which may be inappropriate.
- *Inappropriate expression of requirement:* Some requirements will be amended by requirements engineers when stakeholders also state why the requirement should exist—often, particularly during development of the logical architecture, stakeholders articulate requirements in terms of what they think the solution should be, not the real requirement. This may occur because stakeholders did not know how to state their real requirement, or it may be because they have a hidden agenda of some sort. In either case, the articulation of rationale may assist in exposing the deeper meaning.
- *Communication:* The recording of rationale allows readers at any time to understand why the requirement exists, which will assist with the removal of ambiguity but more importantly will assist in situations where more than one person is involved in the requirements analysis over the course of the project (or, indeed, at any time in the future). Similarly, the rationale communicates to lower level designers the reason why each requirement is necessary and the logic behind performance and verification characteristics assigned to the requirement.
- *Requirements management:* Recall from Chapter 2 that change analysis requires the impact of the change to be analyzed—the rationale for the requirements will be of great assistance when understanding that impact.

A rationale can be provided for each functional requirement and then inherited by the associated performance and verification statements. It is often useful, however, to provide a separate rationale for each performance and verification statement as well. That is, there is a rationale for having the function in the first place, another for the level of performance, and perhaps another for why a certain verification method has been selected.

8.3.7.1 Example of the Utility of Rationale

An example commonly used in systems engineering circles involves the design of a back windscreen of a passenger car, which had a user requirement to be able to withstand the air pressure associated with the car moving at 60 mph. During the design it was established that the car's maximum speed in reverse gear would only be 10 mph, so designers reduced the requirement for the rear windscreen to withstand air pressure from 60 mph to 10 mph. Such a decision seemed to make sense, particularly since there was likely to be a subsequent cost saving in the back windscreen component. However, when the new vehicles were reversed onto a car transporter for delivery from the factory to retail outlets, the rationale for the original user requirement became readily apparent when the rear windscreens were subjected to 60 mph on the transporter. That example maybe a little apocryphal, but it serves to illustrate the point.

There is, of course, another reason as to why the rear windscreen would have to have been specified to withstand air pressure commensurate with the car traveling at 50 mph. Before front windscreens were laminated they were made of hardened glass, which, when hit by a stone, shattered into small fragments rather than shards. As the driver was breaking after the impact, since the front windscreen was now missing, the back windscreen would be subjected to the additional air pressure until the car slowed to a halt. Having both reasons recorded in the rationale in the first instance is essential, because both reasons hold as the vehicle enters utilization, but then during the system life cycle when windscreen technology allows laminated glass to be incorporated, the requirement still holds for the first reason even though the second is no longer present. Consequently, change management throughout the systems life cycle has to ensure that if the windscreen is upgraded from hardened glass to laminated glass, then the rationale is adjusted to the remove the second reason but to retain the first.

8.3.8 Requirements Allocation

Requirements analysis also involves investigation of the relationships between requirements. In addition to the elaboration of requirements, system designers group and allocate functions. As discussed in Chapter 2, allocation has the following three aspects:

- First, functions are allocated to the appropriate logical group of the RBS, which is the aspect under consideration in this activity.
- Second, higher level requirements flow down to lower level requirements—that is, for example, a relevant proportion of the all-up weight of an aircraft will be allocated to its various subsystems. Other examples include, among many more, the allocation of the overall system reliability, availability, survivability, safety, and security.
- Third, there is the logical-to-physical allocation during the development of the subsystems and CIs.

8.4 C3.3—Draft System Requirements Specification

The complete and populated RBS forms the basis for the SyRS, which, when approved, becomes the centerpiece of the system's FBL. At this stage, however, the populated RBS forms the basis for the draft SyRS (Figure 8.1, activity C3.3), because the requirements collected and analyzed so far have yet to be synthesized into an architectural solution. There will therefore be some degree of latitude allowed in certain requirements and priority labels (such as mandatory, important, and desirable), leaving some flexibility for subsequent design.

The SyRS must have its roots in the StRS, as the StRS is the source of endorsed stakeholder requirements. Requirements engineers are responsible for elaborating StRS requirements and producing a set of detailed requirements in the SyRS that, when combined, ensure that this set of StRS requirements are met. Although the SyRS is drawn from the StRS, it is much more detailed and has translated the StRS statements in the stakeholders' language into a more formal specification language.

8.4.1 SyRS Structure

The SyRS may take any one of many forms [7], and the most appropriate form depends on the application. Figure 8.11 illustrates a suggested structure for a SyRS.

8.4.2 SyRS Format

The requirements included in Section 2 of the logical specification must be placed into a structured format. For example, a suggested paper-based format for a specification is illustrated in Figure 8.12, where columns are included to contain most of the information required to be recorded along with the requirements.

Note that it is very difficult to include all the information on a single sheet of paper in a manageable way. The difficulties with paper-based formats for specifications can be overcome with the use of requirements management tools (databases).

8.4.3 Semifunctional/Semiphysical Descriptions

Recall that conceptual design is functional design, focused on what is required, not physically (how) it will be achieved (which is the focus of preliminary

Section 1—System Overview
1.1 Introduction
1.2 System Purpose
1.3 Scope
1.4 Needs, Goals, and Objectives
1.5 Context Diagram
1.6 Stakeholders
1.7 Boundaries
1.8 External Interfaces
1.9 System Constraints

Section 2—Specification Statements
Functional Statements
Performance Statements
Verification Statements
Interface Statements

Section 3—Supporting Artefacts
References
Major Trade Studies
Standards
Glossary/Definitions
FFBDs
RBS Diagrams

Figure 8.11 Suggested structure for an SyRS.

design and detailed design and development). Quite often, however, it is very useful to break (or perhaps just bend) that rule and allow the SyRS to be a little more physical in nature, rather than purely functional. It is important to allow the system specification to be more physical; otherwise it can be too vague, particularly if it is to be used as the major compliance artifact in a competitive tendering process.

ID	Requirement	Attribute 1	Attribute 2	Attribute 3	Attribute 4	Attribute 5	
12.1.1	***Requirement statement***						
12.1.1P	Performance statement						
12.1.1V	Verification statement						
12.1.2	***Requirement statement***						
12.1.2P	Performance statement						
12.1.2V	Verification statement						
12.1.3	***Requirement statement***						

Figure 8.12 Simple example format for the inclusion of requirements in a simple, paper-based specification.

Recall also from Chapter 5 that the mission statement should have in it evidence of its own solution—the same is true for the full functional description in the SyRS. Conceptual designers must be sure that the system they have described in functional terms is able to be realized in physical terms. Before signing off on conceptual design and beginning preliminary design (particularly when going to tender), it is critical that the conceptual designers can validate their functional description to ensure that it is actually achievable and will ultimately manifest in a physical system that will meet the SNR and BNR.

For both those reasons, therefore, it is appropriate for the conceptual design to be couched in what could be called semiphysical terms by describing the system in a generic subsystem, building-block fashion.

For example, in our aircraft replacement example, we know what the typical subsystems of the aircraft are going to be. Unless we propose a radical new design, there will be a cockpit, a passenger cabin (of several classes), galleys, bathrooms, fuselage, avionics, wings, engines, landing gear, and so on. These elements are common on all medium to large passenger aircraft, although they are implemented differently of course.

In our Medical Center example, we know there are going to be separate rooms for each of the GPs along with a shared surgery. We know what a dentist's room will contain as well as the physical therapist's facilities. We know that there will be a waiting room (assuming we have concluded so in a trade study). We know that there will be receptionists. (We just haven't decided yet whether they are shared by all professionals and how many of them there are.) We know that there will be a main entrance, and the privacy requirement has all but mandated a staff/trade entrance. We know that there is a shopping-mall parking lot and much more. If we start to describe those aspects in purely functional terms, we need to take care not to become very vague and ambiguous. Sooner or later, a physical description is required, and it cannot be left to the contractor to make choices that are really the responsibility of the customer.

In our Domestic Security Alarm, we know that the basic building blocks will be those in the system block diagram in Figure 8.13, described as follows:

- *Communications:* The external communications means connects the alarm to the monitoring agent (if one is present) and the resident. We saw already that ACME can conduct a trade study to consider how the current public communications infrastructure allows direct communications with the resident.
- *Power and backup power:* Recall also that consideration of the system interfaces derived the requirement for backup power. ACME can now conduct a trade study to select the appropriate technology. As this will

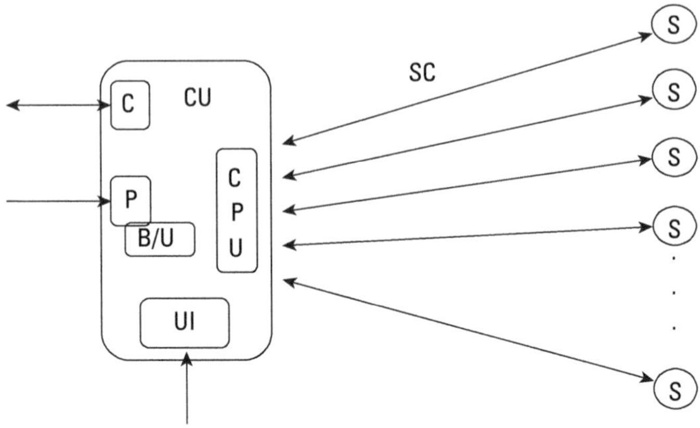

C: Communications
P: Power
B/U: Backup power
CU: Control unit
CPU: Central processing unit
UI: User interface
S: Sensor
SC: Sensor communications

Figure 8.13 A generic system block diagram for a domestic security alarm.

be part of the project scope, it needs to be decided as soon as possible, particularly if a fixed-price contract is to be signed.

- *Control unit:* The control unit will be in a box of some sort. Should it be lockable; should its appearance be polished enough that it can be on show? These and other questions are important considerations before going to tender and signing the subsequent contract. In an in-house development these issues affect the build-to price as well as the company's ability to produce the alarm in-house.
- *Central processing unit:* The processing unit itself does not necessarily have to be chosen in terms of the exact hardware, but ACME needs to consider the price break point for certain functionality and, in particular, for terminating a certain number of sensors.
- *User interface:* The user interface is necessary for setting and disabling the alarm as well as for maintenance and zoning.
- *Sensors:* ACME can conduct a trade study for the sensor technology: infrared versus optical, for example.
- *Sensor communications:* ACME can also decide whether the sensors should be wired, wireless, or both. Again, that decision will affect the project scope, as well as determine the company's ability to produce the alarm in-house.

Using those basic building blocks, therefore, the conceptual designers are more able to make decisions about the risks of certain technologies. Recall from earlier that there were trade studies to be conducted regarding communications (PSTN or SIM card) and user interface (authentication means). ACME could also address sensor technology and sensor communications. In an in-house development, they will also be keen on identifying opportunities for reuse of components from existing products, and the production manager will be interested in ensuring that new components will be able to be manufactured in house.

8.4.4 Technological Growth

Another advantage to developing a generic block diagram is to examine how to accommodate technological growth. That is very difficult if the SyRS description is purely functional and becomes much easier if the description is semiphysical. For example, Figure 8.13 can be used to identify what, if anything of that semiphysical architecture would change over the next five to 10 years. The following are examples of some of the changes that may be considered:

- *Communications:* The external communications has traditionally been the PSTN over a four-wire circuit. That interface inside the control unit will be a small circuit board connected to the main board. If the internal interfaces are designed appropriately, it ought to be relatively straightforward to replace the small communications interface board with another based on a cellular network SIM, for example (at the same time replacing the external line interface with a small cellular antenna). The new communications circuit board will be the same size (if not a little smaller) than the old, so there will be no need to change the control unit casing.
- *Power and backup power:* It is highly unlikely that the power interface to the domestic power supply will change in the next couple of decades. The technology for backup power will change, but that new battery will most likely have a smaller form factor for the same power rating. Again, the control unit casing should be unaffected.
- *Control unit:* The control unit casing will need to accommodate any internal technological changes over a couple of decades. Most technological changes in the internal components will result in smaller components, so the control unit should be sufficient in size for the foreseeable future. If there was some concern, designers could simply specify a larger box that would accommodate any anticipated growth.

- *Central processing unit:* Processing units in the future will likely be smaller and draw less power for the same functionality, so the same form factor will no doubt accommodate future processors that are more powerful.
- *User interface:* The user interface is not likely to change from some form of keypad. However, if it does change to a new technology, the interface to the main control unit will remain the same, and a new user interface component will replace the old.
- *Sensors:* ACME can design a sensor interface protocol that can accommodate a binary response from sensors (detected or not), an image response, or a series of images (streaming video), along with a standard physical interface that will accommodate any future sensor.
- *Sensor communications:* Wired and wireless connections between sensors and the control unit are already greater in capacity than is required for high-resolution transfer of video signals and so will be able to cope with any growth in sensor technology.

8.5 C3.4—Define TPMs

The parametric requirements determined during the preceding stages assist in developing the quantitative criteria (Figure 8.1, activity C3.5), or TPMs, that the system must meet. TPMs are identified early in the system development effort and are continually monitored and tracked throughout the system development as a means of managing the risks associated with system development.

The first step in identifying TPMs is to identify the quantitative parameters that require tracking throughout the project. Once the parameters have been identified, they should be prioritized in terms of their importance as viewed by the customer. A second list of TPMs may be established, prioritized, and maintained by the contractor. As is often the case with subjective issues such as risk, a second perspective is always valuable. Both lists will imply priority to potential suppliers of the system during its design and development. This level of priority will assist the designer in determining the design emphasis to be placed on achieving each of the TPMs and the potential for trading off the performance of one TPM against the performance of another.

Throughout the design process, a list of TPMs and associated metrics should be maintained along with the priority, the "benchmark" objective, and the current level of achievement and projected/estimated performance. It is normal for some TPMs to move down in the priority order or drop from the list entirely as the design proceeds due to the fact that their ultimate achievement is no longer uncertain. It is also normal for activities that were not considered risky during early stages of design to be added or moved up in priority due to previously unforeseen complications. This assists both the customer and the supplier in the identification and mitigation of technical risks during the acquisition phases. Recall that IEEE-STD-1220 [8] states that appropriately selected TPMs can be used to: assess conformance to requirements, assess conformance to levels of technical risk, trigger development of recovery plans for identified deficiencies, and examine marginal cost benefits of performance in excess of requirements.

Bands of acceptable variation from the expected value will also be determined and agreed upon so that unnecessary risk management actions are not instigated at the first sign of a very minor variation. The variation bands will be typically large in the early stages of the design but will become increasingly narrower as the design matures.

8.6 C3.5—Conduct System Requirements Review (SRR)

As discussed in Chapter 3, SRRs (Figure 8.1, activity C3.6) may be conducted periodically throughout conceptual design to verify and approve the various iterations of the sets of system-level requirements. As can be seen from the locations of each review in Figure 8.1, the aim of the SRRs is to monitor and approve progressively the system-level requirements that are developed on the way to the Initial FBL. Progressive reviews allow the requirements analysis effort to continue to lower levels in the logical hierarchy in the RBS by providing validation of the higher levels of abstraction, allowing for a firm start point for the subsequent analysis. The SRR also reviews other information such as manufacturing plans, design schedules, and personnel plans. The number of SRRs depends on the size and complexity of the system.

Figure 8.14 (after Kotonya and Sommerville [9]) illustrates that, while there is a natural flow to the activities, iteration is required. Although a significant number of requirements can be elicited and elaborated early on, the set must be analyzed and negotiated and then allocated to logical groupings so that the work to-date can be validated. Having started at the inner point of the spiral, therefore, the first pass elicits informal requirements and negotiates agreed-on requirements that are then allocated and validated to form an early draft of the set of requirements available for stakeholder review. Requirements definition then continues through another pass of elicitation and elaboration; analysis and negotiation; allocation; and validation to a second review in Figure 8.14. The process continues for as many passes as required. Note that,

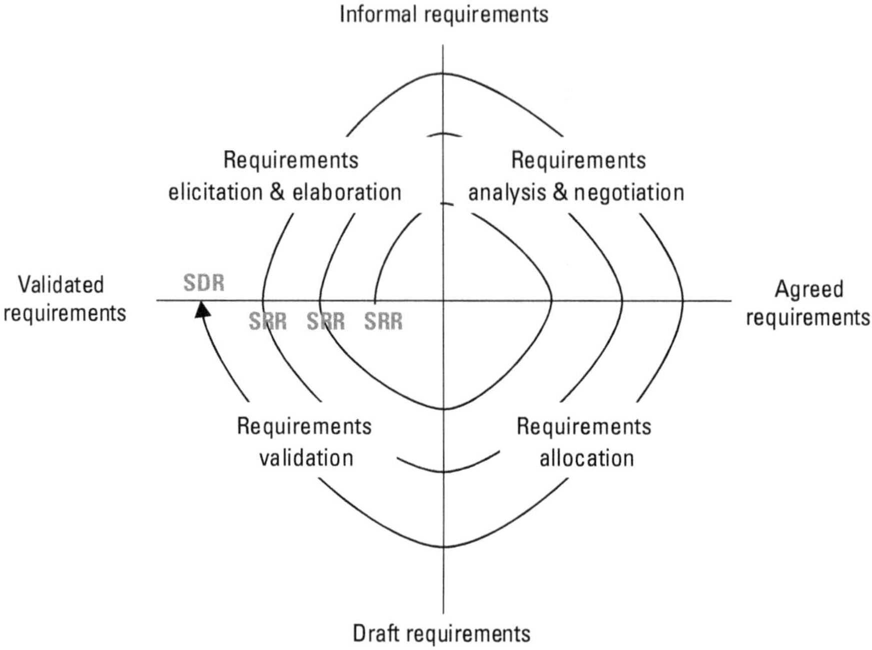

Figure 8.14 Iterative requirements-engineering activities.

at each pass, we have a draft set of requirements that can be validated at an SRR. Once designers are satisfied that the set of requirements is sufficiently complete at the system level, the last review is the SDR.

The number of iterations required to be completed before SDR is described as part of the complete requirements-engineering process which should be articulated in the systems engineering management plan (SEMP), which, in addition to the detail described in this section, should include what activities are to be carried out, who is responsible for them, the inputs and outputs, and what tools are to be used.

8.7 C4—Conduct System-Level Synthesis

System design has now progressed to the stage where some of the system-level design decisions can be made based on the draft system requirements. Synthesis at the conceptual design level establishes a system configuration that is representative of the final system form at the subsystem level. At this point, the configuration is not assumed to be final, as the design may be adjusted later in the design process as the design becomes physical.

In system-level synthesis, potential system solutions are developed (Figure 8.15, activity C4.1). Information is collected (Figure 8.15, activity C4.2) regarding how the potential solutions meet the draft SyRS. The potential solutions are then evaluated (Figure 8.15, activity C4.3) by the acquirer, which requires the development of suitable evaluation criteria as well as an evaluation framework. Any discrepancies in the systems engineering products are noted and fed back to previous activities. The preferred solution is chosen (Figure 8.15, activity C4.4),and the system architecture is generally captured in a SDD [10] or the equivalent, which is used to show compliance against the requirements in the draft SyRS.

The system synthesis activities described in activities C4.1 to C4.4 in Figure 8.15 can also be illustrated by Figure 8.16 in which architectural options

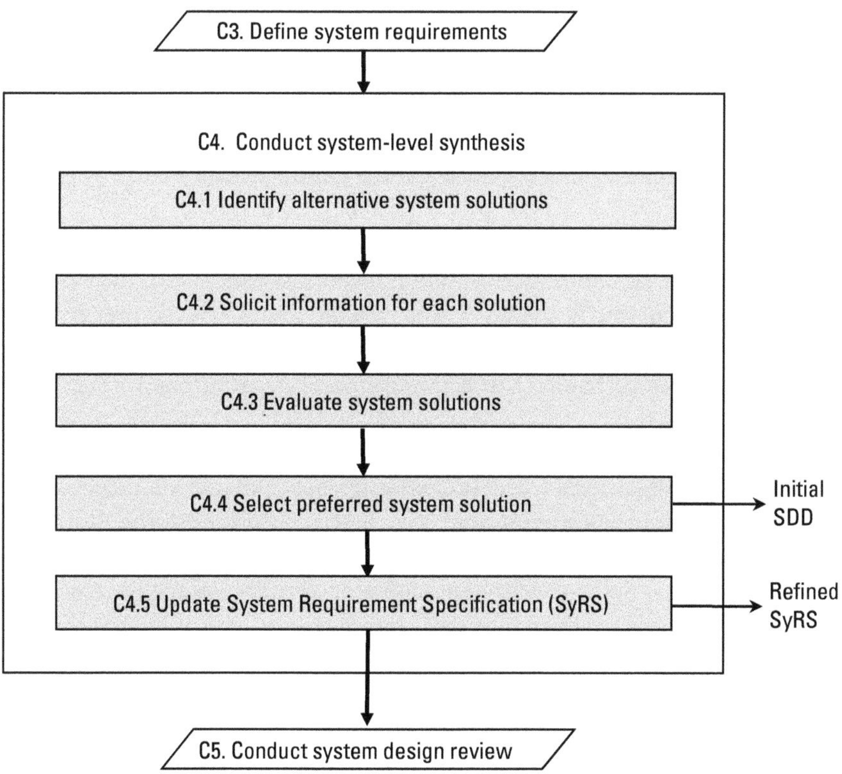

Figure 8.15 A summary of C4—Conduct system-level synthesis.

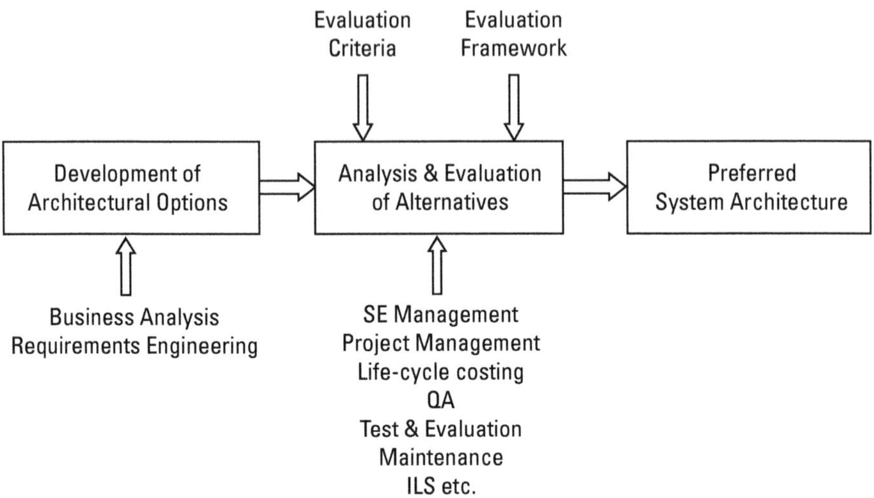

Figure 8.16 System synthesis activities.

are developed; they are evaluated in the context of an evaluation framework using selected criteria with input from the stakeholders involved; and a preferred system architecture is developed.

Although system-level synthesis can be conducted internally by the customer, it often takes the form of a RFT in which tenderers are provided with a draft SyRS and an SOW. Each tender response shows how the solution offered provides a feasible system-level design that meets (complies with) the SyRS and how the work requested in the SOW will be conducted. The acquirer evaluates the tender responses before selecting the preferred system-level solution and inviting the preferred tenderer to negotiations. If negotiations are successful, a contract is signed.

The draft SyRS will contain margins of acceptable performance and priorities providing a sufficiently broad design space to allow a range of solutions to be tendered. As part of tender evaluation (synthesis), the draft SyRS is tested against a range of feasible system solutions and a preferred solution is selected. The draft SyRS requirements now need to be refined to remove the ranges of acceptable performance and to reflect more precisely what is to be achieved by the selected system and the minimum acceptable level of performance. New system-level requirements may need to be added to the SyRS and errors or conflicts in existing requirements need to be clarified. The priority of requirements is often removed so that each requirement statement is essential. As feasible alternatives are presented, the customer may also need to revisit the

BNR/SNR and seek endorsement from the business/stakeholders for proposed scope changes to the draft SyRS if any of the endorsed requirements are not feasible within the approved schedule or cost constraints.

So, the final activity in process (Figure 8.15, activity C4.5) refines the SyRS, updates BNR and SNR, and establishes the initial FBL. The main products of synthesis are the refined SyRS and the broad system solution described in the SDD.

The SyRS is perhaps the most important systems engineering artifact, because it becomes the source of reference for all the subordinate specifications that are produced during later stages of the design process. A solid SyRS forms an effective foundation for the remainder of the design and development effort. Errors or omissions in the SyRS flow into the remaining design effort. The later these errors are discovered, the more expensive and time-consuming the rectification will be. Additionally, the SOW is perhaps the most important project management artifact as the work described in it will be translated into the work packages in the developer's WBS—again, any errors or omissions will have consequences in terms of the project budget and schedule.

8.8 C5—Conduct SDR

At the end of conceptual design, the SDR provides the following from a systems engineering perspective:

- Formal confirmation that the logical design meets the business and stakeholder requirements;
- Formal record of design decisions and acceptance;
- Formalized communication of the intended design approach, as articulated by the SDD or equivalent, to the major players in the design effort;
- Approval of the V&V plans for the system;
- Approval of the SEMP and supporting plans.

In addition to review of the systems engineering effort, the following review activities will most likely be performed at SDR to support project management:

- Confirmation that the system to be procured aligns with the customer's organizational goals;
- The SOW is refined and the work required is better understood;

- The PMP is refined;
- Other early plans such as production plans are refined;
- Cost estimates are refined;
- The schedule is refined and is confirmed to be consistent with the cost and risk goals for the project;
- Confirmation that all required project resources are available.

Figure 8.17 summarizes the activities involved in the conduct of the SDR. As with all reviews, it is essential to establish the entry and exit criteria for the review—that is, it must be clear what must be in place before the review begins and what must have been achieved before the review can be considered to be complete. Consequently, the review cannot begin until the entry criteria have been met (Figure 8.17, activity C5.3). Part of that confirmation is to ensure

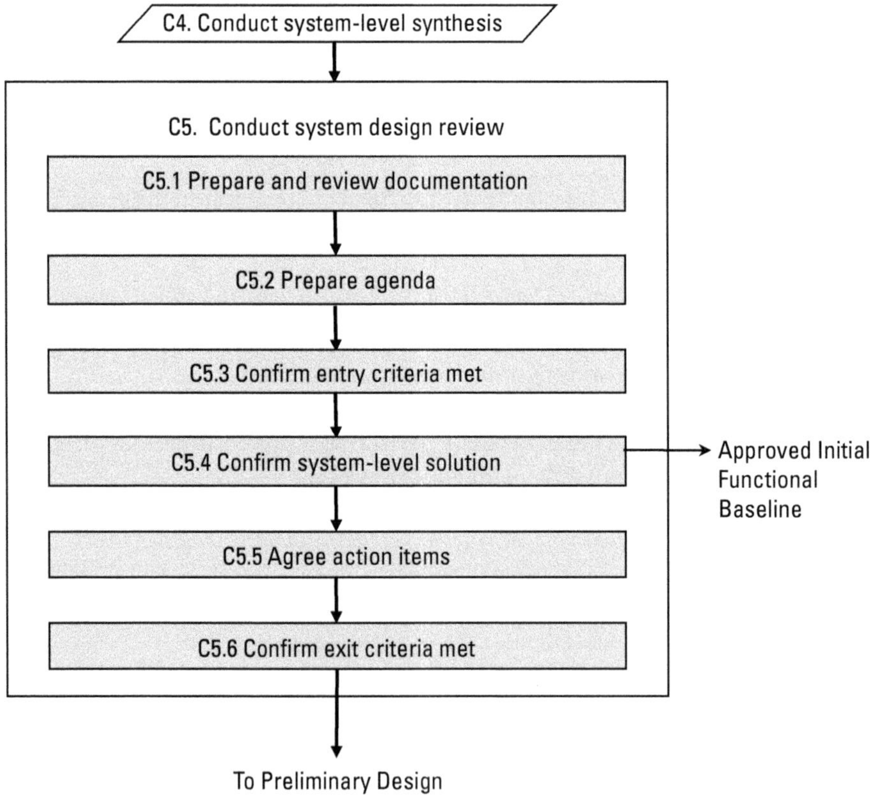

Figure 8.17 A summary of C5—Conduct system design review.

that all of the necessary artifacts have been prepared and reviewed (Figure 8.17, activity C5.1) in advance of the review and that all parties have agreed to the agenda for the review (Figure 8.17, activity C5.2).

The principal purpose of SDR is to confirm that system solution described in the SDD meets the requirements of the refined SyRS (Figure 8.17, activity C5.4). If all is satisfactory, the SyRS is approved and baselined as part of the initial FBL. It is rare that SDR ends with all outstanding issues resolved during the review since many will require some investigation and analysis. Rather than hold up subsequent design, it is common for a number of action items to be agreed on (Figure 8.17, activity C5.5) with agreed-on timelines to account for any outstanding action (providing, of course, that resolution of those issues will not affect the remainder of the design). These outstanding actions are then completed in parallel with the early preliminary design activities and are reviewed for completeness in conjunction with a subsequent review or audit. Before the review can be considered to be complete, the exit criteria must be met (Figure 8.17, activity C5.6).

The boundary between conceptual design and preliminary design is often not clear. From a system design perspective, SDR aims to provide demarcation by approving the SyRS and establishing the initial FBL from which the remainder of the design will follow. For simplicity, we are focused on the SyRS, but we should keep in mind that the initial FBL includes considerably more than the SyRS in terms of systems engineering and project management artifacts. At the very least, there will also be the PMP, the SEMP, the major plans outlined in Chapter 8, along with production plans. Consequently, one of the other major aims of SDR is to mark the transition of responsibility from the customer/acquirer/tasking-activity to the contractor/developer/performing-activity.

8.9 Summary

This chapter addresses the final aspects of conceptual design: the definition of system requirements, which starts with the endorsed SNR and decomposes/derives them into system requirements in the SyRS. The aim of system requirements definition is to determine what the system must do in order to meet the stakeholder needs and requirements—that is, to describe the requirements at the system level and be able to relate the logical design back to the SNR, particularly the StRS. System requirements definition, therefore, uses elaboration to bridge from stakeholder requirements to system requirements. Chapter 9 investigates the writing of those requirements.

Review Questions

1. List and briefly describe the activities involved in the CDM process C3—Define system requirements.
2. Briefly describe the role and importance of performance requirements.
3. Briefly describe the role and importance of verification requirements.
4. Briefly describe the role and importance of assigning attributes to each requirement.
5. Briefly describe the role and use of TPMs.
6. Briefly describe the role and conduct of SRRs.
7. Briefly describe the CDM step C4—Conduct system-level synthesis.
8. Briefly describe the CDM step C5—Conduct SDR.

Endnotes

[1] EIA-632, *EIA Standard—Processes for Engineering a System*, Arlington, VA: Electronic Industries Association, 2003.

[2] Olver, A. M., and M. J. Ryan, *On a Useful Taxonomy of Phases, Modes, and States in Systems Engineering*, Systems Engineering and Test and Evaluation (SETE) Conference, 2014.

[3] Wasson, C. S., *System Phases, Modes, and States—Solutions to Controversial Issues*, presented at INCOSE 2011, 2010.

[4] MIL-STD-498B, *Data Item Description DI-IPSC-81431A*, 2000.

[5] International Council on Systems Engineering (INCOSE), INCOSE-TP-2010-004-04, *Guide to Writing Requirements*, 2024.

[6] Wheatcraft, L. S., M. J. Ryan, and T. E. Katz, *Needs and Requirements Manual: Needs, Requirements, Verification, Validation Across the Lifecycle*, Hoboken, NJ: John Wiley & Sons, 2024.

[7] Sources of information on the content of the SyRS can be found in the following:

Data Item Description (DID), *System/Subsystem Specification (SSS), DI-IPSC-81431*, DD Form 1664, US Department of Defense, April 1989.

ECSS-E-10A, *Space Engineering—System Engineering*, Noordwijk: The Netherlands: European Cooperation for Space Standardization, 1996.

MIL-STD-961E, *Department of Defense Standard Practice for Defense and Program-Unique Specifications—Format and Content*, Washington DC: US Department of Defense, August 1, 2003.

MIL-STD-490A, *Military Standard Specification Practices*, Washington, DC: US Department of Defense, 1985.

[8] IEEE-STD-1220, *IEEE Standard for Application and Management of the Systems Engineering Process*, New York: IEEE Computer Society, 2005.

[9] Kotonya, G., and I. Sommerville, *Requirements Engineering: Processes and Techniques*, Chichester, England: John Wiley & Sons, 2000.

[10] Information on the content of the SDD can be found in Data Item Description (DID), *System/Subsystem Design Description (SSDD), DI-IPSC-81432, DD Form 1664*, US Department of Defense, April 1989.

9

Requirements Writing

9.1 Introduction

The preceding chapters have shown how conceptual design contributes to the derivation of requirements. From the mission statement through to the SyRS, it is essential that we have a well-balanced, complete, functional description of the system. When writing the requirement set and implementing it (especially as part of a contract), the focus now shifts to the individual requirements, each of which has to be written so that it can be legally binding as part of the contract. As discussed in Chapter 2, a good requirement is necessary, singular, correct, unambiguous, feasible, appropriate-to-level, complete, conforming, and verifiable. We also learned in Chapter 2 that a set of requirements should be complete, consistent, comprehensible, feasible, and able to be validated.

This chapter discusses how to write each requirement to achieve those properties—we look at the generic structure for a requirement statement as well as some general guidelines for writing good requirements.

9.2 Some Guidelines for Writing Good Requirements

The INCOSE *Guide to Writing Requirements* provides 42 atomic rules for writing good textual requirements [1]. The rules have been written so that each addresses a single aspect of requirement quality so that automatic quality-checking software can implement the rule as part of a requirements management tool. Here we combine them into logical human-friendly groupings to ease discussion. So, from a human perspective, writing well-formed requirements can be achieved by using the following guidelines [2]:

- Use a style guide;
- Use a standard template for a single sentence;
- Use a glossary to define terms;
- Use correct English expression;
- Use a subject appropriate to the level;
- Use explicit lists;
- Use the active rather than the passive voice;
- Use the definite article;
- Use the same term to mean the same thing;
- Avoid vague words, terms, and expression;
- Avoid superfluous words;
- Avoid the use of conjunctions;
- Avoid unbounded statements;
- Avoid escape clauses;
- Use a convention for logical condition;
- Avoid unnecessary precision;
- Be explicit when necessary;
- Use units, ranges, and tolerances;
- Avoid the use of not;
- Avoid cross-reference to other requirements;
- Avoid the use of pronouns;
- Use diagrams wherever appropriate;
- Introduce logical groups of requirements with informative statements;
- Ensure that each requirement is verifiable.

9.2.1 Use a Style Guide and Template for the Requirement Statement

A project-wide style guide should be used for individual requirements and for sets of requirements. The style guide defines the following:

- Structure of requirement statements;
- Structure of requirement sets;
- Attributes to be appended to the requirements statement.

Each requirement must be a necessary, short, definitive statement. The ability to interpret requirements correctly is aided greatly by the use of a standard format for the requirement statement—consequently, many organizations have a standard template for the structure of a requirement statement. The generic structure of a requirement statement is shown in the following:

[subject] + [priority] + [verb] + [object]

The elements of this most basic form of requirement are described as follows:

- *Subject:* The subject is the entity being specified. In a system-level specification, the subject is almost invariably the system itself. In lower level specifications, the subject could be a subsystem, component, or an actor within the system.
- *Priority:* It is strongly recommended that the priority in all requirements is "shall." Other words such as "should" or "may" could be used to identify requirements that are not mandatory, but it is better practice to always use "shall" and to use the priority attribute of the requirement to identify the level of priority. "Will" should never be used in a contractual document, especially when "shall" is also used.
- *Verb:* The verb defines the action to be undertaken by the subject of the requirement. Examples include "display," "calculate," and "process."
- *Object:* The object is the entity being acted upon by the subject.

Consider, for example, the following:

[The System] + [shall] + [display] + [the Time]

We will see shortly that *System* and *Time* have capitalized first letters because those words have been defined in the Glossary.

In some cases, the subject can only conduct the action of the requirement under certain conditions, so the requirement is expanded to include a [condition] under which the action occurs, as follows:

[condition] + [subject] + [priority] + [verb] + [object]

For example, the example requirement above needs to be expanded further if it is to be useful. The system cannot achieve the requirement when it is turned off, so, we could use a condition as part of the requirement as follows:

[When in the "ON" State,] + [the System] +
[shall] + [display] + [the Time]

Further, some qualification must often be made to the action, so a full requirement might have the following form:

[condition] + [subject] + [priority] + [verb] + [object] +
[qualification]

In our example, then we could add a qualifier as follows to avoid the location display obscuring the detail of the map:

[When in the "ON" state,] + [the System] +
[shall] + [display] + [the Time] +
[without obscuring the Work_Space]

Again, *Work_Space* has capitalized letters because the term has been defined in the Glossary.

There are further extensions of the template, but these should probably be avoided as the requirement sentence becomes too long and runs the strong risk of becoming ambiguous.

Note also the following statement form recommended by INCOSE:

[condition] + [subject] + [priority] + [verb] + [object] +
[qualification]

This is not the only structure. Some would prefer that the qualification is first and the condition in last as follows:

[qualification] + [subject] + [priority] + [verb] + [object] +
[condition]

Others would prefer that, as in the following, the subject always comes first and the condition second:

[subject] + [condition] + [priority] + [verb] + [object] +
[qualification]

The organizational style guide would dictate the format. In many senses, it doesn't matter which one is selected as long as all requirements conform to the same pattern.

9.2.2 Use a Glossary to Define Terms

The English language includes several hundred thousand words, which makes it a rich language with very similar (but subtly different) meanings for many words. However, this richness has a downside when we are seeking precision in legally binding requirement statements. In a requirement set we cannot allow there to be any doubt as to what is referred to by the use of a single English word—every important word must therefore be made explicit.

A useful way to ensure that a common English word has a precise meaning throughout the requirement set is to use a glossary in which the words are defined. Terms from the glossary are capitalized throughout the requirement set to signify that the word or term being used has a specific meaning in the context of the set of statements. It is also helpful to inform the reader of the use of the glossary by noting at the beginning of the specification that any capitalized term takes on the meaning as defined in the glossary.

For example, we saw before the statement:

The System shall display the time.

This is ambiguous because it is not obvious what time zone is being referred to, to what resolution the time must be displayed, in what format, and so on. Because it is then unwieldy to add all these qualifications to the one sentence, it is much simpler to pause, define Time in the glossary (to include the attributes of time zone, resolution, and format) and rewrite the statement as follows:

The System shall display the Time.

Of course, there may be more than one time that needs to be referred to—such as current time, local time, or arrival time. When more than one word is being defined—that is, a term is being defined—it is often useful to use an underscore to signify that the definition is of a term, not of two adjacent words. Consider the following, for example:

The System shall display the Current_Time.

This signifies that Current_Time is a glossary item, whereas the use of the following:

The System shall display the Current Time.

may imply, inappropriately, that *Current* and *Time* are two separate glossary items.

The glossary is also a useful place to define acronyms and abbreviations, which then must be used consistently throughout the requirement set. As with other glossary items, the use of different acronyms and abbreviations for the same item implies that the two items being referred to are different—which leads to ambiguity.

9.2.3 Use Correct English Expression

Just as in normal written English, the use of incorrect grammar, spelling, and punctuation leads to ambiguity. Text-based requirements must therefore be written using correct English expression to ensure clarity in understanding. Consider, for example, the following:

The Weapon_Subsystem shall store the location of each Ordinance.

This is unacceptable, because it contains "*Ordinance*" as the misspelling of "*Ordnance*," which leads to ambiguity—the location to be stored is of the weapons and munitions (ordnance) not the regulations (ordinance). The requirement must be rewritten as follows:

The Weapon_Subsystem shall store the location of each Ordnance.

In this following example, the writers intended to state that the CVR would not emit light but chose the incorrect word:

The *Cockpit_Voice_Recorder shall be unlit.*

This mistake inadvertently specified that the device should have no light fall upon it (be unlit), a completely different requirement.

9.2.4 Use a Subject Appropriate to Level

Our first consideration is that the subject of a requirement statement should be appropriate to the level at which it is stated. System-level requirements should therefore begin with, *The <system> shall. . .*; subsystem-level requirements with *The <subsystem> shall. . .*; component-level statements with *The <component> shall. . .*; and so on.

When writing requirements at any given level, the writer must consider whether the current level is the appropriate level to state the requirement, or whether the requirement be better expressed at a lower or higher level (and, of course, verified at that level). So, requirements at the system level should not be discussing subsystems issues unless there is a good reason to do so.

That also means that most higher level requirements are logical not physical—that is, they do not state a solution, because it would normally not be appropriate to do so at those higher levels.

9.2.5 Use Explicit Lists

When a list of conditions is given, it may not be clear whether all the conditions must hold (a conjunction) or whether any one of them must hold (a disjunction). The wording of the requirement should make this clear. Consider, for example, the following:

> *The Audit _System shall allow the Clerk to change the Status of the Item when: the Clerk originated the Item; the Clerk is the Actionee; and the Clerk is the Reviewer.*

It is not clear in this statement whether the action can be allowed when any one of the conditions hold or when all conditions hold. When all conditions hold, we should write the following:

> *The Audit_System shall allow the Clerk to change the Status of the Item when the following conditions are simultaneously true:*

- *the Clerk originated the Item;*
- *the Clerk is the Actionee; and*
- *the Clerk is the Reviewer.*

When any one of the conditions hold, we should write the following:

> *The Audit_System shall allow the Clerk to change the Status of the Item when one of the following conditions is true:*

- *the Clerk originated the Item;*
- *the Clerk is the Actionee; and*
- *the Clerk is the Reviewer.*

9.2.6 Use the Active Rather Than the Passive Voice

The "voice" of a verb expresses the relationship that the verb conveys between the subject and the object of the sentence. In a sentence written in the active voice, the actor that performs the action of the verb is the subject of the sentence; in the passive tense, the subject is the target of the action. Consider the following two examples, in which the first is in the active voice, and the second is in the passive voice:

The System shall display the Account_Balance.

The Account_Balance shall be displayed by the System.

The requirement statement must make obvious what or who is performing the function—it is the subject of the requirement that is also to be the subject of the verification statement. (If it is not clear who/what is responsible for performing the action, it won't be clear as to who/what is to be tested to ensure that the function is able to be performed.) The difficulty with the passive voice is that the true subject of the requirement is not always obvious.

In the example we have given here, the difference may seem to be one of semantics, because both sentences are most likely to be interpreted the same way. The difficulty with the use of the passive voice arises, because it allows the author to ask for an action without specifying the subject at all—for example, they can simply state the following:

The Account_Balance shall be displayed.

The discipline of writing the statement in the active voice avoids this imprecision because the author has to write the following:

The {. . .. . .} shall display the Account_Balance.

and they can't continue until they determine who/what that subject should be.

Quite often when a passive requirements is rewritten as an active requirement and the subject is inserted, it transpires that the requirement is not appropriate to level because the subject is not the subject of the current level of specification.

9.2.7 Use the Definite Article

Requirement statements should use definite article "the" instead of indefinite article "a." Use of the indefinite article is tantamount to using "any one of," which is not verifiable. For example, if the requirement refers to "a user," it is unclear whether it means any user or one of the defined users for which the system has been designed. This causes further confusion in verification.

For example, a baby is arguably "a user" of baby food, but the system would fail if the test agency sought to verify that a baby could independently perform any action associated with baby food. On the other hand, if the requirement refers to "the User," the reference is explicitly to the nature of the user defined in the glossary.

For example, consider the following requirement:

The System shall allow a student to Enroll on the first encounter, at the first attempt.

This is easily failed if the test agency can find any single person who is unable to enroll. (Perhaps we could recruit one of the babies we used as test subjects in the previous example.) We need to rewrite to be explicit as follows:

The System shall allow the Student to Enroll on the first encounter, at the first attempt.

And we are even more explicit by defining "Student" in the Glossary.

9.2.8 Use the Same Term to Mean the Same Thing

As well as the development of a glossary, it is important to be consistent in the use of words and terms (whether they are included in the glossary or not). In a set of statements that is purported to be precise, the reader must assume that, if the same word/term is used in two different sentences, the author deliberately used that word/term to be the same in both sentences. In a similar vein, if the author uses different words/terms in each statement, the reader must infer a different meaning (why else would the author have chosen to use a different word/term?).

Commonly, however, requirement sets contain statements in which the same term is used to refer to different things, or different words are used to refer to the same thing. Consider, for example, in a subsystem specification, the following three requirement statements:

The Radio shall operate . . . .

The Receiver shall report . . . .

The Terminal shall contain . . . .

They probably all refer to the same subject, and the statements must be modified to use the same word (or, if they are meant to be different, the words must be defined to be so).

Similarly, the following two requirements:

The System shall display the Position of each Vehicle . . . .

The System shall display the Location of the Vehicle . . . .

use different words for the location of the object of the statement, but they probably mean the same thing, and one requirement is therefore most likely redundant.

9.2.9 Avoid Vague Words, Terms, and Expressions

Ambiguity in a requirement statement is also caused by the use of vague words terms and expressions such as: *any, many, some, several, a lot, a few, approximate, nearly, about, almost, close to, significant, flexible, scalable, expandable, usual, typical, sufficient, adequate, appropriate, efficient, effective, reasonable, prompt, fast, quick, routine, common, generic, customary, maximum, minimum, minimal, optimum, nominal, light, heavy, relevant, large, small, easy-to-use, high-speed, medium-sized, best practices,* and *user-friendly.* This is really only listing the common mistakes—there are many others.

The following are some requirements that need to be fixed:

The User _Interface shall be flexible.

The User _Interface shall be user-friendly.

The User _Interface shall require minimal training.

So, let's look at some common mistakes and see how to fix them.

One of the most common uses of vague words is when "all," "any," or "both" are used to refer to a set of things. The use of these words has two problems: first, it is not clear when we say "all" for example, whether we are referring to each one of the elements of the set or the entire set simultaneously. Second, the set is often not defined, and the requirement refers to something such as "all users," which is unverifiable since we would need to confirm that each of an infinite set of users could perform the function.

So, the best practice is to use "each" instead of vague words such as "all," "any," or "both," which are not sufficiently precise to be verifiable. For example, consider the following two statements:

The Record_Subsystem shall store the location of all Line-Items.

The Record_Subsystem shall store the location of any Line-Items.

These are unacceptable because they contain the vague terms "all" and "any." We need to rewrite to as follows:

The Record_Subsystem shall store the location of each Line_Item.

A second common issue relates to use of words that are vague because they can't be measured: such as *prompt, fast, routine, maximum, minimum, optimum, nominal, easy to use, close quickly, high speed, medium-sized, best practices,* and *user-friendly.* The vague term must be replaced by a specific quantity that can be measured. For example, consider the following:

The Radio shall use minimum power.

We cannot measure "minimum," so we need to state a measurable target as follows:

The Radio shall consume less than 50W.

The act of quantification is even more important than making sure that the quantity can be measured. It behooves the designer at this level to think seriously about their requirement—not just say, "It must be as light as possible," "It must be as small as possible," "It must generate as little heat as possible," and so on. At the very least, quantifiable bounds must be stated—or else the requirement cannot be verified.

Some vague words are just simply ambiguous: *significant, typical, sufficient, adequate, appropriate, efficient, effective, reasonable, routine, common, generic, customary,* and *relevant.* The presence of these words normally indicates that the writer has not bothered to consider what is actually required. For example, consider the following:

The Radio shall conform to best practices for spurious emissions.

Best practices are relative, open-to-interpretation, and, arguably, in the eye of the beholder—a specific, measurable level must be stated, and specific reference must be made to the standard that describes best practice, as follows.

The Radio shall limit Spurious_Emissions in accordance with Clause 7 of Standard XYZ.

Note also the use of the Glossary to be specific about what is meant by Spurious_Emissions.

9.2.10 Avoid Superfluous Words

In addition to introducing ambiguity through the use of vague words, confusion in the mind of the reader can also be introduced by using superfluous words—that is, words that are not necessary and take away from the meaning of the statement. For example, we often use the construct ". . . shall be able to . . . ." Unfortunately, this introduces ambiguity into a requirement. Consider, for example, the following:

The Record_Subsystem shall be able to store the location of each Line_Item.

This is unacceptable because "be able to" can imply that the Record_Subsystem does not currently have that capability but might be "able to" do it in the future. So, for example, if the customer inspects the Record_Subsystem and notes that it can't record the location, the developer might be able to say, "Well it doesn't currently have that feature but is able to—all you have to do is buy this plug-in." Such a requirement is therefore ambiguous and cannot be verified. So, we should remove the superfluous words as follows:

The Record_Subsystem shall store the location of each Line_Item.

This form now states that the function is always available. If the requirement is only to be true under certain circumstances, we should add a condition clause (use of "be able to" is a strong indication of the lack of a suitable condition clause in the requirement), as follows:

When in the XY_Mode, the Record_Subsystem shall store the location of each Line_Item.

Another form of superfluous words results when we include additional verbiage associated with describing the purpose of the requirement. That is not to say that we should not consider the purpose—the articulation of the purpose of the requirement is very important but it should be included in the Rationale attribute of the requirement expression, not the requirement statement itself. Consider, for example, the following:

The Record_Subsystem shall store the location of each Line-Item, so that the Line_Item can located easily by Users.

This is unacceptable because it contains an unnecessary description of the purpose of the requirement. The additional words should be removed.

The Record_Subsystem shall store the location of each Line_Item.

Parentheses (brackets) are often indicators of the inclusion of additional verbiage. Consider, for example:, the following:

The Record_Subsystem shall store the location of each Line_Item (so that the Line_Item can located easily by Users).

This is unacceptable because it contains an unnecessary description of the purpose of the requirement. The additional words should be removed as follows.

The Record_Subsystem shall store the location of each Line_Item.

Again, the articulation of the purpose of the requirement is important, but it should be included in the Rationale attribute of the requirement expression, not the requirement statement itself.

9.2.11 Avoid the Use of Conjunctions (Using a Convention for Logical Conditions)

The use of conjunctions is to be avoided so that requirement statements can be singular. We therefore should avoid use of words such as *and, or, then, unless, but, however, whether, meanwhile, whereas,* and *otherwise.* We should also avoid phrases with similar intent such as *as well as, but also,* and *on the other hand.* We should also avoid the oblique (/) symbol (as in *and/or*). The presence of conjunctions in a requirement usually indicates that multiple requirements should be written. Additionally, use of such words creates ambiguity.

For example, looking at the first issue of multiple requirements, consider the following statement:

The System shall display the Location and Identity of the . . . .

This is in fact stating two requirements (through the use of the conjunction "and") that must be split as follows:

The System shall display the Location of the . . . .

The System shall display the Identity of the . . . .

Note also that the requirements must be split because the verification statement written for each requirement will often be different. For example, consider the following requirement statement:

The Vehicle shall have a Maximum_Speed of . . . and a Maximum_Range of . . . .

It must be split as follows:

The Vehicle shall have a Maximum_Speed of . . . .

The Vehicle shall have a Maximum_Range of . . . .

This is because there are two requirements, but also because we will verify the maximum range in a different manner to that we will use to verify maximum speed. (Note that considerable effort is still required to define what is meant by maximum speed and range and the conditions under which each is to be verified.)

As an example of avoiding conjunctions because they introduce ambiguity, consider the following:

The Maintenance_Staff shall fertilize and spray weeds on the Grass_Area.

As previously noted, the presence of the conjunction results in the combining of two actions, which warns us that the requirement should be split into two requirements. However, the conjunction also causes ambiguity as to whether the requirement means that the staff are to do the following:

- Fertilize and then spray the weeds that are on the grass.
- Fertilize the grass areas, and then spray the weeds on the grass areas.

Common sense would seem to imply the latter, but both interpretations are, strictly speaking, correct.

For similar reasons, the conjunction "or" should be avoided because it often hides multiple requirements, but also because it more often leads to ambiguity. Consider, for example, the following requirement:

The System shall provide an Audible or Visual Alarm.

This is ambiguous because it is a moot point whether the designer can choose which alarm to implement, or whether both alarms should exist. However, the operator can select one, or whether both exist and operate, but the system passes its verification test if only one of the alarms alerts the operator.

Before leaving our discussions, we should point out that the words "and" and "or" may be used in requirements statements when used in the context of logical operators, in which case the terms are normally written in all capitals, or italics to show that they are used as operators rather than conjunctions (. . . AND . . .; . . . and . . . OR . . .). Consider, for example, the following:

The User_Interface shall display the Location AND Identity of the Lead_Vehicle.

Note that the logical AND still leads to a little ambiguity because it is assumed that the Location is of the Lead_Vehicle. Strictly speaking, we should write the following:

The User_Interface shall display the Location of the Lead_Vehicle AND the Identity of the Lead_Vehicle.

That still leads to a little ambiguity, because it is not clear whether the AND joins words, phrases or clauses.

A convention for logical conditions uses a formal structure to ensure that it is clear which elements of the requirement are joined by the AND. Consider, the following, for example:

The User_Interface shall display {the Location of the Lead_Vehicle} AND {the Identity of the Lead_Vehicle}.

or:

The User_Interface shall display [the Location of the Lead_Vehicle] AND [the Identity of the Lead_Vehicle].

Alternatively, recognizing that Location and Identity will no doubt need to be declared in subsequent design, the requirement could be simplified, as follows, using the Glossary:

The User_Interface shall display the Lead_Vehicle_Location AND the Lead_Vehicle_Identity.

9.2.12 Avoid Unbounded Statements and Escape Clauses

Another poor practice in requirements writing is to provide loose bounds on the conditions under which the statement is to hold. Consequently, we should avoid phrases such as *and so on, etc.,* and *but not limited to.* If the authors of the requirement statement require all items in a list, each of those items must be specified explicitly. The use of phrases such as . . . *and so on* leads to ambiguity, because it allows different interpretation as to which are the items included in such an open-ended list. For example, consider the following:

The System shall display the Account_Balance, the Account_Holder, and so on.

This should be rewritten as follows:

The System shall display the Account_Balance.

The System shall display the Account_Holder.

It can be followed by additional statements if there are any other items to be displayed.

In addition to being open-ended, requirements authors are often tempted to generalize, using phrases such as *where possible, if practicable, if necessary,* and *as much as possible.* These are escape clauses that, in an acquirer-supplier relationship, may allow the supplier to have an excuse not to bother with a requirement by simply stating that, in their view, it wasn't possible, necessary, or practicable.

9.2.13 Avoid Unnecessary Precision

While we are looking for precision in requirement statements, we must be careful to avoid overspecifying by being too precise or by using absolute values inappropriately. For example, consider the following:

The System shall provide 100% Availability.

This is unacceptable because 100% availability is unachievable. Similarly, consider the following requirement statement:

The Pump_System shall maintain a flow rate of 100 gallons/second.

This is unachievable, because a pump cannot be manufactured with such accuracy. Further, since it is also impossible to manufacture a measuring system with such accuracy, the requirement is also unverifiable. Some latitude must be given as follows (in either absolute or percentage terms):

The Pump_System shall maintain a flow rate of 100±2 gallons/second.

The Pump_System shall maintain a flow rate of 100±2% gallons/second.

Other examples of inappropriate narrowness relate to being unnecessarily prescriptive. Consider, for example, the following requirement for a library system:

The System shall allow Users to search for a catalogued item by ISBN.

This is a poor requirement, because only books are cataloged by ISBN (for example, journals have an ISSN).

9.2.14 Be Explicit Where Necessary

There are some cases, however, where it is necessary to be explicit about some aspect of the system. That is particularly true if it is essential that the system meet some existing constraint, in which case the precise nature of the constraint should be stated. For example, consider the following:

The System shall utilize the YYY operating system.

or:

The System software shall be written in C++.

9.2.15 Use Units, Ranges, and Tolerances

All numbers should have units explicitly stated. For example, consider the following:

The Refrigerator shall maintain an internal temperature of less than or equal to 4 degrees.

This is unacceptable because the units are not specified (i.e., Celsius, Fahrenheit, or Kelvin). The requirement should be rewritten as follows:

The Refrigerator shall maintain an internal temperature of less than or equal to 4 degrees Celsius.

When we describe ranges, there are a few rules, listed as follows:

- If the maximum value is not included in the range, then: ". . . shall be less than . . ." or ". . . < 3."
- If the maximum value of the range is included in the range, then: ". . . shall be less than OR equal to . . ." or ". . . ≤ 3."
- If the minimum value is not included in the range, then: ". . . shall be greater than . . ." or " . . .> 1."
- If the minimum value of the range is included in the range, then: ". . . shall be greater than OR equal to . . ." or " . . .≥ 1."
- If both the maximum and minimum values are not included in the range, then: ". . . shall be greater than . . . AND less than . . ." or ". . . 1 < . . . < 3."
- If both the maximum and minimum values are included in the range, then: ". . . shall be greater than OR equal to . . . AND less than OR equal to . . ." or ". . . 1 ≤ . . . ≤ 3."

When the tolerance is the same on either side of the value, then the following can be used:

- ". . . 10 plus OR minus 2 . . .," or ". . . 10 ± 2 . . . ."

When the tolerances are unequal, then the following can be used:

- ". . . 10 plus 2 OR minus 3 . . . ."

9.2.16 Avoid the Use of "Not"

The presence of "not" in a requirement implies "not ever," which cannot be verified in a finite time. Rewriting the requirement to avoid the use of "not" results in a requirement that is clearer and is verifiable. Consider the following, for example:

The System shall not allow Unauthorized_Access.

This is unacceptable because there is no way of verifying that the system will never allow access to an unauthorized user—and it is made worse by the double negative. An improved requirement would be the following:

The System shall provide access only to Authorized_Users.

Similarly, consider the following:

The System shall not fail.

This is unacceptable because verification of the requirement would require infinite time. A better requirement would be the following:

The System shall have an Availability of greater than or equal to 95%.

9.2.17 Avoid Cross-References Including Pronouns

Another source of ambiguity results when the writer assumes incorrectly that reader has access to additional information. Since each requirement must stand alone as a complete statement, statements cannot cross-reference to other requirements in case those requirements are subsequently moved, modified, or deleted. For example, consider the following two requirements:

The System shall display the Position of the Lead_Vehicle . . . .

The System shall display the Identity of the same vehicle . . . .

The second requirement statement relies on the context provided by the first requirement, through cross-reference by using the word "same." That is probably bad practice even if the sentences remain adjacent. If, however, the two statements are separated, the second does not make sense.

Other cross-referencing words to be avoided are *below, above, aforementioned, aforesaid, following, herein, hereafter,* and *preceding.*

In terms of cross-referencing, we should avoid the use of pronouns, which are words that stand for nouns—examples include *it, this, that, he, she, they, their,* and *them.* Although the set of statements is therefore more stilted to read than general English (where pronouns avoid having to repeat the noun), each statement must stand alone, and the noun must be repeated to avoid ambiguity. (A pronoun is effectively a cross-reference to a noun in another requirement statement.)

9.2.18 Use Diagrams Wherever Appropriate

It is often much easier to describe many aspects of a system (interfaces, for example) by illustration, rather than in a large number of statements—a picture is often worth a thousand words. For example, it is much simpler and more concise to require that the system act as a station on a radio network by referring to a figure that illustrates the net diagram, primary and alternate frequencies, and so on—the alternative is to write several tens of requirement statements that effectively describe the diagram.

As well as saving space, diagrams are useful in avoiding ambiguity and confusion, particularly when the requirements are multidimensional and interdependent. For example, rather than trying to write a statement for every combination of acceptable combination of an aircraft's maximum range and maximum speed (which are related by the amount of fuel that must be carried and the burn rate), a table could be referred to, in which the pairs of acceptable parameters are defined. If the relationship was continuous, a graph may be more useful.

One of the difficulties with the use of diagrams and figures is that additional care must be taken with verification statements. While it might be simpler to write one requirement ". . . in accordance with Table 1 . . . ," it is not generally possible to write a single verification statement that can make the same simple reference.

9.2.19 Introduce Logical Groups of Requirements with Informative Statements

If we follow the rules for writing good requirement statements, the requirement set is invariably very difficult to read, because we have not allowed any of the techniques that lead to good English prose—we have short sentences with no pronouns, no synonyms, and a narrow set of glossary-defined terms. To assist the reader in understanding the requirement set, it is often useful to introduce logical groups of requirements with informative statements that illustrate to the reader the intent of that section of the specification.

These sections are not normally considered normative (that is, they are not legally binding, and we are not going to verify them in the same manner as the legally binding requirement statements), but they do greatly assist in removing potential ambiguity. It should be noted, however, that while these informative statements are not normative, they do communicate the customer's intent and therefore provide a basis for the fitness-for-purpose argument that might inform a claim against common-law warranty should the requirement set not be achieved by the supplier to the satisfaction of the customer.

9.2.20 Invoking Other Documents

There are a number of mechanisms that can be used to refer to content in other documents (other specifications, or standards): "*. . . in accordance with . . .*," "*. . . as specified in . . .* ," or "*. . . as set out in . . . .*" It does not really matter which one is chosen, although "*. . . in accordance with . . .*" is probably the most useful. Whichever is chosen, it is essential that only one is used throughout the requirement set.

9.2.21 Ensure that Each Requirement Is Verifiable

It is essential that authors, at the time of writing, ensure that each functional requirement is verifiable through the discipline associated with writing a verification statement for every functional statement. The following example illustrates the utility of considering verification as we write the requirement statement.

Imagine that we have a ticketing system, and the stakeholders are keen that it can be used easily by the general public without any detailed instructions or training. So, our first attempt at a functional statement might be the following:

The System shall be easy to use to purchase the Ticket.

Clearly, we need to state the requirement better than that—ease of use depends on the extent of available instructions, training, and experience, which are aspects we are trying to avoid. So, with a bit of consultation, we might refine the requirement as follows:

The System shall allow a member of the public to purchase the Ticket without training.

The requirement is now more useful, but it is still too ambiguous—there is still considerable doubt about what "without training" means. If we consider it from a performance point of view, we note that we can be more precise and could write the following:

The System shall allow a member of the public to purchase the Ticket on the first encounter, at the first attempt.

The requirement is better, but it is still too severe. If we consider it from a verification point of view, we note that the requirement could cause significant angst at acceptance testing—for example, the contractor could assert that the requirement should be accepted if one member of the public can be found that can use the system on their first encounter at their first attempt; the customer could equally assert that it will fail if one member can be found that cannot use it at their first attempt. We should therefore allow some latitude in the test to be more realistic, as follows. (After all, there is probably a significant percentage of the public who would never be able to use any new system on their first encounter at their first attempt.)

The System shall allow 90% of a Panel representative of the General_Public to purchase the Ticket on the first encounter, at the first attempt.

This is now a much more useful requirement. The contractor and the customer now have to agree on a panel that is representative of the public (much like jury selection), and then the test must be passed by 90% of that panel, which allows for people who are having an off day and are distracted to perform the task properly. The requirement is much fairer to both parties in the contract and is verifiable.

As an aside, as we move through the requirement set, we may find that there are a number of times that we may want to verify requirements through the use of a panel representative of the general public. In that case, we might define a glossary item *Test_Panel* and rewrite the above requirement (and similar) to use the glossary item as follows:

> *The System shall allow 90% of the Test_Panel to purchase the Ticket on the first encounter, at the first attempt.*

Review Questions

1. List the 24 guidelines for writing good requirements.
2. Outline the generic structure of a good requirement statement and describe each of the elements.
3. Briefly describe why requirement elements should be defined in a Glossary.
4. Briefly describe why correct English expression is required when writing requirement statements.
5. Briefly describe why requirement statements should use the active rather than the passive voice.
6. Briefly describe why requirement statements should avoid vague words, terms, and expressions.
7. Briefly describe why requirements statements should avoid unbounded statements and escape clauses.
8. Briefly describe why requirement statements should avoid the use of conjunctions using a formal convention for logical conditions.
9. Briefly describe why requirements statements should avoid unnecessary precision but be precise when necessary.
10. Briefly describe why requirements statements should avoid the use of "not."
11. Briefly describe why requirement statements should avoid cross-references including pronouns.
12. Briefly describe the utility of introducing logical groups of requirement statements with informative statements.
13. Briefly describe why the following are poor requirements, and rewrite them to be "good" requirements:
    - *The same operator shall be able to see the display from all angles.*
    - *The engine temperature shall be displayed to the driver.*
    - *The system shall be easy to use.*
    - *The data shall be stored on a high-speed magnetic disk or tape drive.*
    - *The user interface shall be flexible and user-friendly.*

- The users shall be able to operate the system with minimal training.
- *The system shall be available 100% of the time.*

Endnotes

[1] International Council on Systems Engineering (INCOSE), INCOSE-TP-2010-004-04, *Guide to Writing Requirements*, 2024.

[2] Including the guidance contained in the following:

IEEE-STD-1233, *IEEE Guide for Developing System Requirements Specifications*, New York: Institute of Electrical and Electronic Engineers (IEEE), 1998.

International Council on Systems Engineering (INCOSE), INCOSE-TP-2010-004-04, *Guide to Writing Requirements*, 2024.

ISO/IEC/IEEE 29148, *Systems and Software Engineering—Lifecycle Processes—Requirements Engineering*, International Standards Organization, 2018.

Acronyms and Abbreviations

ABL	Allocated baseline
ANSI	American National Standards Institute
AT&E	Acceptance test and evaluation
BNR	Business needs and requirements
BRS	Business requirement specification
CCB	Configuration control board
CDR	Critical design review
CI	Configuration item
COI	Critical operational issue
CONOPS	Concept of operations
COTS	Commercial-off-the-shelf
CWBS	Contract work breakdown structure
CVR	Cockpit voice recorder
DID	Data item description
DT&E	Developmental test and evaluation
EIA	Electronics Industries Association
EMC	Electromagnetic compatibility
EMI	Electromagnetic interference
FBL	Functional baseline
FFBD	Functional flow block diagram
FoS	Family of systems
FQR	Formal qualification review

HDBK	Handbook
HMI	Human-machine interface
HR	Human resources
ICD	Interface control document
ICWG	Interface (integration) control working group
IEC	International Electrotechnical Commission
IEEE	Institute of Electrical and Electronics Engineers
IIBA	International Institute of Business Analysis
ILS	Integrated logistics support
INCOSE	International Council on Systems Engineering
IS	Interim standard
ISO	International Organization for Standardization
LC	Life cycle concepts
LCC	Life cycle cost
MBSE	Model-based systems engineering
MIL-HDBK	(US) Military Handbook
MIL-STD	(US) Military Standard
MOE	Measure of effectiveness
MOP	Measure of performance
MOS	Measure of suitability
NCOSE	National Council on Systems Engineering
NRM	(INCOSE) *Needs and Requirements Manual*
OpsCon	Operational concept
OT&E	Operational test and evaluation
OTS	Off-the-shelf
PBL	Product baseline
PDR	Preliminary design review
PLC	Preliminary life cycle concepts
PM	Project management
PMI	Project Management Institute
PMP	Project management plan
PoS	Portfolio of systems

PSTN	Public Switched Telephone Network
QA	Quality assurance
RAAF	Royal Australian Air Force
RBS	Requirements breakdown structure
RF	Tadio frequency
RFT	Request for tender
MP	Risk management plan
SBP	Strategic business plan
SBS	System breakdown structure
SDD	System design document
SDR	System design (definition) review
SE	Systems engineering
SEI	Software Engineering Institute
SEMP	Systems Engineering Management Plan
SNR	Stakeholder needs and requirements
SOI	System of interest
SOP	Standard operating procedure
SoS	System of systems
SOW	Statement of work
SRR	Systems requirements review
SS	System specification
STD	Standard
StRS	Stakeholder requirements specification
SVR	System verification review
SyRS	System requirements specification
SySML	System modeling language
T&E	Test and evaluation
TPM	Technical performance measure
UML	Uniform modeling language
V&V	Verification and validation
WBS	Work breakdown structure
WSOI	Wider SOI

About the Author

Michael Ryan holds BE, MEngSc, and PhD degrees in electrical engineering from the University of New South Wales. He is a Fellow of Engineers Australia (FIEAust), a Chartered Professional Engineer (CPEng) in systems engineering, electrical and ITEE colleges, a senior member of IEEE (SMIEEE), a Fellow of the International Council on Systems Engineering (FINCOSE), a Fellow of the Institute of Managers and Leaders (FIML), and a Fellow of the Royal Society of NSW (FRSN). Since 1981, he has held a number of positions in management and project management and communications and systems engineering. From 1998 to 2020, he was with the University of New South Wales, retiring as a professor. He is currently the director of Capability Associates Pty Ltd. His consulting and teaching interests are in communications and information systems, requirements engineering, systems engineering, project management, and technology management. He is a cochair of the INCOSE Requirements Working Group. Over the last 35 years, he has facilitated mission engineering and needs and requirements development activities for more than 50 major projects. He is the author or coauthor of 14 books, four book chapters, and over 450 technical papers and reports.

Index

Artech House Technology Management and Professional Development Library

Seth Sullivan, Series Editor

Actionable Strategies Through Integrated Performance, Process, Project, and Risk Management, Stephen S. Bonham

Advanced Systems Thinking, Engineering, and Management, Derek K. Hitchins

Applied Systems Engineering, Second Edition, Ian Faulconbridge and Michael Ryan

Applying Total Quality Management to Systems Engineering, Joe Kasser

Building Successful Virtual Teams, Francine Gignac

Critical Chain Project Management, Third Edition, Lawrence P. Leach

Decision Making for Technology Executives: Using Multiple Perspectives to Improve Performance, Harold A. Linstone

Designing the Networked Enterprise, Igor Hawryszkiewycz

Electrical Product Compliance and Safety Engineering, Volume 2, Steli Loznen and Constantin Bolintineanu

The Engineer's Guide to Authentic Leadership, Ronald J. Bennett and Elaine R. Millam

Engineer's and Manager's Guide to Winning Proposals, Donald Helgeson

Engineering and Technology Management Tools and Applications, B. S. Dhillon

Enterprise Release Management: Agile Delivery of a Strategic Change Portfolio, Louis Taborda

The Entrepreneurial Engineer: Starting Your Own High-Tech Company, R. Wayne Fields

Evaluation of R&D Processes: Effectiveness Through Measurements, Lynn W. Ellis

From Engineer to Manager: Mastering the Transition, Second Edition, B. Michael Aucoin

Global High-Tech Marketing: An Introduction for Technical Managers and Engineers, Jules Kadish

How to Become an IT Architect, Cristian Bojinca

Integrated IT Project Management, Kenneth R. Bainey

Introduction to Information-Based High-Tech Services, Eric Viardot

Introduction to Innovation and Technology Transfer, Ian Cooke and Paul Mayes

ISO 9001:2000 Quality Management System Design, Jay Schlickman

IT Project Portfolio Management, Stephen S. Bonham

Managing Complex Technical Projects: A Systems Engineering Approach, Second Edition, Ian Faulconbridge and Michael J. Ryan

Managing Successful High-Tech Product Introduction, Brian P. Senese

Managing Virtual Teams: Practical Techniques for High-Technology Project Managers, Martha Haywood

Mastering Technical Sales: The Sales Engineer's Handbook, Fourth Edition, John Care

The New High-Tech Manager: Six Rules for Success in Changing Times, Kenneth Durham and Bruce Kennedy

The Parameter Space Investigation Method Toolkit, Roman Statnikov and Alexander Statnikov

Planning and Design for High-Tech Web-Based Training, David E. Stone and Constance L. Koskinen

A Practical Guide to Managing Information Security, Steve Purser

Practical Model-Based Systems Engineering, Jose L. Fernandez and Carlos Hernandez

Practical Reliability Data Analysis for Non-Reliability Engineers, Darcy Brooker with Mark Gerrand

The Project Management Communications Toolkit, Second Edition, Carl Pritchard

Project Managment Process Improvement, Robert K. Wysocki

Reengineering Yourself and Your Company: From Engineer to Manager to Leader, Howard Eisner

The Requirements Engineering Handbook, Ralph R. Young

Requirements Practice in Conceptual Design, Second Edition: A Practical Guide, Michael J. Ryan

Running the Successful Hi-Tech Project Office, Eduardo Miranda

Successful Marketing Strategy for High-Tech Firms, Second Edition, Eric Viardot

Successful Proposal Strategies for Small Businesses: Using Knowledge Management to Win Government, Private Sector, and International Contracts, Sixth Edition, Robert S. Frey

Successful Proposal Strategies On-the-Go!, Robert S. Frey

Systems Approach to Engineering Design, Peter H. Sydenham

Systems Engineering Principles and Practice, H. Robert Westerman

Systems Reliability and Failure Prevention, Herbert Hecht

Team Development for High-Tech Project Managers, James Williams

For further information on these and other Artech House titles, including previously considered out-of-print books now available through our In-Print-Forever® (IPF®) program, contact:

Artech House
685 Canton Street
Norwood, MA 02062
Phone: 781-769-9750
Fax: 781-769-6334
e-mail: artech@artechhouse.com

Artech House
16 Sussex Street
London SW1V 4RW UK
Phone: +44 (0)20 7596-8750
Fax: +44 (0)20 7630-0166
e-mail: artech-uk@artechhouse.com

Find us on the World Wide Web at: www.artechhouse.com

www.ingramcontent.com/pod-product-compliance
Ingram Content Group UK Ltd.
Pitfield, Milton Keynes, MK11 3LW, UK
UKHW022206051026
15624UKWH00001B/21

9 781685 691356